Annual Reports in
Computational Chemistry

Annual Reports in Computational Chemistry

Volume 1

Editor
David C. Spellmeyer
Almaden Services Research
IBM Almaden Research Center
San Jose, California
USA

Section Editors
H. Carlson, T.D. Crawford, Y. Martin, C. Simmerling, R. Wheeler, and T. Zielinski

Sponsored by the Division of Computers in Chemistry of
the American Chemical Society

2005

ELSEVIER

Amsterdam • Boston • Heidelberg • London • New York • Oxford • Paris
San Diego • San Francisco • Singapore • Sydney • Tokyo

ELSEVIER B.V.
Radarweg 29
P.O. Box 211, 1000 AE Amsterdam
The Netherlands

ELSEVIER Inc.
525 B Street, Suite 1900
San Diego, CA 92101-4495
USA

ELSEVIER Ltd
The Boulevard, Langford Lane
Kidlington, Oxford OX5 1GB
UK

ELSEVIER Ltd
84 Theobalds Road
London WC1X 8RR
UK

First edition 2005

Library of Congress Cataloging in Publication Data
A catalog record is available from the Library of Congress.

British Library Cataloguing in Publication Data
A catalogue record is available from the British Library.

ISBN: 0-444-51857-6
ISSN: 1574-1400 (Series)

♾ The paper used in this publication meets the requirements of ANSI/NISO Z39.48-1992 (Permanence of Paper). Printed in the United States.

CONTENTS

Contributors ix

Preface xi

1. QUANTUM MECHANICAL METHODS

Section Editor: T. Daniel Crawford, Department of Chemistry,
Virginia Tech University, Blacksburg, Virginia 24061, USA

1. An Introduction to the State of the Art in Quantum Chemistry 3
Frank Jensen
Department of Chemistry, University of Southern Denmark,
DK-5230 Odense, Denmark

2. Time-Dependent Density Functional Theory
in Quantum Chemistry 19
Filipp Furche[1] and Kieron Burke[2]
[1]Institut für Physikalische Chemie, Universität Karlsruhe,
Kaiserstraße 12, 76128 Karlsruhe, Germany
[2]Department of Chemistry and Chemical Biology,
Rutgers University, 610 Taylor Road, Piscataway,
NJ 08854, USA

3. Computational Thermochemistry: A Brief Overview
of Quantum Mechanical Approaches 31
Jan M.L. Martin
Department of Organic Chemistry, Weizmann Institute of Science,
76100 Reḫovot, Israel

4. Bond Breaking in Quantum Chemistry 45
C. David Sherrill
Center for Computational Molecular Science and Technology,
School of Chemistry and Biochemistry, Georgia Institute of
Technology, Atlanta, GA 30332-0400, USA

2. MOLECULAR MODELING METHODS

Section Editor: Carlos Simmerling, Center for Structural Biology, Stony Brook University, Stony Brook, NY 11794, USA

5. A Review of the TIP4P, TIP4P-Ew, TIP5P, and TIP5P-E Water Models 59
Thomas J. Dick and Jeffry D. Madura
Department of Chemistry and Biochemistry, Center for Computational Sciences, Duquesne University, 600 Forbes Avenue, Pittsburgh, PA 15282, USA

6. Molecular Modeling and Atomistic Simulation of Nucleic Acids 75
Thomas E. Cheatham III
Departments of Medicinal Chemistry and of Pharmaceutics and Pharmaceutical Chemistry, College of Pharmacy, University of Utah, 2000 East 30 South Skaggs Hall 201, Salt Lake City, UT 84112, USA

7. Empirical Force Fields for Proteins: Current Status and Future Directions 91
Alexander D. MacKerell Jr.
Department of Pharmaceutical Sciences, School of Pharmacy, University of Maryland, 20 Penn Street, Baltimore, MD 21201, USA

8. Nonequilibrium Approaches to Free Energy Calculations 103
Adrian E. Roitberg
Quantum Theory Project, Department of Chemistry, University of Florida, P.O. Box 118435, Gainesville, FL 32611-8435, USA

9. Calculating Binding Free Energy in Protein–Ligand Interaction 113
Kaushik Raha and Kenneth M. Merz Jr.
Department of Chemistry, The Pennsylvania State University, University Park, PA 16802, USA

3. ADVANCES IN QSAR/QSPR

Section Editor: Yvonne Martin, Abbott Laboratories, Abbott Park, IL 60064, USA

10. Computational Prediction of ADMET Properties: Recent Developments and Future Challenges 133
David E. Clark
Argenta Discovery Ltd, 8/9 Spire Green Centre, Flex Meadow, Harlow, Essex CM19 5TR, UK

4. APPLICATIONS OF COMPUTATIONAL METHODS

Section Editor: Heather Carlson, University of Michigan, College of Pharmacy, 428 Church Street, Ann Arbor, MI 48109-1065, USA

11. Filtering in Drug Discovery 155
Christopher A. Lipinski
Pfizer Global Research and Development, Groton, CT 06340, USA

12. Structure-Based Lead Optimization 169
Diane Joseph-McCarthy
Wyeth Research, 200 CambridgePark Drive, Cambridge, MA 02140, USA

13. Targeting the Kinome with Computational Chemistry 185
Michelle L. Lamb
AstraZeneca R&D Boston, Cancer Research, 35 Gatehouse Drive, Waltham, MA 02451, USA

5. CHEMICAL EDUCATION

Section Editor: Theresa Zielinski, Department of Chemistry, Medical Technology, and Physics, Edison Science Hall, Room E245, Monmouth University, 400 Cedar Avenue, West Long Branch, NJ 07764-1898, USA

14. Status of Research-Based Experiences for First- and Second-Year Undergraduate Students 205
Jeffrey D. Evanseck[1] and Steven M. Firestine[2]
[1]Department of Chemistry and Biochemistry, Center for Computational Sciences, Duquesne University, 600 Forbes Avenue, Pittsburgh, PA 15282, USA
[2]Division of Pharmaceutical Sciences, Duquesne University, 600 Forbes Avenue, Pittsburgh, PA 15282, USA

15. Crossing the Line: Stochastic Models in the Chemistry Classroom 215
Michelle M. Francl
Department of Chemistry, Bryn Mawr College, 101 N. Merion Avenue, Bryn Mawr, PA 19010, USA

16. Simulation of Chemical Concepts, Systems and Processes Using Symbolic Computation Engines: From Computer-Assisted Problem-Solving Approach to Advanced Tools for Research 221
Jonathan Rittenhouse and Mihai Scarlete
Bishop's University, Lennoxville, Que., Canada J1M 1Z7a

6. EMERGING SCIENCE

Section Editor: Ralph Wheeler, Department of Chemistry & Biochemistry, University of Oklahoma, 620 Parrington Oval, Room 208, Norman, OK 73019, USA

17. The Challenges in Developing Molecular Simulations of Fluid Properties for Industrial Applications 239
Raymond D. Mountain and Anne C. Chaka
Computational Chemistry Group, Physical and Chemical Properties Division, Chemical Science and Technology Laboratory, National Institute of Standards and Technology, Gaithersburg, MD 20899-8380, USA

18. Computationally Assisted Protein Design 245
Sheldon Park and Jeffery G. Saven
Department of Chemistry, University of Pennsylvania, 231 S. 34th Street, Philadelphia, PA 19104, USA

Subject Index 255

CONTRIBUTORS

Burke, Kieron 19
Chaka, Anne C. 239
Cheatham III, Thomas E. 75
Clark, David E. 133
Dick, Thomas J. 59
Evanseck, Jeffrey D. 205
Firestine, Steven M. 205
Francl, Michelle M. 215
Furche, Filipp 19
Jensen, Frank 3
Joseph-McCarthy, Diane 169
Lamb, Michelle L. 185
Lipinski, Christopher A. 155
MacKerell Jr., Alexander D. 91
Madura, Jeffry D. 59
Martin, Jan M. L. 31
Merz Jr., Kenneth M. 113
Mountain, Raymond D. 239
Park, Sheldon 245
Raha, Kaushik 113
Rittenhouse, Jonathan 221
Roitberg, Adrian E. 103
Saven, Jeffery G. 245
Scarlete, Mihai 221
Sherrill, C. David 45

PREFACE

Annual Reports in Computational Chemistry is a new periodical focusing on timely reviews of topics important to researchers in the field of computational chemistry. It is published and distributed by Elsevier for the Division of Computers in Chemistry (COMP) of the American Chemical Society. The COMP Executive Committee is pleased to have developed this periodical over the past several years and expect that it will bring great benefit to the Division members and to the scientific community at large. We expect that future volumes of the ARCC will build on the example set by the contributions of the first volume.

The Section Editors have completed the selection and editing of 18 contributions in six sections for the first volume. Sections covered include contributions in the area of Quantum Mechanical Methods (Section Editor: T. Daniel Crawford), Molecular Modeling Methods (Carlos Simmerling), QSAR/QSPR (Yvonne Martin), Applications of Computational Methods (Heather Carlson), Chemical Education (Theresa Zielinski), and Emerging Science (Ralph Wheeler).

The Annual Reports in Computational Chemistry is assembled entirely by volunteers in order to produce a high-quality scientific publication at the lowest cost possible. I would like to thank those people who have contributed to make this first edition possible. The authors of each of this year's contributions and the Section Editors have been gracious in working with the Editor and the publisher during the development of this edition. I would like to thank all the members of the COMP Executive team, past and present, for their commitment to this publication, for pushing and prodding, and for their ideas on topics, contributors, Section Editors, and in general for helping to make this a reality. Peter Grootenhius deserves special thanks for challenging us to publish our first volume in early 2005. Thank you one and all.

I hope that you will find this first edition to be interesting and valuable. We are currently planning the second edition and solicit input from our readers about future topics. Please e-mail me your suggestions and to volunteer as a contributor. I can be reached at arcc_editor@yahoo.com.

Sincerely
David C. Spellmeyer

Section 1
Quantum Mechanical Methods

Section Editor: T. Daniel Crawford
Department of Chemistry
Virginia Tech University
Blacksburg
Virginia 24061
USA

CHAPTER 1

An Introduction to the State of the Art in Quantum Chemistry

Frank Jensen

Department of Chemistry, University of Southern Denmark, DK-5230 Odense, Denmark

Contents

1. Introduction 3
2. Hartree–Fock 5
3. Electron correlation methods 8
   3.1. Configuration interaction and multi-configurational self-consistent field methods 9
   3.2. Many-body perturbation theory 10
   3.3. Coupled cluster methods 10
4. Density functional theory 11
5. Semi-empirical methods 12
6. Basis sets 13
7. Summary 15
References 16

1. INTRODUCTION

The fundamental building blocks in quantum chemistry are nuclei and electrons. The small electronic mass necessitates the use of quantum mechanics for describing the electron distribution, but the nuclear masses are sufficiently heavy that their motion to a good approximation can be described by classical mechanics. The large difference in mass is the basis for the Born–Oppenheimer approximation, where the coupling between the nuclear and electronic motions is neglected. From the electron point of view, the nuclei are thus stationary, and the electronic Schrödinger equation can be solved with the nuclear positions as parameters. A (large) set of such solutions forms a $3N-6$ dimensional potential energy surface (PES) upon which the nuclear motions can be solved subsequently. The multi-dimensional nature of the surface prevents a complete mapping for systems with more than four nuclei, and for larger systems, the effort must therefore be focused on the chemically important low-energy region. Traditionally such investigations have been done by a static approach, by locating minima and first-order saddle points on the PES [1]. Minima describe stable molecules, while first-order saddle points relate to the chemical transformation of

ANNUAL REPORTS IN COMPUTATIONAL CHEMISTRY, VOLUME 1
ISSN: 1574-1400 DOI 10.1016/S1574-1400(05)01001-7

one species to another *via* transition state theory. More recently, the PES has also been explored by direct dynamics, where Newton's equations for the nuclear motions are solved using energies and derivatives generated on-the-fly, as required by the dynamics [2]. In the present context, we will only be concerned with methods for solving the electronic Schrödinger equation and not with methods for exploring the resulting PES.

Solving the electronic Schrödinger equation is difficult for two main reasons:

- The electrons are indistinguishable and the differential equation couples all the electronic coordinates.
- The interaction between electrons is only a factor of Z (nuclear charge) less than the interaction between the nucleus and the electrons.

The standard approach for solving multi-variable differential equations is to find a set of coordinates where the variables can be separated and solve them one at a time. This is not possible for the electronic Schrödinger equation with more than one electron and the relatively large electron–electron interaction compared to the nucleus–electron interaction prevents a central-field approximation. Neglect of the electron–electron interaction leads to a wave function composed of hydrogen-like orbitals, but this is too poor a model to be useful. A qualitatively correct description can be obtained by a mean-field approximation, where the average electron–electron interaction is included, and within the wave function approach, this is known as the Hartree–Fock (HF) method. In order to improve the computational efficiency, various approximations to the HF equations can be made, with the reduction in fundamental accuracy being (partly) made up for by parameterization against experimental data. Such methods can collectively be called semi-empirical methods. Alternatively, the inherent deficiencies due to the mean-field approximation can be reduced by adding many-body corrections and these are called electron correlation methods [3].

Density functional theory (DFT) may be considered as an alternative formulation of quantum mechanics, where the electron density is the fundamental variable, rather than the electron coordinates [4]. DFT can also be considered as an improvement of the HF model, where the many-body correlation is modeled as a function of the electron density. DFT is analogous to the HF method a pseudo one-particle model, leading to a computationally efficient way of determining the electronic structure for large systems. While DFT has been a widely used tool for several decades in solid-state physics, it was only after the introduction of so-called gradient-corrected functionals in the early 1990s that the accuracy improved sufficiently to become a useful tool in computational chemistry.

An integrated element in practical calculations is the expansion of the orbitals (one-particle wave functions) in a set of known functions, the basis set. Only Gaussian-type basis functions will be considered in the present case, as these are used almost universally for application purposes. An ideal basis set should give a good accuracy for a small number of functions, be computationally efficient and allow a systematic way of extrapolating to the basis set limit. Unfortunately, different methods have different basis set requirements and it is not possible to find a single basis set optimum for all purposes.

In the following, we will briefly review the theoretical background for methods aimed at solving the electronic Schrödinger equation and present some highlight of the recent research within each area.

2. HARTREE–FOCK

The electronic Schrödinger equation in abbreviated form can be written as

$$\mathbf{H}\Psi = E\Psi$$

where the Hamilton operator **H** contains four terms corresponding to the electron kinetic energy, the nuclear–electron attraction, the electron–electron repulsion and the nuclear–nuclear repulsion. The latter is an additive constant within the Born–Oppenheimer approximation.

$$\mathbf{H} = \mathbf{T}_e + \mathbf{V}_{ne} + \mathbf{V}_{ee} + \mathbf{V}_{nn}$$

The variational principle states that an approximate wave function will always have an energy higher than the exact wave function and the (energetically) best wave function can thus be determined by minimizing the energy. In an independent particle picture (HF), each electron is described by an orbital ϕ and the whole wave function is a product of such orbitals. The antisymmetry requirement means that the wave function is not simply a product of orbitals, but rather a linear combination of such products, which conveniently can be written in the form of a (Slater) determinant. The HF energy is given by integrals of molecular orbitals over the $\mathbf{T}_e$ and $\mathbf{V}_{ne}$ operators, often collected in a one-electron operator **h**, and over the two-electron $\mathbf{V}_{ee}$ operator. The latter gives rise to two contributions: a Coulomb term corresponding to the classical interaction between two charge clouds and an exchange term arising from the wave function antisymmetry. Setting the first derivative of the energy with respect to the molecular orbitals to zero gives the HF equation, which has the same form as the Schrödinger equation, except that it is now at the orbital (one-electron) level.

$$\mathbf{F}\phi = \varepsilon\phi$$

The Fock operator **F** describes the motion of one electron in the field of all the nuclei and the mean-field of all the other electrons. Since the latter is given by the orbitals, the HF equation depends on its own solutions and must be solved iteratively.

In practical applications, the molecular orbitals are expanded in a basis set, thereby transforming the HF equations into the Roothaan–Hall equations, which can be written as a (generalized) matrix eigenvalue problem.

$$\mathbf{FC} = \varepsilon \mathbf{SC}$$

The variational parameters are the molecular orbital coefficients contained in the **C** matrix which, together with the basis functions, determine the shape of the molecular orbitals. The dimension of the matrix equation is the number of the basis functions, a quantity that is under user control. The elements in the **F** matrix contain integrals of the one- and two-electron operators over basis functions, multiplied with products of the molecular orbital coefficients collected into a density matrix **D**. The iterative sequence corresponds to diagonalization of the Fock matrix to give an updated density matrix, which is used for constructing the next Fock matrix, etc. The iteration is started with a suitable guess of the density matrix and continued until the difference between two consecutive density matrices are within a suitable (small) threshold. At this point, the solution corresponds to a self-consistent field (SCF), i.e., the calculated electric field generated by the electrons is consistently with the electron distribution.

There are two major computational problems in a HF calculation, calculating the two-electron integrals and solving the HF (Roothaan–Hall) equations. The calculation of a single two-electron integral is computationally quite easy, but the number of such integrals is approximately $1/8M^4$ for a basis set containing M functions. A medium-sized basis set will typically have 15–20 functions for each atom, and already a 100-atom system may result in several thousand basis functions, and thus potentially $\sim 10^{12}$ two-electron integrals. In a traditional implementation, these integrals are calculated and stored on disk, requiring $\sim$10 TB of disk space. A straightforward implementation of the HF model is, therefore, an N^4 method, increasing the system size by a factor of two will increase the computational time and storage by a factor of 16, and this effectively limits the application to systems with less than $\sim$30 atoms.

Two developments have been essential for reducing the scaling and thereby pushing the limit for feasible calculations. The first is the introduction of the direct SCF method with differential update of the Fock matrix [5]. The Fock matrix can be written as a one-electron contribution **h**, which is independent of the density matrix, and a contraction of the density matrix **D** with a two-electron tensor **G**. The change in the Fock matrix is thus given by the change in the density matrix.

$$\mathbf{F}_i = \mathbf{h} + \mathbf{D}_i\mathbf{G}$$

$$\Delta \mathbf{F}_i = \Delta \mathbf{D}_i \mathbf{G}$$

Rather than calculating all the two-electron integrals prior to solving the Roothaan–Hall equation, the integrals can be recalculated in each iteration. While this avoids the requirement for massive amounts of disk storage, it potentially increases the computational time with a factor close to the number of iterations. The availability of the density matrix elements, however, means that not all integrals have to be calculated, only those that will be multiplied with sufficiently large density matrix elements are required. The density matrix elements between atoms that are spatially far apart will be close to zero and this effectively reduces the method scaling from N^4 to N^2 for large systems. Furthermore, as the iterative solution proceeds, the change in the density matrix (hopefully) becomes smaller and smaller and the integral screening therefore becomes more and more efficient.

For large systems, the dominating integrals are those describing the Coulomb interaction between electrons, leading to an overall N^2 scaling. In fast-multipole methods, the Coulomb contribution is not calculated by two-electron integrals, but is replaced with the interaction between two electron densities [6]. The latter can be calculated in a more efficient fashion by partitioning the physical space into boxes and evaluating the interaction between densities within the boxes as interactions between multipoles located at the center of the boxes. The required multipole order and box size depends on the distance between boxes for a given final accuracy, and distant interactions can, therefore, be calculated with a coarser granulation than the near-field contribution. For sufficiently large systems, this leads to a computational complexity of order N, i.e., the computational effort only increases linearly with the system size.

Solution of the Roothaan–Hall equation by repeated diagonalization of the Fock matrix requires a computational time proportional to the cube of the matrix dimension. For small systems, the matrix diagonalization time is insignificant compared to the construction of the Fock matrix, but for large systems, the N^3 diagonalization becomes the dominating step. In order to achieve a true linear scaling implementation, the HF energy can be written as a function directly of the density matrix elements and standard optimization techniques like conjugate gradient methods can be used for minimization [7].

When discussing the method scaling, i.e., the increase in computational time with system size, the focus is on the most demanding computational step in the large system limit. More important for practical calculations is the size of the corresponding prefactor, i.e., Time = Prefactor $*$ N^n. A low-order scaling method with a large prefactor will ultimately be more efficient than a higher order method with a small prefactor, but this region may not be attainable within the limitation of the available computational resources. Figure 1 illustrates the computational time as a function of the system size for three methods with linear (N), quadratic (N^2) and quartic (N^4) scaling.

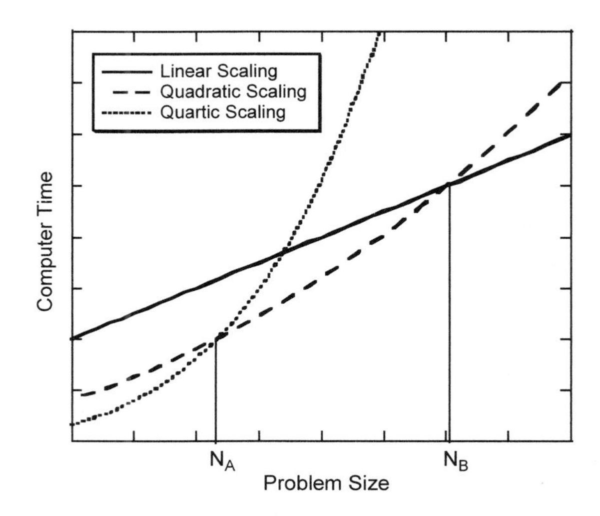

Fig. 1. Illustrating the regimes where different methods are most efficient.

As seen from Fig. 1, the N^4 method is the most efficient for systems smaller than N_A, the best method for systems between N_A and N_B is the N^2 method, while the linear scaling method becomes the preferred choice for systems larger than N_B. For the HF model, the switch between a conventional N^4 and an N^2 (direct) SCF method occurs for so small systems that direct methods often are the default option for all systems. The switch for when the linear scaling method becomes favorable, however, occurs for systems that currently are near the feasibility limit, i.e., a few hundred atoms.

In the above, it has been assumed implicitly that the iterative Roothaan–Hall procedure converges, i.e., the difference between two consecutive density matrices decreases during the iterative process, but this is by no means guaranteed. A closely related problem is that the optimization converges on a local, rather than the global, minimum. Close-lying states often mean that the density matrix switches between different state descriptions during the iterative procedure, leading to a non-convergent behavior. The direct inversion in the iterative subspace (DIIS) has been a favorite tool for improving both the convergence and reducing the number of iterations at the same time [8], while second-order methods can be used to force convergence, albeit at a significant increase in computational time [9].

3. ELECTRON CORRELATION METHODS

The HF model only accounts for the average electron–electron interaction and thus neglects the correlation between electrons. Since HF is the energetically best single determinant wave function, correlated methods must necessarily involve more than one Slater determinant. This also means that the mental picture of each electron residing in a separate orbital must be abandoned. Rather, one must accept a picture with a range of orbitals having a fractional number of electrons. The HF model has N_{elec} orbitals with occupation numbers being exactly 1, while correlated methods have N_{elec} (natural) orbitals with occupation numbers close to 1, and the remaining having occupation numbers close to 0. The total amount of electrons moved from occupied to empty HF orbitals is a measure of how important electron correlation is for the particular system.

There are three main methods for calculating the correlation energy: configurational interaction, many-body perturbation theory and coupled cluster [10].

3.1. Configuration interaction and multi-configurational self-consistent field methods

The configuration interaction (CI) method relies on writing a multi-determinant trial wave function and using the variational principle for determining the weight of each determinant in the total wave function. The additional determinants beyond HF are generated by 'exciting' electrons from occupied to empty orbitals, and they can be characterized by the excitation level, i.e., singly (S), doubly (D), triply (T), quadruply (Q), etc. excited determinants relative to the HF configuration. The orbitals are taken from a HF calculation and kept fixed. If they also are optimized, the method is called multi-configurational self-consistent field (MCSCF), and is primarily used for obtaining a correct qualitative description for systems where the single determinant HF wave function is insufficient. The full CI method includes all excitations and represents an exact solution within the limitations of the basis set. The number of excited determinants, however, increases factorially with the number of electrons and basis functions, and is therefore only possible for small systems. For practical applications the excitation level must be truncated, leading to the CISD, CISDT, CISTDQ, etc. models. Due to the orbital optimization in the HF model, CIS is identical to HF for ground state energies. The CISD method scales as N^6 with the system size, CISDT has a scaling of N^8, while CISDTQ is an N^{10} order process, and only the CISD method is sufficiently efficient computationally to be generally useful.

A major disadvantage of truncated CI methods is the lack of size consistency, i.e., calculating the energy of two non-interacting systems do not give the same results as adding the energies from two separate calculations. CI is the oldest method for including electron correlation, but the lack of size-consistency means that there have been relatively few developments in recent years. The full CI is useful as an absolute reference for evaluating new approximate methods and improvements in computational algorithms and computer hardware have pushed the limit to $\sim 10^{10}$ excited determinants. The factorial dependence on size, however, means that this number of determinants is reached already for $\sim$10 electrons and $\sim$30 basis functions.

3.2. Many-body perturbation theory

The most popular many-body perturbation theory is the Møller–Plesset (MP) form, where the Hamilton operator is partitioned into a sum of Fock operators and a remaining 'fluctuation' potential. The Fock operator part can be solved exactly, with the solutions being excited Slater determinants generated from the HF wave function. By assigning a perturbation parameter to the fluctuating potential, the corrections to the energy and wave function can be determined to various orders. The HF result is recovered at first order and electron correlation thus starts at order two (MP2). Higher order methods become increasingly complex, but MP3 and MP4 results can be obtained routinely for quite large systems, and implementation of the MP5 and MP6 contributions have also been reported [11]. MP2 is the least expensive method for including electron correlation, having a scaling of (only) N^5 with the system size and a quite small prefactor. It furthermore recovers a substantial fraction of the correlation energy, often $\sim$80% of the amount possible within the limitations of the basis set. Each successive term in the perturbation series has a computational complexity one order higher, i.e., MP3 is an N^6 method, MP4 is N^7, MP5 is N^8, etc. The MP form of perturbation theory is size consistent.

Although an explicit implementation of perturbation corrections beyond MP6 has not been reported, the results to any order can be generated from a full CI calculation (equivalent to MP∞). Analysis of such results has shown that the perturbation series for many systems is divergent, i.e., inclusion of higher order terms does not necessarily lead to more accurate results [12]. The MP2 result may still provide a reasonable estimate of the correlation effect, but there is no guarantee that the MP3 or MP4 results will be more accurate. For this reason the development of MP methods beyond MP2 have to a large extent been abandoned.

3.3. Coupled cluster methods

In the coupled cluster (CC) approach, the excited Slater determinants are generated by an exponential parameterization, in contrast to the CI method, where the excitations enter in a linear fashion [13]. The exponential parameterization ensures that the method is size consistent, but also leads to sets of non-linear coupled cluster equations, which must be solved by iterative methods. CCSD is a coupled cluster model where only single and double excitation operators are considered, but due to the exponential parameterization, higher order excited determinants are accounted for implicitly. If the triple excitation operator is also included, the CCSDT model arises, addition of the quadruply excitation operator leads to the CCSDTQ model, etc. The CCSD model is an N^6 method, analogous to CISD, CCSDT has an N^8 complexity and CCSDTQ is an N^{10} process. The iterative nature of the coupled cluster equations means that CCSD is computationally more expensive than CISD, although they both are N^6 methods, i.e., CCSD has a larger prefactor than CISD.

The CCSD model neglects the important connected triple excitations, but the CCSDT model is computationally so complex that it can only be used for small systems. CCSD(T) is a popular hybrid method, where the result from a CCSD calculation is augmented with the T-contribution calculated by fourth-order perturbation theory, and it has a computational scaling of N^7. While CCSD(T) results have been shown to give a very accurate description even for difficult systems, the high scaling and large prefactor limits the applicability to relatively small systems. Several attempts have been made at reducing the scaling of coupled cluster methods, e.g., Cholesky decomposition [14], reformulating the coupled cluster equations as matrix equations [15] and using localized orbitals [16]. It is unclear, however, what the increase in the resulting prefactors is, i.e., at what system size will such methods become favorable relative to more traditional methods.

4. DENSITY FUNCTIONAL THEORY

DFT rests on the Hohenberg–Kohn theorem, which states that there is a unique one-to-one correspondence between the ground state electron density and the energy of a system [17]. In the Kohn–Sham version of DFT, the density is written as an antisymmetric product of orbitals, analogous to the HF model [18]. The one-electron term and the Coulomb interaction between electrons are identical to those in the HF model, but the exchange and correlation contributions are incorporated as functionals of the electron density. DFT shares the same computational complexity as HF, i.e., a formal N^4 method, which can be reduced to a linear scaling method by, e.g., fast-multipole methods and direct optimization of the Kohn–Sham equations. The problem of SCF convergence is more severe for DFT methods than for HF, but the use of fractional occupation numbers have recently been shown to improve the convergence towards the global minimum [19].

Although it can be proven that there is a unique exchange-correlation functional, its mathematical form is unknown. There are currently three relatively well-defined levels of DFT:

1. *Local spin density approximation* (LSDA). For the special case of a uniform electron density it is possible to derive explicitly that the exchange energy can

be calculated as a constant times the integral of the density raised to the power of 4/3. The correlation energy can be calculated by Monte Carlo methods and mathematical fitting functions have been devised for providing an analytical form of the correlation energy as a function of the electron density [20].

2. *Generalized gradient approximation* (GGA). These methods express the exchange and correlation functionals as a function of the electron density and its first derivative. Several different functional forms have been proposed, some of which employ empirical parameters determined by fitting to experimental data [21].
3. *Hybrid methods*. These methods employ a mix of the LSDA and GGA functional forms and mixes in the 'exact' exchange energy as calculated by HF theory [22]. The appropriate mixing of the three terms is determined by fitting to experimental data.

LSDA methods have been used since the dawn of scientific computation [23]. For chemical applications, the accuracy is similar or inferior to HF, and HF has been preferred in the chemical community, but LSDA has played a major role in solid-state physics. The introduction of gradient corrected methods in the early 1990s significantly improved the accuracy and moved DFT into mainstream computational chemistry. Development of new exchange-correlation functionals is a very active research area, and has been pursued both with respect to increasing the number of fitting parameters and by including exact exchange, orbital-dependent quantities and second derivatives of the electron density [24]. Following the initial rapid improvements in performance by introduction of gradient corrections and exact exchange, there has been relatively little progress in recent years. The main problem in DFT is the non-systematic way the results can be improved toward the exact result, but its strength is that it can deliver results of a good accuracy at a computationally very favorable price. The accompanying review by Furche and Burke in this volume provides a review of the most recent developments for time-dependent DFT methods.

5. SEMI-EMPIRICAL METHODS

The major computational effort in the HF model is calculating two-electron integrals over basis functions. In order to improve the computational efficiency, and thus allowing treatment of larger systems, it is necessary both to limit the size of the basis set and neglect certain classes of integrals. Semi-empirical methods are characterized by using only a minimum valence basis set, i.e., core electrons are accounted for by reducing the effective nuclear charge.

All integrals extending over more than two nuclear centers are furthermore neglected. Since the HF model in the full implementation is only capable of modest accuracy, such drastic approximations are expected to lead to a poor model. The success of semi-empirical methods relies on not actually calculating the surviving terms, but turning them into parameters and fitting these to experimental data. The best known of these methods is the MNDO, AM1 and PM3 family [25], primarily used for geometries and energetics, although the ZINDO method has been widely used for spectroscopic purposes [26]. The AM1/PM3 methods rely on extensive parameterization against experimental data, primarily structures and stabilities, and have been developed for elements important for organic chemistry, since this is where most of the required experimental information is available.

The fundamental computational problem is the same as for the HF methods, formation of a Fock-type matrix and iterative solution of the Roothaan–Hall equations. The Fock matrix construct is formally an N^2 process, since only two-center terms are included explicitly and the prefactor is much smaller than for HF. This means that matrix diagonalization methods for solving the variational problem rapidly becomes dominating and linear scaling methods for solving the HF equations becomes important even for relatively small systems [27]. More recently, the MNDO method has been extended to include d-orbitals [28], the PM3 parameterization has been extended to transition metals [29] and a new PM5 method has been introduced [30].

6. BASIS SETS

The use of nuclear centered basis functions allows a formal way of approaching the basis set limit, by including more and more functions, and of increasingly higher angular momentum. Since a complete (infinite) basis set is computationally infeasible, error cancellation is the key to achieving good results with modest-sized basis sets. A good basis set is characterized by having a balanced composition, i.e., a proper number of functions and angular momenta. Unfortunately, different methods have different requirements for achieving a basis set balance. HF and DFT methods only need to represent the electron density, and have a (fast) exponential convergence towards the basis set limit as a function of the highest angular momentum functions included in the basis set [31]. Electron correlation methods, on the other hand, display a (slow) inverse power convergence towards the basis set limit [32]. The slow basis set convergence can to some extent be solved by including the interelectronic distance as a variable, although such methods are still primarily used for calibration studies [33]. The difference in convergence rate indicates that a balanced basis set for HF and DFT will contain more low-angular momentum functions than basis sets tailored for electron correlation. For calculation of molecular properties, like polarizabilities or nuclear chemical shifts, the basis set may be augmented by special diffuse or tight functions in order to improve the performance.

The smallest basis set, called a minimum basis set, has only the necessary number of functions for containing the electrons in the isolated atom. For hydrogen this is one s-function, for a first row element like carbon it is two s- and one p-function, for a second row element it is three s- and two p-functions, etc. An improved description can be obtained by doubling (DZ), tripling (TZ), quadrupling (QZ), etc. the number of valence functions. The core orbitals are described by a single set of functions, as they are essentially independent of the molecular environment. In order to describe the distortion of the electron density upon bond formation, and for describing electron correlation, the basis set must be augmented with higher angular momentum functions. The number and nature of these polarization functions should balance the number of s- and p-functions. In order to increase the computational efficiency, the number of variational parameters can be reduced by contracting some of the basis functions. Typically the functions describing the core electrons are contracted into fixed linear combinations, and the DZ, TZ, QZ,... classification refers to the number of contracted functions. A given basis set can thus be classified by:

- the highest angular momentum functions included;
- the number of lower angular momentum functions;
- the degree of contraction (i.e., how many primitive functions are used for each contracted function); and
- the basis set exponents.

Many different basis sets have been proposed over the years [34]. The early ones were typically determined from HF calculations, and especially the 6-31G basis set has been popular in routine applications. Dunning and co-workers have proposed a series of correlation-consistent basis set of double, triple, quadruple, etc. quality, with the acronym cc-pVXZ ($X =$ D, T, Q, 5, 6) [35]. These basis sets have been developed specifically for recovering electrons correlation, using correlation energies for determining the balance between the different types of basis functions and their exponents. These basis sets can be augmented by diffuse functions (aug-cc-pVXZ) for improving the quality of the results for properties depending on the wave function tail [36], and with additional tight functions (cc-pCVXZ) for recovering the core and core-valence correlation energy [37].

More recently, the concept of using energetic criteria for determining basis set composition have been used for designing basis sets tailored for DFT type calculations, denoted 'polarization consistent' in analogy with the correlation-consistent basis sets for electron correlation [38]. The level of polarization beyond the isolated atom is denoted with a number, i.e., a pc-1 type basis set is of DZ quality, a pc-2 type basis set is of TZ quality, etc.

In the limit of a complete basis set and recovering all the electron correlation, the Schrödinger equation is solved exactly. By performing a series of calculations with different methods and increasingly larger basis set, the full-CI-complete-basis-set limit can be estimated. Several groups are active in designing a computationally efficient way of estimating this 'exact' limit [39] and the accompanying review in this volume by Martin discusses this in more detail.

7. SUMMARY

While computer hardware continues to closely follow Moore's law (doubling the performance–price ratio every 18 months), the introduction of new algorithms over the years has given at least the same amount of improvements. For HF and DFT methods, the scaling with system size appears to have been solved, and these methods are well suited for running in parallel on inexpensive cluster-type computers. There is little doubt that systems containing up to thousands of atoms will be attempted in the near future. Unfortunately, there is at present no clear picture of how the current exchange-correlation functionals can be improved for achieving a better accuracy of DFT methods.

Semi-empirical methods have been resurrected after lying dormant for almost a decade, and are being parameterized for more elements, and attempts are being made at developing DFT analogues of semi-empirical methods. Already with the current technology, systems with 10,000 atoms are possible on single CPU machines. Semi-empirical methods currently hold the best promise for performing direct dynamics for systems with thousands of atoms, but the fundamental accuracy is still somewhat lower than desired.

The two major problems in wave function based electron correlation methods are the agonizing slow convergence with respect to basis set size and the high scaling of computer time with system size. Methods using the interelectronic distance as a variable hold promises for improving the basis set convergence, but have so far primarily been used for calibration purposes. The few attempts of designing methods with reduced scaling with system size have so far had very little influence, presumably because the break-even point in terms of computer time is well beyond the current feasibilities. The high scaling of these methods is at variance with the fundamental physical interaction being pair-wise, and achieving even a N^2 scaling would be a major breakthrough. Algorithmic improvements are also required before these methods can be used efficiently on massively parallel computers.

REFERENCES

[1] H. B. Schlegel, Exploring potential energy surfaces for chemical reactions: an overview of some practical methods, *J. Comput. Chem.*, 2003, **24**, 1514–1527.
[2] X. Li, J. M. Millam and H. B. Schlegel, Ab initio molecular dynamics studies of the photodissociation of formaldehyde, $H_2CO \rightarrow H_2 + CO$: direct classical trajectory calculations by MP2 and density functional theory, *J. Chem. Phys.*, 2000, **113**, 10062–10067.
[3] T. Helgaker, P. Jørgensen and J. Olsen, *Molecular Electronic-Structure Theory*, Wiley, Chichester, 2000.
[4] W. Koch and M. C. Holthausen, *A Chemist's Guide to Density Functional Theory*, Wiley-VCH, New York, 2000.
[5] J. Almloef, K. Faegri, Jr. and K. Korsell, Principles for a direct SCF approach to LCAO–MO ab initio calculations, *J. Comput. Chem.*, 1982, **3**, 385–399.
[6] M. C. Strain, G. E. Scuseria and M. J. Frisch, Achieving linear scaling for the electronic quantum Coulomb problem, *Science*, 1996, **271**, 51–53.
[7] H. Larsen, J. Olsen, P. Jorgensen and T. Helgaker, Direct optimization of the atomic-orbital density matrix using the conjugate-gradient method with a multilevel preconditioner, *J. Chem. Phys.*, 2001, **115**, 9685–9697; X. Li, J. M. Millam, G. E. Scuseria, M. J. Frisch and H. B. Schlegel, Density matrix search using direct inversion in the iterative subspace as a linear scaling alternative to diagonalization in electronic structure calculations, *J. Chem. Phys.*, 2003, **119**, 7651–7658.
[8] T. P. Hamilton and P. Pulay, Direct inversion in the iterative subspace (DIIS) optimization of open-shell, excited-state, and small multiconfiguration SCF wave functions, *J. Chem. Phys.*, 1986, **84**, 5728–5734.
[9] C. Kollmar, Convergence optimization of restricted open-shell self-consistent field calculations, *Int. J. Quant. Chem.*, 1997, **62**, 617–637.
[10] F. Jensen, *Introduction to Computational Chemistry*, Wiley, Chichester, 1999.
[11] Z. He and D. Cremer, Sixth-order many-body perturbation theory. III. Correlation energies of size-extensive MP6 methods, *Int. J. Quantum Chem.*, 1996, **59**, 57–69.
[12] J. Olsen, O. Christiansen, H. Koch and P. Jørgensen, Surprising cases of divergent behavior in Møller–Plesset perturbation theory, *J. Chem. Phys.*, 1996, **105**, 5082–5090.
[13] T. D. Crawford and H. F. Schaefer, An introduction to coupled cluster theory for computational chemists, *Rev. Comp. Chem.*, 2000, **14**, 33–136.
[14] H. Koch and A. Sanchez de Meras, Size-intensive decomposition of orbital energy denominators, *J. Chem. Phys.*, 2000, **113**, 508–513.
[15] G. E. Scuseria and P. Y. Ayala, Linear scaling coupled cluster and perturbation theories in the atomic orbital basis, *J. Chem. Phys.*, 1999, **111**, 8330–8343.
[16] M. Schuetz and F. R. Manby, Linear scaling local coupled cluster theory with density fitting. Part I: 4-external integrals, *Phys. Chem. Chem. Phys.*, 2003, **5**, 3349–3358.
[17] P. Hohenberg and W. Kohn, Inhomogeneous electron gas, *Phys. Rev.*, 1964, **136**, B864–B871.
[18] W. Kohn and L. J. Sham, Self-consistent equations including exchange and correlation effects, *Phys. Rev.*, 1965, **140**, A1133–A1138.
[19] E. Cances, K. N. Kudin, G. E. Scuseria and G. Turinici, Quadratically convergent algorithm for fractional occupation numbers in density functional theory, *J. Chem. Phys.*, 2003, **118**, 5364–5368.
[20] S. H. Vosko, L. Wilk and M. Nusair, Accurate spin-dependent electron liquid correlation energies for local spin density calculations: a critical analysis, *Can. J. Phys.*, 1980, **58**, 1200–1211.
[21] A. D. Becke, Density-functional exchange-energy approximation with correct asymptotic behavior, *Phys. Rev. A*, 1988, **38**, 3098–4000; J. P. Perdew, K. Burke

and M. Ernzerhof, Generalized gradient approximation made simple, *Phys. Rev. Lett.*, 1996, **77**, 3865–3868.
[22] A. D. Becke, Density-functional thermochemistry. III. The role of exact exchange, *J. Chem. Phys.*, 1993, **98**, 5648–5652.
[23] J. C. Slater, A simplification of the Hartree–Fock method, *Phys. Rev.*, 1951, **81**, 385–390.
[24] J. Tao, J. P. Perdew, V. N. Staroverov and G. E. Scuseria, Climbing the density functional ladder: nonempirical meta-generalized gradient approximation designed for molecules and solids, *Phys. Rev. Lett.*, 2003, **91**, 146401/1–146401/4.
[25] J. J. P. Stewart, Optimization of parameters for semiempirical methods. I. Method, *J. Comput. Chem.*, 1989, **10**, 209–220.
[26] D. R. Kanis, M. A. Ratner, T. J. Marks and M. C. Zerner, Nonlinear optical characteristics of novel inorganic chromophores using the Zindo formalism, *Chem. Mater.*, 1991, **3**, 19–22.
[27] A. D. Daniels and G. E. Scuseria, What is the best alternative to diagonalization of the Hamiltonian in large scale semiempirical calculations?, *J. Chem. Phys.*, 1999, **110**, 1321–1328.
[28] W. Thiel and A. A. Voityuk, Extension of MNDO to d orbitals: parameters and results for the second-row elements and for the zinc group, *J. Phys. Chem.*, 1996, **100**, 616–626.
[29] T. R. Cundari, J. Deng and W. Fu, PM3(tm) parameterization using genetic algorithms, *Int. J. Quant. Chem.*, 2000, **77**, 421–432.
[30] J. J. P. Stewart, Comparison of the accuracy of semiempirical and some DFT methods for predicting heats of formation, *J. Mol. Mod.*, 2004, **10**, 6–12.
[31] F. Jensen, The basis set convergence of the Hartree–Fock Energy for H_3^+, Li_2 and N_2, *Theor. Chem. Acc.*, 2000, **104**, 484–490.
[32] W. Kutzelnigg and J. D. Morgan, III, Rates of convergence of the partial-wave expansions of atomic correlation energies, *J. Chem. Phys.*, 1992, **96**, 4484–4508.
[33] J. Noga, P. Valiron and W. Klopper, The accuracy of atomization energies from explicitly correlated coupled-cluster calculations, *J. Chem. Phys.*, 2001, **115**, 2022–2032.
[34] E. R. Davidson and D. Feller, Basis set selection for molecular calculations, *Chem. Rev.*, 1986, **86**, 681–696.
[35] A. K. Wilson, T. van Mourik and T. H. Dunning, Jr., Gaussian basis sets for use in correlated molecular calculations VI. Sextuple zeta correlation consistent basis sets for boron through neon, *THEOCHEM*, 1996, **388**, 339–349.
[36] R. A. Kendall, T. H. Dunning, Jr. and R. J. Harrison, Electron affinities of the first-row atoms revisited. Systematic basis sets and wave functions, *J. Chem. Phys.*, 1992, **96**, 6796–6806.
[37] K. A. Peterson and T. H. Dunning, Jr., Accurate correlation consistent basis sets for molecular core-valence correlation effects: the second row atoms Al–Ar, and the first row atoms B–Ne revisited, *J. Chem. Phys.*, 2002, **117**, 10548–10560.
[38] F. Jensen, Polarization consistent basis sets. II. Estimating the Kohn–Sham basis set limit, *J. Chem. Phys.*, 2002, **116**, 7372–7379.
[39] L. A. Curtiss, P. C. Redfern, K. Raghavachari and J. A. Pople, Gaussian-3X (G3X) theory using coupled cluster and Brueckner energies, *Chem. Phys. Lett.*, 2002, **359**, 390–396; A. D. Boese, M. Oren, O. Atasoylu, J. M. L. Martin, M. Kallay and J. Gauss, W3 theory: robust computational thermochemistry in the kJ/mol accuracy range, *J. Chem. Phys.*, 2004, **120**, 4129–4141.

CHAPTER 2

Time-Dependent Density Functional Theory in Quantum Chemistry

Filipp Furche[1] and Kieron Burke[2]

[1] *Institut für Physikalische Chemie, Universität Karlsruhe, Kaiserstraße 12, 76128 Karlsruhe, Germany*
[2] *Department of Chemistry and Chemical Biology, Rutgers University, 610 Taylor Road, Piscataway, NJ 08854, USA*

Contents

1. Background 20
2. Electronic excitations 20
3. Computational aspects 21
4. Performance 22
   4.1. Vertical excitation and CD spectra 22
   4.2. Excited state structure and dynamics 24
5. Qualitative limitations of present functionals 25
   5.1. Inaccurate ground state KS potentials 25
   5.2. Adiabatic approximation 25
   5.3. Multiple excitations 25
   5.4. Extended systems 26
   5.5. Charge transfer problems 26
6. Promising developments 26
   6.1. Exact exchange 26
   6.2. Beyond the adiabatic approximation 27
   6.3. TD current DFT 27
7. Outlook 27
Acknowledgements 28
References 28

Time-dependent density functional theory (TDDFT) is increasingly popular for predicting excited state and response properties of molecules and clusters. We review the present state of the art, focusing on recent developments for excited states. We cover the formalism, computational and algorithmic aspects, and the limitations of present technology. We close with some promising developments. Extensive reviews on many aspects of TDDFT exist [1–5], and no pretence at comprehensive coverage is made here; instead, we rely heavily on our own work and that of our collaborators.

ANNUAL REPORTS IN COMPUTATIONAL CHEMISTRY, VOLUME 1
ISSN: 1574-1400 DOI 10.1016/S1574-1400(05)01002-9

1. BACKGROUND

Standard (i.e., ground state) density functional theory (DFT) is derived from traditional wavefunction-based quantum mechanics. The Hohenberg–Kohn (HK) theorem is a simple rewriting of the Rayleigh–Ritz variational principle [6]. Time-dependent DFT is based on a different theorem [7], which is a simple consequence of the *time-dependent* Schrödinger equation. For a given initial wavefunction and particle interaction, a time-dependent one-electron density $\rho(\mathbf{r}\,t)$ can be generated by at most one time-dependent external (i.e., one-body) potential. By starting in a non-degenerate ground state, the dependence on the initial wavefunction can be absorbed into the density dependence, by virtue of HK.

We define a set of time-dependent Kohn–Sham (TDKS) equations that reproduce $\rho(\mathbf{r}\,t)$, from a TDKS potential. This consists of the external potential, the Hartree potential, and the unknown time-dependent exchange correlation (XC) potential $v_{XC}[\rho](\mathbf{r}\,t)$. This is a much more sophisticated object than the ground state $v_{XC}[\rho](\mathbf{r})$, as it encapsulates all the quantum mechanics of all electronic systems subjected to all possible time-dependent perturbations.

2. ELECTRONIC EXCITATIONS

To extract electronic excitations, apply a weak electric field, and ask how the system responds, as in standard perturbation theory. We do not need the entire $v_{XC}[\rho](\mathbf{r}\,t)$, but only its value close to the ground state. This is captured in the XC kernel, $f_{XC}(\mathbf{r}\,\mathbf{r}';t-t') = \delta v_{XC}[\rho](\mathbf{r}\,t)/\delta\rho(\mathbf{r}'\,t')$. This is a *new* functional introduced by the time dependence. Its Fourier transform, $f_{XC}(\mathbf{r}\,\mathbf{r}';\omega)$, reduces to the ground state value as $\omega \to 0$. The standard *adiabatic* approximation ignores the frequency dependence and uses the second derivative of the ground state XC energy functional. Typical examples are the local density approximation (LDA), generalized gradient approximation (GGA), and hybrids, such as B3LYP.

Several practical routes have been adopted for extracting excitation energies from TDDFT response theory. In 1995, Casida converted the optical response problem into the solution of an eigenvalue problem (EVP) [2] whose indices are the single-particle transitions of the ground state Kohn–Sham potential:

$$\left[\begin{pmatrix} \mathbf{A} & \mathbf{B} \\ \mathbf{B} & \mathbf{A} \end{pmatrix} - \Omega_n \begin{pmatrix} \mathbf{1} & \mathbf{0} \\ \mathbf{0} & -\mathbf{1} \end{pmatrix}\right]\begin{pmatrix} X_n \\ Y_n \end{pmatrix} = 0$$

The form of this EVP is well known from time-dependent Hartree–Fock (TDHF) theory. $(\mathbf{A}+\mathbf{B})$ corresponds to the electric and $(\mathbf{A}-\mathbf{B})$ to the magnetic Hessian of the electronic ground state energy. The dominant contributions to the $\mathbf{A}$ matrix are the Kohn–Sham transition frequencies along the diagonal.

(Explicit expressions for **A** and **B** may be found, e.g., in Ref. [8].) The transition vectors $(X_n\ Y_n)$ correspond to collective eigenmodes of the TDKS density matrix with eigenfrequencies Ω_n. The Hartree and XC kernels produce both diagonal and off-diagonal contributions to **A** and **B**, correcting the transitions between occupied and unoccupied levels of the ground state KS potential into the true transitions of the system. If the different Kohn–Sham transitions do not couple strongly to one another, a useful approximation is to take only the diagonal elements of **A**. One can view the KS transition as being corrected by an integral over $f_{xc}(\mathbf{r}\ \mathbf{r}'; \omega)$ on the transition matrix elements, and the KS oscillator strengths will be good approximations to the true ones [9].

Alternatively, many physicists propagate the TDKS equations in real time, usually on a real-space grid inside a large sphere. They calculate the time-dependent dipole moment of their system, whose Fourier-transform yields the optical response.

3. COMPUTATIONAL ASPECTS

The response theory outlined above can be re-cast in variational form [8]. To this end, one defines a Lagrangian L which is stationary with respect to *all* its parameters at the excited state energies. L depends on the ground state KS molecular orbitals (MOs), on the excitation vector, and three Lagrange multipliers. This is convenient for excited state property calculations, because the Hellmann–Feynman theorem holds for L. The LCAO (linear combination of atomic orbitals)-MO expansion reduces the computation of excited state energies and properties to a finite-dimensional optimization problem for L, which can be handled algebraically. The stationarity conditions for L lead to the following problems which have to be solved subsequently in an excited state calculation.

(i) *Ground state KS equations in a finite basis*. Results are the ground state KS MOs and their eigenvalues. Computational strategies to solve this problem have been developed over decades, e.g., direct SCF (self-consistent field), RI (resolution of the identity), and linear scaling methods. Efficient excited state methods take advantage of this technology as much as possible.

(ii) *The finite-dimensional TDKS EVP* (*Casida's equations*) [10,11]. Results are the excitation energies and transition vectors. They are used (a) to compute transition moments and (b) to analyze the character of a transition in terms of occupied and virtual MOs. Complete solution of the TDKS EVP for all excited states leads to a prohibitive $O(N^6)$ scaling of CPU time and $O(N^5)$ I/O (N is the dimension of the one-particle basis). In most applications only the lowest states are of interest; iterative diagonalization methods such as the Davidson method are therefore the first choice [12–14]. In these iterative procedures, the time-determining step is a single matrix-vector operation per excited state and iteration, which can be cast into a form closely resembling a ground state Fock matrix construction [15]. In this way, a single-point excitation energy can be computed with similar effort as a single-point ground state energy. Block algorithms lead to additional savings if several states are computed at the same time [16]. Sometimes, the Tamm–Dancoff approximation is used [3], which amounts to constraining Y to zero in the variation of L.

(iii) *The 'Z-vector equation'*. Results are the TDKS 'relaxed' excited state density and energy-weighted density matrices. Excited state properties such as dipole moments and atomic populations can be computed from the excited state density matrix; analytical gradients of the excited state energy with respect to the nuclear positions require the energy-weighted density matrix as well. Using iterative methods similar to those above, the cost for computing the Z-vector is again in the range of the cost for a single-point ground state energy. Geometry optimizations for excited states are therefore not significantly more expensive than for ground states.

Flexible Gaussian basis sets developed for ground states are usually suited for excited state calculations. The smallest recommendable basis sets are of split valence quality and have polarization functions on all atoms except H, e.g., SV(P) or 6-31G*. Especially in larger systems, these basis sets can give useful accuracy, e.g., for simulating UV spectra (see below). However, excitation energies are typically overestimated by 0.2–0.5 eV, and individual oscillator strengths may be qualitatively correct only. A useful (but not sufficient) indicator of the quality is the deviation between the oscillator strengths computed in the length and the velocity gauge, which approaches zero in the basis set limit. Triple-zeta valence basis sets with two sets of polarization functions, e.g., cc-pVTZ or TZVPP, usually lead to basis set errors well below the functional error; larger basis sets are used to benchmark. Higher excitations and Rydberg states may require additional diffuse functions.

4. PERFORMANCE

4.1. Vertical excitation and CD spectra

So far, simulation and assignment of vertical electronic absorption spectra has been the main task of TDDFT calculations in chemistry. Most benchmark studies agree that low-lying valence excitations are predicted with errors of *ca.* 0.4 eV by LDA and GGA functionals [10,17,18]. Hybrid functionals can be more accurate, but display a less systematic error pattern. Traditional methods such as TDHF or configuration interaction singles (CIS) often produce errors of 1–2 eV at comparable or higher computational cost. Bearing in mind that UV-VIS spectra of larger molecules are mostly low-resolution spectra recorded in solution, and in view of the relatively low cost of a TDDFT calculation, errors in the range of 0.4 eV are acceptable for many purposes.

Calculated oscillator strengths may be severely in error for individual states, but the global shape of the calculated spectra is often accurate. Because semi-local functionals often predict the onset of the continuum to be 1–2 eV too low (due to the lack of derivative discontinuity), this is especially true for excitations in the continuum (excitation energy $>$ |HOMO energy|) [19]. Rotatory strengths which determine electronic circular dichroism (CD) spectra can be computed from magnetic transition moments in the density matrix based approach to TDDF response theory [20]. The simulated CD spectra predict the absolute configuration of chiral compounds in a simple and mostly reliable way. In particular, TDDFT also works well for inherently chiral chromophores [21] and transition metal compounds [22] where semi-empirical methods tend to fail.

Successful applications of TDDFT vertical excitation and CD spectra have been reported in various areas of chemistry, including metal clusters, fullerenes, aromatic compounds, porphyrins and corrins, and many other organic chromophores. As an example, we show the simulated and measured CD spectra of the chiral fullerene C_{76} in Fig. 1 [16]. We used the Becke-Perdew86 GGA together with the RI-J approximation and an SVP basis set augmented with diffuse s functions; a uniform blue shift of 0.4 eV was applied to all excitation energies to correct systematic errors of the calculation and solvent effects. The computed spectrum reproduces the main features of the experimental spectrum; even the intensities are in the right range. The absolute configuration of C_{76} can be

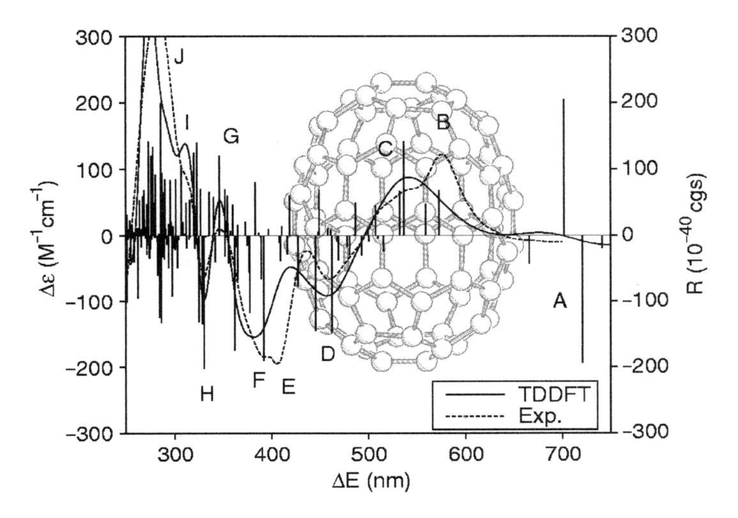

Fig. 1. The simulated CD spectrum of fullerene C_{76} compared to experiment.

determined in this way, because the measured spectrum can be assigned to one of the two enantiomers whose CD spectra differ by their sign only. The simulated spectrum involves 240 excited states; its calculation took 30 h on a single processor 1.2 GHz Athlon PC using TURBOMOLE V5-4 [23].

4.2. Excited state structure and dynamics

An adequate description of most photophysical and photochemical properties requires information on excited potential energy surfaces beyond vertical excitation energies. Early benchmark studies indicated at least qualitative agreement of excited potential surfaces calculated using TDDFT and correlated wavefunction methods [24,25]. An increasing number of excited state reaction path calculations using TDDFT have been reported. A limitation of most studies is that the reaction paths do not correspond to minimum energy paths (MEPs), i.e., the internal degrees of freedom other than the reaction coordinate are not optimized.

Analytical gradients of the excited state energy with respect to the nuclear positions are a basic prerequisite for systematic studies of excited state potential energy surfaces even in small systems. Implementations have become available only recently [2,26,27]. While errors in adiabatic excitation energies are similar to errors in vertical excitation energies, the calculated excited state structures, dipole moments, and vibrational frequencies are relatively accurate, with errors in the range of those observed in ground state calculations. The traditional CIS method, which has almost exclusively been used for excited state optimizations in larger systems, is comparable in cost, but significantly less accurate. Moreover, the KS reference is much less sensitive to stability problems than the HF reference, which is an important advantage especially if the ground and excited state structures differ strongly.

Individual excited states of larger molecules can be selectively investigated by pump-probe experiments. The resulting time-dependent absorption, fluorescence, IR, and resonance Raman spectra can be assigned by TDDFT excited state calculations. First applications show that calculated vibrational frequencies are accurate enough to determine the excited state structure by comparison with experiment [28]. The combination of TDDFT and transient spectroscopy methods offers a promising strategy for excited state structure elucidation in larger systems. Computed normal modes of excited states can be used to study the vibronic structure of UV spectra within the Franck–Condon and Herzberg–Teller approximation [29]. For a detailed understanding of photochemical reactions beyond MEPs, excited state nuclear dynamics simulations including non-adiabatic couplings are necessary. The first steps towards this ambitious goal have already been made [30,31].

5. QUALITATIVE LIMITATIONS OF PRESENT FUNCTIONALS

Next we discuss situations where today's approximations in TDDFT produce much larger errors, or entirely miss important aspects of the optical response.

5.1. Inaccurate ground state KS potentials

It had been well known for many years that the XC potentials of LDA and GGA are inaccurate. At large distances, they decay exponentially rather than as the correct $-1/r$. This can be a severe problem for TDDFT, since the orbital energies can be very sensitive to the details of the potential. This is not a problem if only low-lying valence excitations of large molecules are required, but the energy of low-lying diffuse states is often considerably underestimated, while higher Rydberg states are completely missing in the bound spectrum [32].

There now exist several schemes for imposing the correct asymptotic decay of the XC potential [33]. But such potentials are not the functional derivative of any XC energy. While this has no direct effect on vertical excitation energies, other excited state properties are not well defined. Exact exchange DFT methodology is developing rapidly (see Section 5.2), which does not suffer from this problem. Furthermore, when correctly interpreted, even the physicists' TDLDA calculations recover the correct oscillator strength despite these difficulties [19].

5.2. Adiabatic approximation

The frequency dependence of the XC kernel is ignored in most calculations. A simple approximation is to use the ω-dependent XC kernel of the uniform gas [34]. However, any collective motion of the electrons that does not deform the density, e.g., an overall boost, should not excite the electrons, but a frequency-dependent kernel violates this exact condition (whereas adiabatic approximations do not) [35].

5.3. Multiple excitations

In principle, the exact electronic response functions contain all levels of excitation. But Casida's equations span the space of KS single-particle excitations only, and this is unchanged by a frequency-independent XC kernel, i.e., within the adiabatic approximation.

5.4. Extended systems

Unlike ground state DFT, there are non-trivial complications when TDDFT is applied to bulk systems. These arise because the XC kernel has long-range contributions, comparable to the Hartree $1/|\mathbf{r} - \mathbf{r}'|$. However, our usual local and semi-local approximations yield XC kernels that are of the form $\delta^{(3)}(\mathbf{r} - \mathbf{r}')$, or derivatives thereof. Thus they have *little effect* on the calculated optical response of extended systems.

5.5. Charge transfer problems

Charge transfer (CT) excitations are notoriously predicted too low in energy, sometimes by more than 1 eV [36]. In chain-like systems such as polyenes, polyacenes, or other conjugated polymers, the error in CT excitation energies increases with the chain length [37,38]. In the limit of complete charge separation, this can be related to the lack of derivative discontinuities in semi-local functionals [36]. To correct CT excitation energies, methods have been suggested that estimate the derivative discontinuity from a ΔSCF calculation [36,39]. The validity of this approach depends on assumptions such as complete charge separations that may rarely be justified in real systems.

6. PROMISING DEVELOPMENTS

Here we discuss several promising paths to overcome present limitations.

6.1. Exact exchange

Many problems are related to spurious self-interaction, which affects energies and potentials computed with semi-local functionals. The self-interaction free exact exchange functional leads to a potential with the correct $-1/r$ tail, greatly improving the description of Rydberg states [40,41]. Moreover, the absence of self-interaction is a prerequisite for a correct derivative discontinuity, as has been demonstrated numerically. The use of exact exchange potentials improves the description of optical properties of conjugated polymers. Unfortunately, exchange alone is not enough. So far, calculations employing the full frequency-dependent exchange kernel have been reported for solids only [42]. Excitation energies of valence states obtained with exchange-only potentials plus ALDA kernel are not systematically better than those from GGA calculations. Moreover, the neglect of correlation effects generally leads to an overestimation of the energy of ionic states, as is well known, e.g., from TDHF. Adding an LDA or GGA correlation potential to the x-only potential leads to marginal improvements only, because the error compensation between approximate exchange and correlation is lost. In practice, one often resorts to hybrid functionals, which contain a (relatively small) fraction of exact exchange only. Thus, moderately diffuse states and certain CT excitations can still be handled [43]. A more fundamental solution may require correlation functionals compatible with exact exchange.

6.2. Beyond the adiabatic approximation

Higher-order excitations are accounted for by dramatic frequency dependence in f_{xc}, and building it into the kernel allows one to recover, e.g., a double excitation close to a single. In fact, the usual adiabatic approximation simply combines both into one peak, which will be a good approximation to the total oscillator strength [44,45].

Over the last year, it has been shown that incorporation of the essential terms of the polarizability from the Bethe–Salpeter equation (i.e., an orbital-dependent functional) recovers excellent excitonic peak shifts in semiconductors [46,47]. Chemists with long molecules should be aware of this, as the standard methodology misses these effects.

6.3. TD current DFT

The Runge–Gross theorem in fact establishes that the potential is a functional of the current density, $\mathbf{j}(\mathbf{r})$. This approach allowed Vignale and Kohn [48] to construct a gradient expansion in $\mathbf{j}(\mathbf{r})$ that goes beyond the adiabatic approximation without violating exact conditions for boosts. This formulation leads naturally to ultra non-local functionals that can shift exciton peaks and correct polarizability problems for polymers [49] and solids [50], but no accurate universally applicable approximation is yet available [51,52].

7. OUTLOOK

TDDFT in its present incarnation works remarkably well for many systems and properties. The number of papers is growing exponentially. While most are focused on extracting electronic transitions, there are many other promising applications. For example, atoms and molecules in intense laser fields can be handled with this formalism. Recently, it has been shown that scattering cross-sections can also be extracted [53].

This is a golden age of TDDFT in quantum chemistry, in which we are right now discovering which systems and properties can be handled routinely, where our favorite approximations fail, and how to fix these failures. We anticipate several more exciting years.

ACKNOWLEDGEMENTS

This work was supported by the Center for Functional Nanostructures (CFN) of the Deutsche Forschungsgemeinschaft (DFG) within project C2.1, and by NSF under grant number CHE-9875091 and DOE under grant number DE-FG02-01ER45928.

REFERENCES

[1] M. A. L. Marques and E. K. U. Gross, Time-dependent density functional theory, *Annu. Rev. Phys. Chem.*, 2004, **55**, 427–455.
[2] R. Van Leeuwen, Key concepts in time-dependent density functional theory, *Int. J. Mod. Phys. B*, 2001, **15**, 1969–2023.
[3] G. Onida, L. Reining and A. Rubio, Electronic excitations: density-functional versus many-body Green's-function approaches, *Rev. Mod. Phys.*, 2002, **74**, 601–659.
[4] N. T. Maitra, A. Wasserman and K. Burke, What is time-dependent density functional theory? Successes and challenges. In *Electron Correlations and Materials Properties 2* (eds A. Gonis, N. Kioussis and M. Ciftan), Kluwer/Plenum, New York, 2003, pp. 285–298.
[5] F. Furche and D. Rappoport, Density functional methods for excited states: equilibrium structure and electronic spectra. In *Computational Photochemistry* (ed. M. Olivucci), Elsevier, Amsterdam, 2005, in press.
[6] P. Hohenberg and W. Kohn, Inhomogeneous electron gas, *Phys. Rev.*, 1964, **136**, B864–B871.
[7] E. Runge and E. K. U. Gross, Density-functional theory for time-dependent systems, *Phys. Rev. Lett.*, 1984, **52**, 997–1000.
[8] F. Furche and R. Ahlrichs, Adiabatic time-dependent density functional methods for excited state properties, *J. Chem. Phys*, 2002, **117**, 7433–7447.
[9] H. Appel, E. K. U. Gross and K. Burke, Excitations in time-dependent density-functional theory, *Phys. Rev. Lett.*, 2003, **90**, 043005.
[10] M. E. Casida, Time-dependent density functional response theory for molecules. In *Recent Advances in Density Functional Methods* (ed. D. P. Chong), World Scientific, Singapore, 1995, pp. 155–193.
[11] R. Bauernschmitt and R. Ahlrichs, Treatment of electronic excitations within the adiabatic approximation of time dependent density functional theory, *Chem. Phys. Lett.*, 1996, **256**, 454–464.
[12] J. Olsen, H. J. A. Jensen and P. Jørgensen, Solution of the large matrix equations which occur in response theory, *J. Comput. Phys.*, 1988, **74**, 265–282.
[13] R. E. Stratmann, G. E. Scuseria and M. J. Frisch, An efficient implementation of time-dependent density-functional theory for the calculation of excitation energies of large molecules, *J. Chem. Phys.*, 1998, **109**, 8218–8224.
[14] V. Chernyak, M. F. Schulz, S. Mukamel, S. Tretiak and E. V. Tsiper, Krylov-space algorithms for time-dependent Hartree–Fock and density functional computations, *J. Chem. Phys.*, 2000, **113**, 36–43.

[15] H. Weiss, R. Ahlrichs and M. Häser, A direct algorithm for self-consistent-field linear response theory and application to C60: excitation energies, oscillator strengths, and frequency-dependent polarizabilities, *J. Chem. Phys.*, 1993, **99**, 1262–1270.
[16] F. Furche, Dissertation, Universität Karlsruhe, 2002.
[17] S. Hirata and M. Head-Gordon, Time-dependent density functional theory within the Tamm–Dancoff approximation, *Chem. Phys. Lett.*, 1999, **314**, 291–299.
[18] M. Parac and S. Grimme, Comparison of multireference Moller–Plesset theory and time-dependent methods for the calculation of vertical excitation energies of molecules, *J. Chem. Phys. A*, 2002, **101**, 6844–6850.
[19] A. Wasserman, N. T. Maitra and K. Burke, Accurate Rydberg excitations from the local density approximation, *Phys. Rev. Lett.*, 2003, **91**, 263001.
[20] F. Furche, On the density matrix based approach to time-dependent density functional theory, *J. Chem. Phys.*, 2001, **114**, 5982–5992.
[21] F. Furche, R. Ahlrichs, C. Wachsmann, E. Weber, A. Sobanski, F. Vögtle and S. Grimme, Circular dichroism of helicenes investigated by time-dependent density functional theory, *J. Am. Chem. Soc.*, 2000, **122**, 1717–1724.
[22] J. Autschbach, F. E. Jorge and T. Ziegler, Density functional calculations on electronic circular dichroism spectra of chiral transition metal complexes, *Inorg. Chem.*, 2003, **42**, 2867–2877.
[23] R. Ahlrichs, M. Bär, M. Häser, H. Horn and C. Kölmel, Electronic structure calculations on workstation computers: the program system Turbomole, *Chem. Phys. Lett.*, 1989, **162**, 165–169, See also: http://www.turbomole.com.
[24] M. E. Casida, K. C. Casida and D. R. Salahub, Excited-state potential energy curves from time-dependent density-functional theory: a cross section of formaldehyde's $^1A_1$ manifold, *Int. J. Quantum Chem.*, 1998, **70**, 933–941.
[25] A. L. Sobolewski and W. Domcke, Ab initio potential-energy functions for excited state intramolecular proton transfer: a comparative study of *o*-hydroxybenzaldehyde, salicylic acid and 7-hydroxy-1-indanone, *Phys. Chem. Chem. Phys.*, 1999, **1**, 3065–3072.
[26] C. Van Caillie and R. D. Amos, Geometric derivatives of density functional theory excitation energies using gradient-corrected functionals, *Chem. Phys. Lett.*, 2000, **317**, 159–164.
[27] J. Hutter, Excited state nuclear forces from the Tamm–Dancoff approximation to time-dependent density functional theory within the plane wave basis set framework, *J. Chem. Phys.*, 2003, **118**, 3928–3934.
[28] D. Rappoport and F. Furche, Photoinduced intramolecular charge transfer in 4-(dimethyl)aminobenzonitrile – a theoretical perspective, *J. Am. Chem. Soc.*, 2004, **126**, 1277–1284.
[29] M. Dierksen and S. Grimme, Density functional calculations of the vibronic structure of electronic absorption spectra, *J. Chem. Phys.*, 2004, **120**, 3544–3554.
[30] V. Chernyak and S. Mukamel, Density-matrix representation of nonadiabatic couplings in time-dependent density functional (TDDFT) theories, *J. Chem. Phys.*, 2000, **112**, 3572–3579.
[31] U. F. Rohrig, I. Frank, J. Hutter, A. Laio, J. VandeVondele and U. Röthlisberger, QM/MM Car-Parrinello molecular dynamics study of the solvent effects on the ground state and on the first excited singlet state of acetone in water, *Chem. Phys. Chem.*, 2003, **4**, 1177–1182.
[32] M. E. Casida, C. Jamorski, K. C. Casida and D. R. Salahub, Molecular excitation energies to high-lying bound states from time-dependent density-functional response theory: characterization and correction of the time-dependent local density approximation ionization threshold, *J. Chem. Phys.*, 1998, **108**, 4439–4449.
[33] M. E. Casida and D. R. Salahub, Asymptotic correction approach to improving approximate exchange-correlation potentials: time-dependent density-functional

theory calculations of molecular excitation spectra, *J. Chem. Phys.*, 2000, **113**, 8918–8935.
[34] E. K. U. Gross and W. Kohn, Local density-functional theory of frequency-dependent linear response, *Phys. Rev. Lett.*, 1985, **55**, 2850–2852; 1986, **57**, 923 (E).
[35] J. F. Dobson, Harmonic potential theorem, *Phys. Rev. Lett.*, 1994, **73**, 2244–2247.
[36] A. Dreuw, J. L. Weisman and M. Head-Gordon, Long-range charge-transfer excited states in time-dependent density functional theory require non-local exchange, *J. Chem. Phys.*, 2003, **119**, 2943–2946.
[37] A. Pogantsch, G. Heimel and E. Zojer, Quantitative prediction of optical excitations in conjugated organic oligomers: a density functional theory study, *J. Chem. Phys.*, 2002, **117**, 5921–5928.
[38] M. Parac and S. Grimme, Substantial errors from time-dependent density functional theory for the calculation of excited states of large π systems, *Chem. Phys. Chem.*, 2003, **4**, 292–295.
[39] D. J. Tozer, Relationship between long-range charge-transfer excitation energy error and integer discontinuity in Kohn–Sham theory, *J. Chem. Phys.*, 2003, **119**, 12697–12699.
[40] S. Hirata, S. Ivanov, I. Grabowski and R. J. Bartlett, Time-dependent density functional theory employing optimized effective potentials, *J. Chem. Phys.*, 2002, **116**, 6468–6481.
[41] F. Della Sala and A. Görling, Excitation energies of molecules by time-dependent density functional theory based on effective exact exchange Kohn–Sham potentials, *Int. J. Quantum Chem.*, 2003, **91**, 131–138.
[42] Y.-H. Kim and A. Görling, Excitonic optical spectrum of semiconductors obtained by time-dependent density functional theory with the exact-exchange kernel, *Phys. Rev. Lett.*, 2002, **89**, 096402.
[43] C. Adamo, G. E. Scuseria and V. Barone, Accurate excitation energies from time-dependent density functional theory: assessing the PBE0 model, *J. Chem. Phys.*, 1999, **111**, 2889–2899.
[44] N. T. Maitra, F. Zhang, R. J. Cave and K. Burke, Double excitations in time-dependent density functional theory linear response, *J. Chem. Phys.*, 2004, **120**, 5932–5937.
[45] R. J. Cave, F. Zhang, N. T. Maitra and K. Burke, A dressed TDDFT treatment of the 2^1A_g states of butadiene and hexatriene, *Chem. Phys. Lett.*, 2004, **389**, 39–42.
[46] A. Marini, R. Del Sole and A. Rubio, Bound excitons in time-dependent density-functional theory: optical and energy-loss spectra, *Phys. Rev. Lett.*, 2003, **91**, 256402.
[47] F. Sottile, V. Olevano and L. Reining, Parameter-free calculation of response functions in time-dependent density-functional theory, *Phys. Rev. Lett.*, 2003, **91**, 056402.
[48] G. Vignale and W. Kohn, Current-dependent exchange-correlation potential for dynamical linear response theory, *Phys. Rev. Lett.*, 1996, **77**, 2037–2040.
[49] M. van Faassen, P. L. de Boeij, R. van Leeuwen, J. A. Berger and J. G. Snijders, Ultranonlocality in time-dependent current-density-functional theory: application to conjugated polymers, *Phys. Rev. Lett.*, 2002, **88**, 186401.
[50] P. L. de Boeij, F. Kootstra, J. A. Berger, R. van Leeuwen and J. G. Snijders, Current density functional theory for optical spectra: a polarization functional, *J. Chem. Phys.*, 2001, **115**, 1995–1999.
[51] C. A. Ullrich and K. Burke, Excitation energies from time-dependent density-functional theory beyond the adiabatic approximation, *J. Chem. Phys.*, 2004, **121**, 28–35.
[52] M. van Faassen and P. L. de Boeij, Excitation energies for a benchmark set of molecules obtained within time-dependent current-density functional theory using the Vignale–Kohn functional, *J. Chem. Phys.*, 2004, **120**, 8353–8363.
[53] A. Wasserman, N. T. Maitra and K. Burke, Scattering amplitudes from time-dependent density functional theory, *J. Chem. Phys.*, 2004, submitted for publication.

CHAPTER 3

Computational Thermochemistry: A Brief Overview of Quantum Mechanical Approaches

Jan M.L. Martin

Department of Organic Chemistry, Weizmann Institute of Science, 76100 Rehovot, Israel

Contents

1.	Introduction	31
2.	Semiempirical methods	31
3.	Density functional methods	32
4.	*Ab initio* thermochemistry: preliminaries	33
5.	Use of isodesmic and isogyric reactions	34
6.	Empirically corrected methods: G1, G2, G3 theory	34
7.	Hybrid extrapolation/correction methods: CBS-*n*	36
8.	Nonempirical extrapolation approaches: W*n* theory	37
9.	Explicitly correlated methods	39
10.	Conclusions	39
	References	40

1. INTRODUCTION

The field of computational thermochemistry – the computational prediction of thermochemical properties – has come of age in the last decade and a half. A very wide variety of tools has become available – and is often leaving the novice confused. The present contribution seeks to offer a compact overview of the available quantum mechanics-based approaches to the problem.

Within the limited scope of this review, it is plainly impossible to offer a detailed explanation of the underlying electronic structure methods – for that, the reader is referred to other contributions in this journal or to a specialized text such as, e.g., the textbooks by Cramer [1] or Jensen [2]. Nor will it be possible to cover every detail of every computational thermochemistry method, but more detailed discussions can be found in two recent review volumes [3,4].

2. SEMIEMPIRICAL METHODS

The performance of semiempirical MO methods for thermochemistry has fairly recently been reviewed by Thiel [5]. Since such methods (especially on massively

ANNUAL REPORTS IN COMPUTATIONAL CHEMISTRY, VOLUME 1
ISSN: 1574-1400 DOI 10.1016/S1574-1400(05)01003-0

parallel machines) can handle systems from hundreds to thousands of atoms, they offer an alternative for cases beyond the reach of all other methods (*pace* molecular mechanics). Semiempirical methods rely on very heavy parametrization (on the order of a dozen empirically adjusted parameters *per chemical element*) of otherwise quite unflexible theoretical models (valence-only minimal basis set Hartree–Fock with various neglect-of-overlap approximations and empirical integral formulas). As a result, they end up yielding respectable performance for common molecules similar to those used in parametrization, and wildly erratic results for chemically outlandish systems. This translates into a very wide error distribution.

3. DENSITY FUNCTIONAL METHODS

While density functional methods can still be applied to rather large systems, the picture for thermochemistry is considerably brighter than with semiempirical methods.

Basis sets in DFT can be as flexible as required, and some small-molecule DFT programs eliminate basis set incompleteness error entirely by working in a basis of numerical orbitals [6]. This leaves the form of the exchange-correlation functional as the main fount of uncertainty and/or empiricism in DFT calculations.

Arguably, the greatest milestone for DFT thermochemistry was the development of the very popular B3LYP functional [7]. This functional is a linear combination of the local density approximation, gradient corrections to the same, and 'exact' Hartree–Fock exchange: the three mixing coefficients were parametrized against the total atomization energies of a set of 55 small molecules. Some next-generation functionals [8,9] indeed contain as many as 10–16 empirical parameters.

While this is a lesser order of empiricism than semiempirical methods that may involve as many parameters for a single *element*, it exposes these functionals to the criticism of being merely very sophisticated 'chemical interpolation methods'. Although parametrization bias cannot be eliminated rigorously, the problem can be greatly mitigated by using ever larger and more diverse parametrization sets, by additionally involving properties other than energetics, and by generating additional parametrization data, not (reliably) available from experiment, by rigorous *ab initio* calculations. The approach of the Handy group [9] combines all of the above, and the HCTH/407 functional of Boese and Handy [10] may well be the most reliable GGA (generalized gradient approximation, i.e., involving only the density and its gradient) functional for thermochemistry available. Best performance overall appears to be achieved [11] by the hybrid GGA functional B97-1, a reparametrization [9] of Becke's 1997 functional [8]. Still marginally better results may be achieved by the τ-HCTH hybrid functional [12], which includes a term involving the kinetic energy density. Its 'pure DFT', counterpart, the τ-HCTH meta-GGA functional, somewhat outperforms HCTH/407, as may another (less parametric) functional, TPSS [13].

Yet it should be noted that even in the case of B97-1, a mean absolute error of 5.6 kcal/mol is obtained for binding energies of 215 neutral molecules [12].

Although basis set convergence in DFT is fairly rapid, its importance has often been underestimated. Boese *et al.* [11] recently investigated the role of the basis set in some detail and found that diffuse-function augmented spdf basis sets generally are close enough to the DFT limit. The recently developed 'polarization consistent' basis sets of Jensen [14] allow essentially reaching the Kohn–Sham limit for first-row systems.

One important deficiency of most contemporary functionals is in the area of transition state structures and reaction barrier heights. Barriers are commonly seriously underestimated and sometimes reactions are erroneously predicted to be barrierless [15,16]. It was noted repeatedly [16–18] that DFT reaction barriers can be greatly improved by increasing the percentage of Hartree–Fock exchange, from the 15–25% region optimal for most equilibrium properties, to the 40–50% region; however, this improvement goes at the expense of all other properties [19]. Very recently, Boese and Martin [19] found that, if terms dependent on the kinetic energy density are additionally admitted, the dependence of performance for equilibrium properties on the percentage of Hartree–Fock exchange is strongly reduced, and thus they were able to develop a new functional (termed BMK) that combines B3LYP-like or better performance for equilibrium properties with accurate barrier heights.

4. *AB INITIO* THERMOCHEMISTRY: PRELIMINARIES

By and large, *ab initio* thermochemistry can be seen as a two-dimensional convergence problem, with one dimension being the one-particle basis set and the other the *n*-particle correlation treatment. (Additional dimensions result from the relativistic treatment and Born–Oppenheimer corrections.) The greatest strength of *ab initio* thermochemistry is that the computational problem is rigorously defined, and that a clear pathway to improving one's predictions exists (even though it may be prohibitive in terms of computing resources needed).

Its greatest weakness is the exceedingly slow convergence of the correlation energy with the one-particle basis set, which in practice limits rigorous *ab initio* methods to quite small systems. A number of approximate *ab initio* thermochemistry schemes have been proposed over the years: we will discuss them in order of increasing rigor.

5. USE OF ISODESMIC AND ISOGYRIC REACTIONS

One way to accelerate convergence of computed thermochemical properties with the level of theory is to not calculate them directly at all, but to obtain them indirectly from (experimental or benchmark *ab initio*) data for some reference compounds and the reaction energy for some chemical transformation involving only the reference compounds and the compound of interest. The greater the chemical similarity between the reactant and the product side, the more rapidly the computed reaction energy will converge with the level of theory.

Reactions in which the number of unpaired electrons is conserved are said to be *isogyric* [20]. Reactions in which the number of formal bonds of each type is also conserved are said to be *isodesmic* [21]. If in addition, the number of carbon atoms in each hybridization state is conserved, the reaction is said to be *homodesmotic* [22,23].

Isodesmic reaction energies converge quite fast with the level of theory (particularly since the correlation energies of both sides of the equation are quite similar), and homodesmotic ones even more so. But clearly, unless accurate data are available for all species in the reaction except one, neither isodesmic nor homodesmotic reactions will be very helpful, except perhaps as interpretative tools.

6. EMPIRICALLY CORRECTED METHODS: G1, G2, G3 THEORY

Compare, e.g., the basis set of the direct dissociation reaction

$$N_2 \rightarrow 2N$$

with that of the isogyric reaction

$$N_2 + 6H \rightarrow 2N + 3H_2$$

The former reaction has a particularly slow basis set convergence behavior. Almlöf's pioneering 1989 paper [24] showed that even *i* functions contribute 0.5 kcal/mol to the dissociation energy, while expanding the basis set from 5s4p3d2f1g to 6s5p4d3f2g1h adds no less than 2.3 kcal/mol.

By contrast, the energy of the latter reaction is reasonably close to convergence even at the QCISD(T)/6-311+G(2df,p) level of theory. In effect, one is using the difference between the known binding energy of H_2 and the error at that level of theory as a correction term.

This was one of two elements of the rationale behind the 'G1 theory' of Pople and co-workers [25]. It gets expressed in a 'high-level correction' term:

$$\Delta E(\text{HLC}) = -0.19n_\alpha - 5.95n_\beta \qquad \text{(millihartree)}$$

where n_α and n_β represent the numbers of spin-up and spin-down electrons, respectively. The other element was an additivity approximation:

$$\begin{aligned}E[\text{QCISD(T)/6-311+G(2df, p)}] &\approx E[\text{QCISD(T)/6-311G(d, p)}] \\ &+ E[\text{MP4/6-311+G(d, p)}] - E[\text{MP4/6-311G(d, p)}] \\ &+ E[\text{MP4/6-311G(2df, p)}] - E[\text{MP4/6-311G(d, p)}]\end{aligned}$$

There are two underlying 'weakness of coupling' assumptions here: between the effects of adding diffuse functions and additional polarization functions on one hand, and between expansion of the basis set and improvement in the electron correlation treatment on the other. (Evidence exists [26] that, in practice, this latter assumption holds less well than usually assumed.)

Reference geometries are obtained at the MP2/6-31G* level; zero-point vibrational energies and thermal corrections are obtained from HF/6-31G* vibrational frequencies scaled by 0.8929.

Curtiss *et al.* introduced some refinements, leading to the G2 theory [27] which is still fairly commonly used. Here (a) an additional basis set additivity step is considered at the MP2 level, namely

$$\begin{aligned}&E[\text{MP2/6-311+G(3df, 2p)}] - E[\text{MP2/6-311+G(d, p)}] \\ &\quad - E[\text{MP2/6-311G(2df, p)}] + E[\text{MP2/6-311G(d, p)}]\end{aligned}$$

and (b) the two coefficients in the 'high-level correction' are turned into empirical parameters. For the original G2 test set [27], 55 atomization energies, 25 electron affinities, 38 ionization potentials, and 7 proton affinities, this approach yields a mean absolute error of 1.21 kcal/mol – close to the target of 'chemical accuracy' and well beyond anything possible with present-day density functional methods. Note that subvalence correlation is not considered at all and neither is atomic spin–orbit splitting, although a paper [28] on the extension of G2 theory to third-row main group molecules recommends spin–orbit splitting be considered there. (For molecules in nondegenerate ground states, this is easily done as a sum of atomic corrections.)

In G3 theory [29], some further improvements are introduced: (a) the 6-311+G(3df,2p) basis set is replaced by a still more extended 'G3large' set; (b) inner-shell correlation is accounted for at the MP2/G3large level; (c) different 'high-level corrections' are used for atoms and molecules; and (d) atomic spin–orbit splitting is taken into account. A minor variant called G3X [30] currently represents the most sophisticated G*n* method: here, more reliable B3LYP/6-31G(2df,p) reference geometries and frequencies (the latter scaled by 0.9854) are employed, and for second-row molecules, an additional SCF step is done with the G3large basis set with an added g function.

As pointed out by Truhlar [31], multiplicative corrections lend themselves much better to being applied to all points on a potential surface than additive correction. In this spirit, and following earlier work by Truhlar and co-workers [31], the G3S [32] and later the G3SX [30] methods were proposed.

For the more extended G2-97 test set [33], the mean absolute error drops from 1.48 kcal/mol for G2 theory to 1.01 kcal/mol for G3 theory: the RMS error drops from 1.93 to 1.45 kcal/mol [29]. Mean absolute (RMS) errors for the even larger G3/99 test set [29] range from 1.07(1.54) kcal/mol for G3 and 1.08(1.55) kcal/mol for G3S to 1.02(1.35) kcal/mol for G3X and 0.95(1.38) kcal/mol for G3SX [30].

Several extensions were considered that are somewhat more robust for species with significant nondynamical correlation [34] and for radicals [35]. Also, reduced-cost approximations like G3(MP2), G3X(MP2), and G3SX(MP3) have been proposed [29,30], in which the MP4 and QCISD(T) steps were eliminated. This entails some sacrifices in accuracy, but the mean absolute (RMS) error of 1.04(1.54) kcal/mol of G3SX(MP3) surprisingly approaches that of standard G3 theory [30].

Martin [36] recognized that σ and π bonds tend to have qualitatively different convergence behaviors, and proposed a three-term correction

$$\mathrm{HLC} = n_\sigma A_\sigma + n_\pi A_\pi + n_{\mathrm{LP}} A_{\mathrm{LP}}$$

where n_σ, n_π, and n_{LP} represent the number of sigma bonds, pi bonds, and lone pairs in the system, respectively. Mean absolute errors for small molecule samples can be brought down into the 0.5 kcal/mol range for spdfg basis sets, but the approach offers no practical advantage over nonempirical extrapolation-based methods like W1 theory (see below) and has been abandoned.

7. HYBRID EXTRAPOLATION/CORRECTION METHODS: CBS-*n*

Schwartz [37] considered the basis set convergence of the correlation energy for a helium-like atom, at second order in many-body perturbation theory (MBPT-2 or MP2). In a singlet state, $E(L) = E_\infty + A/L^3 + \mathrm{O}(L^{-4})$, where L represents the highest angular momentum present in the basis set. For a triplet state, $E(L) = E_\infty + A/L^5 + \mathrm{O}(L^{-6})$. Hill [38] and Kutzelnigg and Morgan [39] extended this treatment to arbitrary electron pairs in a spherically symmetric system.

The CBS-*n* methods of Petersson and co-workers rely on a combination of (a) additivity approximations for higher order correlation effects; (b) Schwartz-type extrapolations for individual MP2 pair correlation energies; and (c) empirical corrections (which are numerically much less important here). A hierarchy of methods has evolved over the years, with the CBS-4M method [40] as an inexpensive low-end option, CBS-QCI/APNO (presently only applicable to first-row systems) at the top [41], and CBS-QB3 [42] as the method of choice in most cases.

Over a somewhat modified G2/97 test set [40], CBS-QB3 and G3 theory have roughly comparable mean absolute errors (1.10 and 0.94 kcal/mol, respectively) and RMS errors (1.45 and 1.33 kcal/mol, respectively). Our experience [43] suggests that CBS-QB3 is somewhat less prone to outliers than ordinary G3 theory (not necessarily G3X theory).

8. NONEMPIRICAL EXTRAPOLATION APPROACHES: W*n* THEORY

Our primary goal in designing the Weizmann-*n* (W*n*) approaches [43–45] was not to have 'chemical accuracy' (± 1 kcal/mol) on average, but to achieve ± 1 kcal/mol worst-case and ± 1 kJ/mol typically. It is impossible to reach this goal without a substantial markup in cost compared to G*n* and CBS-*n* methods.

Secondary design goals were total absence of parameters derived from experiment, black-box character similar to G*n* and CBS-*n* (i.e., operator decisions only being required in exceptional cases), and sufficient cost-effectiveness for application to benzene-sized molecules on commodity workstations.

For first- and second-row compounds, the following components can be expected to contribute at the 1 kJ/mol or greater level to ground-state binding energies: (1) Hartree–Fock; (2) valence CCSD correlation energy; (3) valence (T) correlation energy; (4) inner-shell correlation energy; (5) scalar relativistic effects; (6) first-order spin–orbit coupling; and (7) anharmonicity in the zero-point vibrational energy. A fairly small set of 28 molecules (the W2-1 set) was selected for which (a) atomization energies were known experimentally to great accuracy (better than 1 kJ/mol, better than 0.1 kcal/mol in most cases); (b) anharmonic zero-point corrections were available from experiment or large-scale *ab initio* calculations; and (c) severe nondynamical correlation effects are absent.

All of the above components were then subjected to exhaustive benchmark calculations, and in each component, approximations were then introduced in a controlled fashion until the point of unacceptable deterioration was found.

The valence steps in W2 theory [43] are based on CCSD(T) calculations with spdf ($L = 3$) and spdfg ($L = 4$) basis sets, and a CCSD calculation with an spdfgh ($L = 5$) basis set.

The SCF component was originally extrapolated geometrically [46] from all three SCF energies, but this technique has the annoying property that the result becomes dependent on whether one extrapolates on the energy difference or on the component total energies. Inspired by the SCF extrapolation in the CBS methods, we, therefore, later adopted [47] a simple $A + B/L^5$ extrapolation from the $L = 4$ and $L = 5$ energies. We also note that the basis sets employed have been enhanced with diffuse functions for C through Ne and P through Ar, and that high-exponent d and f functions have been added to second-row elements to accommodate inner polarization on these elements [48].

The CCSD correlation energy is extrapolated from the $L = 4$ and $L = 5$ energies using a simple $A + B/L^3$ formula; the (*T*) component converges faster, and can be extrapolated from the $L = 3$ and $L = 4$ results using the same formula (thus obviating the need for a CCSD(T)/spdfgh calculation).

Extensive investigation into the nature of inner-shell correlation revealed [43] that (a) the contribution of connected triple excitations may reach up to 50% of the differential contribution to binding energies; (b) at least an spdf basis set with high-exponent d and f functions is required. The MP2/G3large approach used in G3 theory yields reasonable results in the first row because of an error compensation between basis set superposition error (leading to an overestimate) and neglect of higher order correlation effects; in the second row, BSSE becomes too large and poor results are obtained [49].

For first- and second-row molecules, it turned out that expectation values of the ACPF wave function with an spdf basis set were quite adequate. However, Douglas–Kroll [50] CCSD(T) calculations in a similar basis set will be even more reliable, and are also applicable to heavier elements. (They do presuppose the availability of a Douglas–Kroll code.)

For systems in degenerate ground states, spin–orbit coupling constants were computed at the CISD level with an spdf basis set. (It was found important to include 2s, 2p correlation in second-row elements for these calculations.)

The totally nonempirical method thus obtained achieved a mean absolute error of 0.23 kcal/mol over the W2-1 set and a maximum error (for O_2) of 0.64 kcal/mol.

A more economical 'W1 theory' was proposed for applications on larger systems. Here all basis sets were reduced by one angular momentum, i.e., the valence contributions are now based on CCSD(T)/spd, CCSD(T)/spdf, and CCSD/spdfg calculations. The use of the $A + B/L^3$ formula led to systematic overshooting here: in a reluctant compromise, we decided to make the exponent an adjustable parameter, but adjusted it to W2 results rather than experiment.

W1 and W2 theory have been validated in some detail in Ref. [47] and a number of cost-effective approximations proposed in Ref. [44].

The main limitation of W2 theory (other than, inevitably, cost) is applicability to systems with severe nondynamical correlation. A general coupled cluster code developed by Kállay *et al.* [51] enabled us to investigate [45] correlation effects beyond CCSD(T). Our conclusions were: (a) higher order connected triple excitations systematically reduce the binding energy (generally away from experiment); (b) connected quadruple excitations systematically increase binding energies, and the excellent performance of CCSD(T) for many systems is the result of error compensation; (c) the connected quadruples contribution converges very rapidly with the basis set, such that even an spd basis set can give useful answers; (d) connected quintuple and higher excitations are unimportant for all except the most pathological systems (like C_2 or BN in their respective lowest singlet states). We thus proposed [45] a W3 theory, which is essentially W2 theory with approximate post-CCSD(T) correlation effects added. Mean absolute errors for the W2-1 set were only marginally improved (from 0.24 to 0.20 kcal/mol): however, for some molecules that used to be beyond the reach of W2 theory, errors are dramatically reduced (e.g., O_3, from 3.01 to 0.38 kcal/mol). For a new parametrization set which now includes a number of 'problem molecules' like O_3, NO_2, and N_2O, mean absolute error is reduced from 0.36 to 0.16 kcal/mol and RMS error from 0.72 to 0.23 kcal/mol.

A number of avenues for increasing accuracy even further were explored, to no significant avail. Possible remaining error sources include higher order correlation effects in the inner shell correlations, deviations from Born–Oppenheimer (primarily for hydrides), and higher order relativistic effects (primarily for heavier systems).

The focal-point approach [52] has a somewhat similar philosophy to W1/W2/W3 theory, but was never intended to be a black-box method. In fact, it has been combined [53] with the explicitly correlated approaches discussed below, and thus straddles the fence between them and W*n*-type methods.

9. EXPLICITLY CORRELATED METHODS

A frontal assault on the slow convergence of pair correlation energies involves the use of explicitly correlated basis functions, i.e., basis functions that involve interelectronic distances [54]. The idea goes back all the way to Hylleraas' historical paper on the He ground-state wavefunction [55], but the obstacles to general application on systems of practical interest are formidable and are only recently beginning to be surmounted. Practical implementation for more general molecular systems is only now starting to become a reality.

In the R12 approach [56,57], terms linear in the interelectronic distances are admitted to the wavefunction. Particularly CCSD(T)-R12 [57] is an extremely powerful method for systems where its formidable computational requirements can be fulfilled. It has been shown by Kutzelnigg and Morgan [39] that pair correlation energies in such a method asymptotically converge as L^{-7} rather than L^{-3}: this means that *de facto* basis set limit results can be achieved using practical-sized basis sets, especially when extrapolations are employed.

10. CONCLUSIONS

In the overview given above, one can discern some parallel trends. The first is from methods applicable – with present-day workstation computers or modest clusters thereof – to 1000-atom systems, to those restricted to few-atom molecules. The second is from methods involving as many as a dozen parameters derived from experiment *per chemical element*, over second-generation DFT functionals that involve less than a score of such parameters total, over compound thermochemistry methods with a few such parameters, to purely nonempirical methods such as W*n* theory and CCSD(T)-R12. Sensitivity to unusual bonding types and nondynamical correlation effects is less neatly hierarchical, as the relatively low sensitivity of DFT methods to the latter bucks the general trend.

The common engineering truism 'better, cheaper, faster – pick any two' could be rephrased in the context of computational thermochemistry as: 'more accurate, less empirical, less computationally demanding – pick any two'.

REFERENCES

[1] C. J. Cramer, *Essentials of Computational Chemistry*, Wiley, New York, NY, 2002.

[2] F. Jensen, *Introduction to Computational Chemistry*, Wiley, Chichester, UK, 1999, Companion website: http://bogense.chem.sdu.dk/~icc/.

[3] K. K. Irikura and D. J. Frurip (eds), *Computational Thermochemistry. Prediction and Estimation of Molecular Thermodynamics*, 7th edition, American Chemical Society, Washington, DC, 1998.

[4] J. Cioslowski (ed.), *Quantum Mechanical Prediction of Thermochemical Data*, Kluwer Scientific Publishing, Dordrecht, The Netherlands, 2001.

[5] W. Thiel, Thermochemistry from semiempirical molecular orbital theory. In *Computational thermochemistry. Prediction and Estimation of Molecular Thermodynamics* (eds K. K. Irikura and D. J. Frurip), 7th edition, American Chemical Society, Washington, DC, 1998, pp. 142–161.

[6] A. D. Becke and R. M. Dickson, Numerical solution of Schrödinger's equation in polyatomic molecules, *J. Chem. Phys.*, 1990, **92**, 3610–3612.

[7] A. D. Becke, Density-functional thermochemistry. III. The role of exact exchange, *J. Chem. Phys.*, 1993, **98**, 5648–5652.

[8] A. D. Becke, Density-functional thermochemistry. V. Systematic optimization of exchange-correlation functionals, *J. Chem. Phys.*, 1997, **107**, 8554–8560.

[9] F. Hamprecht, A. J. Cohen, D. J. Tozer and N. C. Handy, Development and assessment of new exchange-correlation functionals, *J. Chem. Phys.*, 1998, **109**, 6264–6271.

[10] A. D. Boese and N. C. Handy, A new parametrization of exchange-correlation generalized gradient approximation functionals, *J. Chem. Phys.*, 2001, **114**, 5497–5503.

[11] A. D. Boese, J. M. L. Martin and N. C. Handy, The role of the basis set: assessing density functional theory, *J. Chem. Phys.*, 2003, **119**, 3005–3014.

[12] A. D. Boese and N. C. Handy, New exchange-correlation density functionals: the role of the kinetic-energy density, *J. Chem. Phys.*, 2002, **116**, 9559–9569.

[13] J. M. Tao, J. P. Perdew, V. N. Staroverov and G. E. Scuseria, Climbing the density functional ladder: nonempirical meta-generalized gradient approximation designed for molecules and solids, *Phys. Rev. Lett.*, 2003, **91**, 146401.

[14] F. Jensen, Polarization consistent basis sets: principles, *J. Chem. Phys.*, 2001, **115**, 9113–9125; F. Jensen, Polarization consistent basis sets. II. Estimating the Kohn–Sham basis set limit, *J. Chem. Phys.*, 2002, **116**, 7372–7379; F. Jensen, Polarization consistent basis sets. III. The importance of diffuse functions, *J. Chem. Phys.*, 2002, **117**, 9234–9240.

[15] J. Baker, J. Andzelm, M. Muir and P. R. Taylor, $OH + H_2 \rightarrow H_2O + H$. The importance of 'exact exchange' in density functional theory, *Chem. Phys. Lett.*, 1995, **237**, 53–60.

[16] J. L. Durant, Evaluation of transition state properties by density functional theory, *Chem. Phys. Lett.*, 1996, **256**, 595–602.
[17] B. J. Lynch, P. L. Fast, M. Harris and D. G. Truhlar, Adiabatic connection for kinetics, *J. Phys. Chem. A*, 2000, **104**, 4811–4815.
[18] J. K. Kang and C. B. Musgrave, Prediction of transition state barriers and enthalpies of reaction by a new hybrid density-functional approximation, *J. Chem. Phys.*, 2001, **115**, 11040–11051.
[19] A. D. Boese and J. M. L. Martin, Development of novel density functionals for thermochemical kinetics: BMK and B97-K, *J. Chem. Phys.*, 2004, **121**, 3045–3416.
[20] W. J. Hehre, L. Radom, P. Schleyer, R. von and J. A. Pople, *Ab initio Molecular Orbital Theory*, Wiley Interscience, New York, NY, 1986.
[21] To the best of our knowledge, the concept was introduced in W. J. Hehre, R. Ditchfield, L. Radom and J. A. Pople, Molecular orbital theory of the electronic structure of organic compounds. V. Molecular theory of bond separation, *J. Am. Chem. Soc.*, 1970, **92**, 4796–4801.
[22] P. George, M. Trachtman, C. W. Bock and A. M. Brett, Alternative approach to the problem of assessing stabilization energies in cyclic conjugated hydrocarbons, *Theor. Chim. Acta*, 1975, **38**, 121–129.
[23] For more detailed definitions, see also V. I. Minkin, Glossary of terms used in theoretical organic chemistry (IUPAC Recommendations 1999), *Pure Appl. Chem.*, 1999, **71**, 1919–1981, also available online at http://www.iupac.org/reports/1999/7110minkin/.
[24] J. Almlöf, B. J. DeLeeuw, P. R. Taylor, C. W. Bauschlicher, Jr. and P. E. M. Siegbahn, The dissociation energy of N_2, *Int. J. Quantum Chem. Symp.*, 1989, **23**, 345–354.
[25] J. A. Pople, M. Head-Gordon, D. J. Fox, K. Raghavachari and L. A. Curtiss, Gaussian-1 theory: a general procedure for prediction of molecular energies, *J. Chem. Phys.*, 1989, **90**, 5622–5629; L. A. Curtiss, C. Jones, G. W. Trucks, K. Raghavachari and J. A. Pople, Gaussian-1 theory of molecular energies for second-row compounds, *J. Chem. Phys.*, 1990, **93**, 2537–2545.
[26] J. M. L. Martin, Coupling between the convergence behavior of basis set and electron correlation – a quantitative study, *Theor. Chem. Acc.*, 1997, **97**, 227–231.
[27] L. A. Curtiss, K. Raghavachari, G. W. Trucks and J. A. Pople, Gaussian-2 theory for molecular energies of first- and second-row compounds, *J. Chem. Phys.*, 1991, **94**, 7221–7230.
[28] L. A. Curtiss, M. P. McGrath, J.-P. Blaudeau, N. E. Davies, R. C. Binning, Jr. and L. Radom, Extension of Gaussian-2 theory to molecules containing third-row atoms Ga–Kr, *J. Chem. Phys.*, 1995, **103**, 6104–6113.
[29] L. A. Curtiss, K. Raghavachari, P. C. Redfern, V. Rassolov and J. A. Pople, Gaussian-3 (G3) theory for molecules containing first and second-row atoms, *J. Chem. Phys.*, 1998, **109**, 7764–7776; L. A. Curtiss, P. C. Redfern, V. Rassolov, G. Kedziora and J. A. Pople, Extension of Gaussian-3 theory to molecules containing third-row atoms K, Ca, Ga–Kr, *J. Chem. Phys.*, 2001, **114**, 9287–9295.
[30] L. A. Curtiss, P. C. Redfern, K. Raghavachari and J. A. Pople, Gaussian-3X (G3X) theory: use of improved geometries, zero-point energies, and Hartree–Fock basis sets, *J. Chem. Phys.*, 2001, **114**, 108–117.
[31] P. L. Fast, M. L. Sanchez and D. G. Truhlar, Multi-coefficient Gaussian-3 method for calculating potential energy surfaces, *Chem. Phys. Lett.*, 1999, **306**, 407–410.
[32] L. A. Curtiss, K. Raghavachari, P. C. Redfern and J. A. Pople, Gaussian-3 theory using scaled energies, *J. Chem. Phys.*, 2000, **112**, 1125–1132.
[33] L. A. Curtiss, K. Raghavachari, P. C. Redfern and J. A. Pople, Assessment of Gaussian-2 and density functional theories for the computation of enthalpies of formation, *J. Chem. Phys.*, 1997, **106**, 1063–1079; L. A. Curtiss, P. C. Redfern, K. Raghavachari and J. A. Pople, Assessment of Gaussian-2 and density functional

theories for the computation of ionization potentials and electron affinities, *J. Chem. Phys.*, 1998, **109**, 42–55.
[34] L. A. Curtiss, P. C. Redfern, K. Raghavachari and J. A. Pople, Gaussian-3X (G3X) theory using coupled cluster and Brueckner energies, *Chem. Phys. Lett.*, 2002, **359**, 390–396.
[35] D. J. Henry, M. B. Sullivan and L. Radom, G3-RAD and G3X-RAD: modified Gaussian-3 (G3) and Gaussian-3X (G3X) procedures for radical thermochemistry, *J. Chem. Phys.*, 2003, **118**, 4849–4860.
[36] J. M. L. Martin, On the performance of large Gaussian basis sets for the computation of total atomization energies, *J. Chem. Phys.*, 1992, **97**, 5012–5018; J. M. L. Martin, On the performance of correlation consistent basis sets for the calculation of total atomization energies, geometries, and harmonic frequencies, geometries, and harmonic frequencies, *J. Chem. Phys.*, 1994, **100**, 8186–8193.
[37] C. Schwartz, Estimating convergence rates of variational calculations. In *Methods in Computational Physics* (ed. B. J. Alder), Academic Press, New York, NY, 1963, Vol. 2, pp. 241–266.
[38] R. N. Hill, Rates of convergence and error estimation formulas for the Rayleigh-Ritz variational method, *J. Chem. Phys.*, 1985, **83**, 1173–1196.
[39] W. Kutzelnigg and J. D. Morgan, III, Rates of convergence of the partial-wave expansions of atomic correlation energies, *J. Chem. Phys.*, 1992, **96**, 4484–4508, (Erratum, 1992, **97**, 8821).
[40] J. A. Montgomery, M. J. Frisch, J. W. Ochterski and G. A. Petersson, A complete basis set model chemistry. VII. Use of the minimum population localization method, *J. Chem. Phys.*, 2000, **112**, 6532–6542.
[41] J. A. Montgomery, J. W. Ochterski and G. A. Petersson, A complete basis set model chemistry. IV. An improved atomic pair natural orbital method, *J. Chem. Phys.*, 1994, **101**, 5900–5909.
[42] J. A. Montgomery, M. J. Frisch, J. W. Ochterski and G. A. Petersson, A complete basis set model chemistry. VI. Use of density functional geometries and frequencies, *J. Chem. Phys.*, 1999, **110**, 2822–2827.
[43] J. M. L. Martin and G. De Oliveira, Towards standard methods for benchmark quality ab initio thermochemistry – W1 and W2 theory, *J. Chem. Phys.*, 1999, **111**, 1843–1856.
[44] J. M. L. Martin and S. Parthiban, Variants: thermochemistry in the kJ/mol accuracy range. In *Quantum Mechanical Prediction of Thermochemical Data* (ed. J. Cioslowski), Kluwer Scientific Publishing, Dordrecht, The Netherlands, 2001, pp. 31–65.
[45] A. D. Boese, M. Oren, O. Atasoylu, J. M. L. Martin, M. Kállay and J. Gauss, W3 theory: robust computational thermochemistry in the kJ/mol accuracy range, *J. Chem. Phys.*, 2004, **120**, 4129–4141.
[46] D. Feller, Application of systematic sequences of wave functions to the water dimer, *J. Chem. Phys.*, 1992, **96**, 6104–6114.
[47] S. Parthiban and J. M. L. Martin, Assessment of W1 and W2 theories for the computation of electron affinities, ionization potentials, heats of formation, and proton affinities, *J. Chem. Phys.*, 2001, **114**, 6014–6029.
[48] C. W. Bauschlicher and H. Partridge, *Chem. Phys. Lett.*, 1995, **240**, 533; J. M. L. Martin, Basis set convergence study of the atomization energy, geometry, and anharmonic force field of SO_2: the importance of inner polarization functions, *J. Chem. Phys.*, 1998, **108**, 2791–2800.
[49] Paraphrasing T. S. Eliot's *Murder in a Cathedral*, The last temptation is the greatest treason/to get the right result for the wrong reason.
[50] M. Douglas and N. M. Kroll, Quantum electrodynamical corrections to the fine structure of helium, *Ann. Phys. (NY)*, 1974, **82**, 89–155; B. A. Hess, Relativistic electronic-structure calculations employing a two-component no-pair formalism with external-field projection operators, *Phys. Rev. A*, 1986, **33**, 3742–3748.

[51] M. Kállay and P. R. Surján, Computing coupled-cluster wave functions with arbitrary excitations, *J. Chem. Phys.*, 2000, **113**, 1359–1365; M. Kállay and P. R. Surján, Higher excitations in coupled-cluster theory, *J. Chem. Phys.*, 2001, **115**, 2945–2954; N. Kállay, P. G. Szalay and P. R. Surján, A general state-selective multireference coupled-cluster algorithm, *J. Chem. Phys.*, 2002, **117**, 980–990.
[52] A. G. Császár, W. D. Allen and H. F. Schaefer, III, In pursuit of the ab initio limit for conformational energy prototypes, *J. Chem. Phys.*, 1998, **108**, 9751–9764.
[53] See e.g. K. Aarset, A. G. Csaszar, E. L. Sibert, W. D. Allen, H. F. Schaefer, III, W. Klopper and J. Noga, Anharmonic force field, vibrational energies, and barrier to inversion of SiH_3^-, *J. Chem. Phys.*, 2000, **112**, 4053–4063.
[54] For a review see W. Klopper, R_{12} methods, Gaussian geminals. In *Modern Methods and Algorithms of Quantum Chemistry* (ed. J. Grotendorst), 2nd edition, John von Neumann Institute for Computing, Jülich, Germany, 2000, pp. 181–229, Available online at http://www.fz-juelich.de/nic-series/Volume3/klopper.pdf.
[55] E. A. Hylleraas, Neue berechnung der energie des heliums im grundzustande, sowie des tiefsten terms von ortho-helium, *Z. Phys.*, 1929, **54**, 347–366; T. U. Helgaker and W. Klopper, Perspective on new calculation of the helium ground state energy as well as of the lowest term value of ortho-helium, *Theor. Chem. Acc.*, 2000, **103**, 180–181.
[56] This approach was first proposed in W. Kutzelnigg, R_{12}-Dependent terms in the wavefunction as closed sums of partial-wave amplitudes for large L, *Theor. Chim. Acta*, 1985, **68**, 445–469.
[57] W. Klopper, Highly accurate coupled-cluster singlet and triplet pair energies from explicitly correlated calculations in comparison with extrapolation techniques, *Mol. Phys.*, 2001, **99**, 481–507, and references therein.

CHAPTER 4

Bond Breaking in Quantum Chemistry

C. David Sherrill

Center for Computational Molecular Science and Technology, School of Chemistry and Biochemistry, Georgia Institute of Technology, Atlanta, GA 30332-0400, USA

Contents

1. Introduction 45
2. The challenge of breaking bonds 46
   2.1. Difficulties for Hartree–Fock molecular orbital theory 46
   2.2. The multiconfigurational self-consistent-field method 46
   2.3. The complete-active-space self-consistent-field method 47
   2.4. The generalized valence bond method 47
3. Failure of standard single-reference methods 48
   3.1. Methods based on a single, restricted Hartree–Fock reference 48
   3.2. Methods based on a single, unrestricted Hartree–Fock reference 50
4. Methods improving upon MCSCF/CASSCF 51
   4.1. Multireference configuration interaction 51
   4.2. Multireference perturbation theory 51
   4.3. Multireference coupled-cluster theory 52
5. New perspectives 53
   5.1. Approximations to CASSCF 53
   5.2. Spin-flip methods 53
   5.3. Improved coupled-cluster methods 53
   5.4. Methods discarding the potential energy surface 54
6. Conclusions 54
Acknowledgements 54
References 54

1. INTRODUCTION

Ab initio quantum chemical methods can predict the equilibrium properties of small molecules in the gas phase to near-spectroscopic accuracy. However, a detailed understanding of a chemical reaction often requires knowledge of its dynamics, which in turn requires knowledge of the potential energy surface. The computation of potential energy surfaces to high accuracy is, unfortunately, much more challenging theoretically than the computation of equilibrium properties: if any bonds are formed or broken, then near degeneracies arise among the electron configurations, invalidating the assumptions of most standard methods. Although we focus on bond breaking (or, equivalently, bond formation), these

ANNUAL REPORTS IN COMPUTATIONAL CHEMISTRY, VOLUME 1
ISSN: 1574-1400 DOI 10.1016/S1574-1400(05)01004-2

same difficulties can occur for diradicals or metals of the first transition row. This chapter provides an introduction to the bond-breaking problem for nonspecialists and describes the reliability of different theoretical methods in such situations.

2. THE CHALLENGE OF BREAKING BONDS

2.1. Difficulties for Hartree–Fock molecular orbital theory

Hartree–Fock theory fails for bond-breaking reactions; energies at the dissociation limit are far too high. Let us consider the dissociation of H_2 into two H atoms. Both electrons (alpha and beta) are placed in the same σ bonding orbital. Because the σ orbital is delocalized over both H atoms, the single-determinant wave function contains terms corresponding not only to $H^{\cdot} + H^{\cdot}$, but also to $H^{-} + H^{+}$. These ionic terms are unphysical at large distances and increase the energy dramatically.

By allowing the alpha and beta electrons to occupy different orbitals, the wave function can avoid the unphysical ionic terms. This approach is called unrestricted Hartree–Fock (UHF), as opposed to the usual, restricted Hartree–Fock (RHF) method. The UHF potential energy curve is qualitatively correct, but often quantitatively poor. It also has the rather undesirable property that it is not an eigenfunction of the electronic spin operator S^2. At the dissociation limit for the singlet ground state of H_2, the UHF wave function is a 50/50 mixture of singlet and triplet functions; hence, predictions of any spin-dependent properties will be wrong. Next, we discuss alternative solutions to the bond-breaking problem that avoid this spin contamination.

2.2. The multiconfigurational self-consistent-field method

Consider again the case of H_2 dissociation. As the internuclear distance grows, the two H atoms become so far apart that it no longer matters if the H 1s orbitals combine in phase or out of phase, and the σ and σ^* orbitals should become degenerate. At this point, the configurations $(\sigma)^2$ and $(\sigma^*)^2$ should contribute equally to the wave function, and the orbitals should be determined as those appropriate for both configurations. The importance of the $(\sigma^*)^2$ configuration is technically referred to as 'nondynamical electron correlation', and it contrasts with the usual situation in which other determinants serve mainly to help describe the correlated motion of electrons at close range, the 'dynamical electron correlation'.

A multiconfigurational self-consistent-field (MCSCF) wave function is written as a linear combination of *N*-electron functions (Slater determinants or configuration state functions) appropriate to provide a qualitatively correct description of the electronic state, including relevant near-degenerate electron configurations.

The coefficients of this expansion are solved variationally, according to the usual configuration interaction (CI) method. At the same time, the orbitals are determined variationally as those which minimize the energy of this small CI wave function. A review of the MCSCF method has been given recently by Schmidt and Gordon [1].

2.3. The complete-active-space self-consistent-field method

Although MCSCF provides a proper framework for obtaining a qualitatively correct wave function for bond-breaking reactions, it has some disadvantages. Orbitals can be hard to converge for an arbitrary MCSCF wave function [2], and the selection of the 'important' configurations can be difficult for nonspecialists. An approach that helps solve both these problems is complete-active-space self-consistent field (CASSCF) [3], also called full optimized reaction space (FORS) [4]. In a CASSCF procedure, one selects a subset of orbitals that is the most important for the breaking and forming of bonds for a given reaction. Within this 'active space', all possible determinants of the correct symmetry are constructed. This gives an exact, 'full CI' (FCI) treatment of electron correlation within the space of active orbitals. The orbitals are determined as those which minimize the energy of this active-space FCI expansion.

CASSCF is the most widely used quantum chemical method for bond-breaking reactions. CASSCF wave functions are usually easier to converge than general MCSCF wave functions, and the ambiguity of selecting individual configurations is removed. On the other hand, one still must select a set of active orbitals, and the cost of the CASSCF scales factorially with the number of active electrons and orbitals. Quite frequently, severe compromises are made in the number of active orbitals to make the computations feasible. An intermediate method, which allows a larger active space but still allows the user to avoid choosing individual configurations, is the restricted-active-space self-consistent-field (RASSCF) method of Malmqvist *et al.* [5]. In that approach, the orbitals are divided into several subsets, and the configurations are chosen according to how many electrons are placed in each of the orbital subsets. For example, one might approximate a CASSCF wave function by discarding determinants which are more than, say, quadruply substituted relative to the Hartree–Fock reference.

2.4. The generalized valence bond method

The problematic ionic terms that make the RHF energy too high for large separations can also be avoided by using nonorthogonal orbitals in the valence bond approach. Modern versions of this formalism, called generalized valence bond (GVB) methods, have been developed by Goddard and coworkers [6].

The generalized valence bond perfect pairing (GVB-PP) approximation includes pairwise substitutions within the valence space and is analogous to a constrained MCSCF wave function.

3. FAILURE OF STANDARD SINGLE-REFERENCE METHODS

3.1. Methods based on a single, restricted Hartree–Fock reference

With the need to choose configurations and/or active spaces, the MCSCF and CASSCF methods are clearly less 'black box' than other *ab initio* methods. Is this extra complexity truly necessary to solve the bond-breaking problem? After all, any of the correlated, post-Hartree–Fock theories also include both critical configurations, $\cdots(\sigma)^2$ and $\cdots(\sigma^*)^2$, necessary to break a single bond, and many other configurations besides.

The configuration interaction singles and doubles (CISD) wave function includes the Hartree–Fock reference determinant and all determinants which differ from it by the substitution of no more than two orbitals. The $\cdots(\sigma^*)^2$ configuration is included in CISD because it represents a double substitution (one for alpha, one for beta) from the $\cdots(\sigma)^2$ configuration. The other determinants included in CISD are primarily important in describing the short-range, correlated motion of electrons (dynamical correlation) relative to the $\cdots(\sigma)^2$ configuration. We note that dynamical correlation is *not* adequately described with respect to the $\cdots(\sigma^*)^2$ configuration, since that would require double substitutions relative to $\cdots(\sigma^*)^2$, which would be *quadruple* substitutions relative to $\cdots(\sigma)^2$. This imbalanced treatment of dynamical correlation for the near-degenerate configurations also occurs to varying degrees for the other single-reference methods introduced below.

The second-order Møller–Plesset perturbation theory (MP2) wave function includes the same doubly substituted determinants as CISD, but it neglects the coupling between different electron pairs. Coupled-cluster theory with single and double substitutions (CCSD), on the other hand, includes this coupling as well as approximate descriptions of triple, quadruple, and higher-order substitutions as products of singles and doubles [7]. This approach can be improved with an approximate, perturbative treatment of irreducible triple substitutions to yield the CCSD(T) method, which is the most reliable of the commonly used single-reference methods and is often referred to as a 'gold standard' of quantum chemistry. Finally, we will also consider density functional theory (DFT), which dispenses with the wave function and solves instead for the electron density. This method has become very popular among computational chemists in the past decade. Here we will employ the widely used B3LYP hybrid functional, which often gives accurate predictions for molecular properties.

Given that the RHF reference function is qualitatively incorrect, let us consider what ought to be the easiest possible test for the single-reference methods, that of breaking a single bond to hydrogen. Figure 1 displays potential energy curves [8] generated by these methods, along with the exact (FCI) curve for BH in the large, aug-cc-pVQZ basis set. It should be mentioned that DFT energies such as those from the B3LYP functional are not directly comparable to FCI for a finite basis set; in typical cases, including this one, the DFT energies are actually lower than the FCI energies, because DFT methods converge more rapidly with respect to basis set than wave function-based methods. Nevertheless, the comparison of the B3LYP curve and the FCI curve should remain instructive, and the aug-cc-pVQZ basis should be large enough to minimize the basis set errors in the FCI curve.

As discussed previously, the RHF energy rises far too rapidly at large separations. B3LYP, although it includes an approximate treatment of electron correlation, nevertheless fails in essentially the same way as RHF: the dissociation energy is tremendously overestimated. We also note a catastrophic failure for MP2. The MP2 energy diverges to negative infinity because it includes denominators that approach zero as the energies of the σ and σ^* orbitals become degenerate. It is noteworthy that the CCSD(T) method, a gold standard in other contexts, also fails dramatically. At large distances, the energy levels off to a value much lower than the true asymptote. This is due to the perturbative treatment of triple substitutions, and we have already seen from MP2 that nondegenerate perturbation theory fails totally for this problem. The CISD energy is too large near dissociation because it lacks dynamical correlation for the $(\sigma^*)^2$ configuration, as

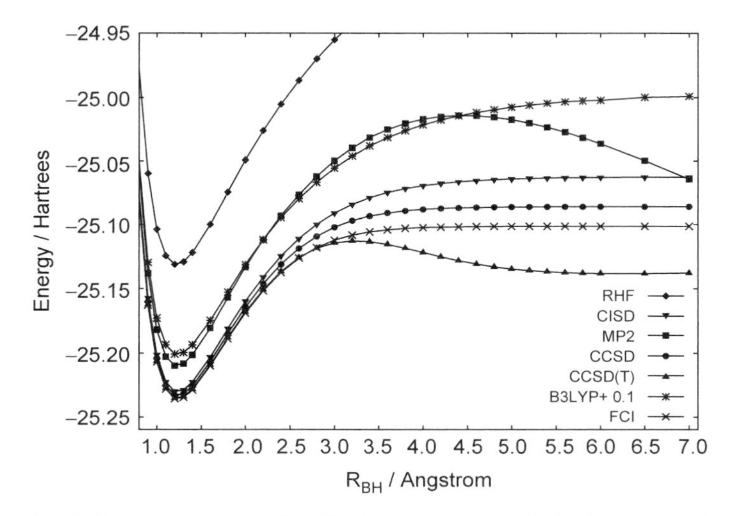

Fig. 1. Potential energy curves for BH in an aug-cc-pVQZ basis set for correlated methods based on an RHF reference function. Data adapted from Ref. [8]. B3LYP curve shifted up by 0.1 hartree.

discussed previously. Only CCSD displays a reasonable potential energy curve for BH, but even that curve is not as parallel to the exact curve as one might wish, featuring a nonparallelity error (the difference between the maximum and minimum errors along the curve) of 8 kcal mol^{-1}.

3.2. Methods based on a single, unrestricted Hartree–Fock reference

Having seen that correlated methods based on an RHF reference are unsatisfactory even for the simple bond-breaking case of BH, let us consider what improvement is afforded by switching to a UHF reference, which at least gives qualitatively correct potential energy curves. We note once again that the price we pay for this alternative is that the wave function is massively contaminated by higher-multiplicity spin states.

Figure 2 displays potential energy curves for correlated methods using a UHF reference [8,9]. The curves are greatly improved over those generated with an RHF reference, and none of them fail catastrophically. However, energies rise too rapidly at intermediate distances. This is perhaps most obvious for UMP2, and the problem becomes less severe for UCCSD and then UCCSD(T). It is also clear that the UHF potential well is much too shallow compared to that for FCI. The nonparallelity errors are 28 (UHF), 17 (UMP2), 5 (UCCSD), 3 [UCCSD(T)], and 6 kcal mol^{-1} (UB3LYP). Although these errors may be acceptable for some applications, they are surprisingly large for such a simple test case. Errors in more typical bond-breaking reactions will be at least this large, and usually significantly larger.

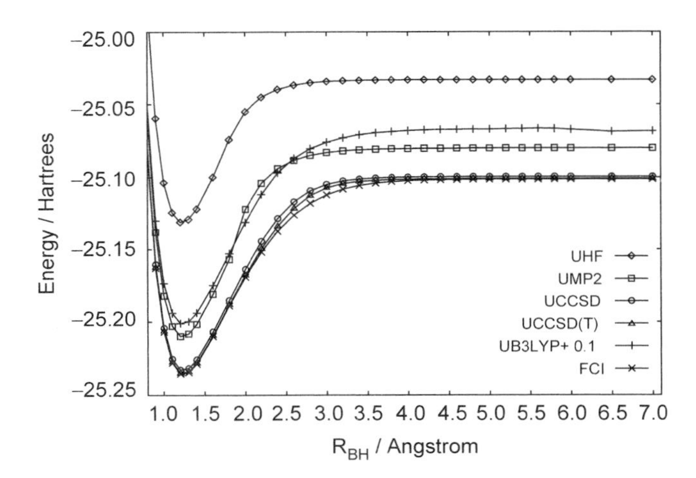

Fig. 2. Potential energy curves for BH in an aug-cc-pVQZ basis set for correlated methods based on a UHF reference function. Data adapted from Ref. [8]. UB3LYP curve shifted up by 0.1 hartree.

The CASSCF method, by including near-degenerate electron configurations, is greatly superior to either RHF or UHF. However, it still has a large error compared to FCI and the error is larger near equilibrium ($\sim$30 kcal mol^{-1} for BH with a valence active space, and a nonparallelity error of about 9 kcal mol^{-1}). CASSCF lacks dynamical electron correlation, which is more important at shorter distances, where the electrons are closer to each other. In the next section, we discuss bond-breaking methods which include dynamical correlation.

4. METHODS IMPROVING UPON MCSCF/CASSCF

As seen above, there are various ways to improve a Hartree–Fock wave function to account for electron correlation, including configuration interaction, perturbation theory, and coupled-cluster theory. There are analogous ways to add dynamical correlation to multiconfigurational reference wave functions (although typically the formalism becomes much more complex).

4.1. Multireference configuration interaction

CISD fails at large distances because it includes only those determinants necessary for dynamical correlation of the leading Hartree–Fock $\cdots$ $(\sigma)^2$ configuration, and not those necessary for dynamical correlation of the degenerate $\cdots$ $(\sigma^*)^2$ configuration. The simplest solution is to include all determinants that can be reached by single and double substitutions from *either* of these configurations. This is an example of a multireference CISD (MR-CISD) wave function [10]. The computational cost of MR-CISD is proportional to the usual cost of CISD (N^6, where N is proportional to the size of the molecule), multiplied by the number of references. For more than a few references, this cost becomes impractical. The most complete type of MR-CISD is the 'second-order CI', in which one takes as references all possible determinants that can be formed in the active space (i.e., the same set of determinants used in a CASSCF wave function). One way to reduce the cost of multireference configuration interaction (MRCI) is to act single and double substitution operators onto the MCSCF wave function taken as a whole, with fixed internal CI coefficients. This internally contracted MRCI introduces only a small error while leading to significant savings in computer time [11].

4.2. Multireference perturbation theory

There are many different ways to add dynamical electron correlation to a multiconfigurational reference function by perturbation theory, as discussed in several recent reviews [12–14]. We may distinguish between methods designed to handle a single electronic state and those designed to handle multiple states (as might be important when two electronic states of the same symmetry become close energetically). The former, single-state methods are generalizations of the usual, single-reference many-body perturbation theory methods, adapted for multiconfigurational reference wave functions. The complete-active-space second-order perturbation theory (CASPT2) method of Roos and coworkers [12] is an example of this approach, and it is the most popular way to improve upon CASSCF. The latter, multistate methods employ quasidegenerate perturbation theory, in which an effective Hamiltonian is constructed and then diagonalized [14].

4.3. Multireference coupled-cluster theory

Considering the great success of coupled-cluster methods for equilibrium properties, multireference coupled-cluster theories (MRCCs) [15] hold great promise for bond-breaking reactions. Such methods have been pursued for some time, but because of the great complexity of the equations, efficient computer programs implementing these theories are not yet widely available. However, recent results are encouraging [16].

To compare MR-CISD against CASPT2, Fig. 3 displays the errors *vs.* FCI for bond breaking in BH [17]. Also included are the errors for UCCSD(T), which performed best among the single-reference methods. Both MR-CISD and CASPT2 have much smoother error curves than UCCSD(T), which is consistent

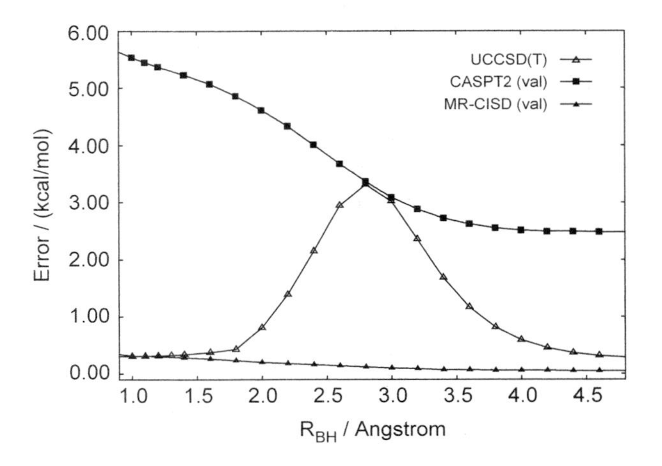

Fig. 3. Errors (*vs.* full CI) for potential energy curves of BH in a cc-pVQZ basis. Valence orbitals were chosen as active for the CASPT2 and MR-CISD wave functions. The MR-CISD is a second-order CI. Data from Ref. [17].

with the great improvement in the underlying reference function. The MR-CISD wave function used here is a second-order CI, and its performance is truly impressive; the nonparallelity error is only 0.3 kcal mol^{-1}. Although the UCCSD(T) and CASPT2 error curves look very different from each other, the nonparallelity errors are about the same at 3 kcal mol^{-1}. However, we note that MR-CISD and CASPT2 should work as well for more difficult bond-breaking reactions, whereas the quality of UCCSD(T) degrades rapidly. For example, in the dissociation of C_2, the nonparallelity error of UCCSD(T) grows to over 20 kcal mol^{-1} [18].

5. NEW PERSPECTIVES

5.1. Approximations to CASSCF

We have already mentioned RASSCF [5], which uses a limited CI instead of an FCI in the active space, as a less expensive alternative to CASSCF. Other works have explored coupled-cluster approximations to the active-space FCI [19] or less expensive choices of orbitals [20].

5.2. Spin-flip methods

Krylov has introduced a new family of single-reference methods based on the idea that triplet states are often easier to model across a potential energy surface than singlets. Starting from a Hartree–Fock determinant for a high-spin triplet state, 'spin-flipped' determinants are generated in which one of the alpha electrons is flipped back to beta, to yield the target value of $M_s = 0$. The simplest such method, spin-flip self-consistent field (SF-SCF), includes all spin-flipped single substitutions relative to the high-spin triplet reference. This approach includes the determinants most critical for a qualitatively correct description of breaking a single covalent bond [21]. This original prescription was recently improved to ensure that the wave functions are spin eigenfunctions, in the spin-complete SF-SCF method [22]. More complete treatments of electron correlation have also been explored [23].

5.3. Improved coupled-cluster methods

Piecuch and coworkers have pursued simple, state-selective energy corrections that improve the usual ground or excited state coupled-cluster methods in bond-breaking reactions. These 'method of moments' coupled-cluster methods (and the related renormalized and completely renormalized variants) were recently reviewed [24]. More robust alternatives to the (T) correction in CCSD(T) have also been considered [25].

5.4. Methods discarding the potential energy surface

Mapping complete potential energy surfaces for molecules with more than a few atoms becomes impossible due to the very large number ($3N-6$) of dimensions. In *ab initio* molecular dynamics (AIMD) methods, one avoids explicit computations of entire potential surfaces and only computes energies and gradients for those nuclear configurations which are accessed along a trajectory of a molecular dynamics simulation.

A more extreme solution is to discard the Born–Oppenheimer approximation entirely and to solve for nuclear and electronic motion simultaneously. It is also possible to retain the Born–Oppenheimer approximation for some nuclei, and couple the electrons only to light nuclei like hydrogen which might tunnel. Over the past few years there has been significant interest in such methods [26], which might be useful in modeling proton transfer reactions in biochemistry. However, some fundamental difficulties remain to be solved [27].

6. CONCLUSIONS

The long-recognized bond-breaking problem has received significantly greater interest in the past few years. FCI studies have been critical in understanding the failures of single-reference methods and in analyzing the performance of new methods. Recent theoretical advances show great promise.

ACKNOWLEDGEMENTS

The author acknowledges a National Science Foundation CAREER Award (grant no. CHE-0094088) and a Camille and Henry Dreyfus New Faculty Award. The Center for Computational Molecular Science and Technology is funded through a Shared University Research (SUR) grant from IBM and by Georgia Tech.

REFERENCES

[1] M. W. Schmidt and M. S. Gordon, The construction and interpretation of MCSCF wavefunctions, *Annu. Rev. Phys. Chem.*, 1998, **98**, 233–266.

[2] (a) H.-J. Werner and P. J. Knowles, A second order multiconfiguration SCF procedure with optimum convergence, *J. Chem. Phys.*, 1985, **82**, 5053–5063; (b) H. J. Aa. Jensen and H. Ågren, A direct, restricted-step, second-order MC SCF program for large scale ab initio calculations, *Chem. Phys.*, 1986, **104**, 229–250.

[3] B. O. Roos, The complete active space SCF method in a Fock-matrix-based super-CI formulation, *Int. J. Quantum Chem.*, 1980, **14**, 175–189.

[4] K. Ruedenberg, M. W. Schmidt, M. M. Gilbert and S. T. Elbert, Are atoms intrinsic to molecular electronic wavefunctions? I. The FORS model, *Chem. Phys.*, 1982, **71**, 41–49.

[5] P.-Å. Malmqvist, A. Rendell and B. O. Roos, The restricted active space self-consistent-field method, implemented with a split graph unitary group approach, *J. Phys. Chem.*, 1990, **94**, 5477–5482.
[6] (a) F. W. Bobrowicz and W. A. Goddard, The self-consistent field equations for generalized valence bond and open-shell Hartree–Fock wave functions. In *Methods of Electronic Structure Theory* (ed. H. F. Schaefer), Plenum Press, New York, 1977, pp. 79–127; (b) F. Faglioni and W. A. Goddard, GVB-RP: a reliable MCSCF wave function for large systems. *Int. J. Quantum Chem.*, 1999, **73**, 1–22.
[7] (a) R. J. Bartlett and J. F. Stanton, Applications of post-Hartree–Fock methods: a tutorial (eds K. B. Lipkowitz and D. B. Boyd), Reviews in Computational Chemistry, VCH, New York, 1994, Vol. 5, pp. 65–169; (b) T. D. Crawford and H. F. Schaefer, An introduction to coupled cluster theory for computational chemists (eds K. B. Lipkowitz and D. B. Boyd), Reviews in Computational Chemistry, VCH, New York, 2000, Vol. 14, pp. 33–136.
[8] A. Dutta and C. D. Sherrill, Full configuration interaction potential energy curves for breaking bonds to hydrogen: an assessment of single-reference correlation methods, *J. Chem. Phys.*, 2003, **118**, 1610–1619.
[9] In Ref. [8], the very lowest UHF solution was followed. Here, we show only UHF solutions maintaining spatial symmetry.
[10] C. D. Sherrill and H. F. Schaefer, The configuration interaction method: advances in highly correlated approaches, *Adv. Quantum Chem.*, 1999, **34**, 143–269.
[11] H.-J. Werner and P. J. Knowles, An efficient internally contracted multi-configuration-reference configuration interaction method, *J. Chem. Phys.*, 1988, **88**, 5803–5814.
[12] K. Andersson and B. O. Roos, Multiconfigurational second-order perturbation theory. In *Modern Electronic Structure Theory* (ed. D. R. Yarkony), Advanced Series in Physical Chemistry, World Scientific, Singapore, 1995, Vol. 2, pp. 55–109.
[13] M. R. Hoffmann, Quasidegenerate perturbation theory using effective Hamiltonians. In *Modern Electronic Structure Theory* (ed. D. R. Yarkony), Advanced Series in Physical Chemistry, World Scientific, Singapore, 1995, Vol. 2, pp. 1166–1190.
[14] E. R. Davidson and A. A. Jarzecki, Multi-reference perturbation theory. In *Recent Advances in Multireference Methods* (ed. K. Hirao), Recent Advances in Computational Chemistry, World Scientific, Singapore, 1999, Vol. 4, pp. 31–64.
[15] D. Mukherjee and S. Pal, Use of cluster-expansion methods in the open-shell correlation-problem, *Adv. Quantum Chem.*, 1989, **20**, 291–373.
[16] (a) P. Ghosh, S. Chattopadhyay, D. Jana and D. Mukherjee, State-specific multi-reference perturbation theories with relaxed coefficients: molecular applications, *Int. J. Mol. Sci.*, 2002, **3**, 733–754; (b) M. Tobita, S. A. Perera, M. Musial, R. J. Bartlett, M. Nooijen and J. S. Lee, Critical comparison of single-reference and multi-reference coupled-cluster methods: geometry harmonic frequencies, and excitation energies of N_2O_2, *J. Chem. Phys.*, 2003, **119**, 10713–10723; (c) X. Z. Li and J. Paldus, *N*-reference, *M*-state coupled-cluster method: merging the state-universal and reduced multireference coupled-cluster theories, *J. Chem. Phys.*, 2003, **119**, 5334–5345.
[17] M. L. Abrams and C. D. Sherrill, An assessment of the accuracy of multireference configuration interaction (MRCI) and complete-active-space second-order perturbation theory (CASPT2) for breaking bonds to hydrogen, *J. Phys. Chem. A*, 2003, **107**, 5611–5616.
[18] M. L. Abrams and C. D. Sherrill, Full configuration potential energy curves for the X $^1\Sigma_g^+$, B $^1\Delta_g$, and B′ $^1\Sigma_g^+$ states of C_2: a challenge for approximate methods, *J. Chem. Phys.*, 2004, **121**, 9211–9219.
[19] (a) A. I. Krylov, C. D. Sherrill, E. F. C. Byrd and M. Head-Gordon, Size-consistent wave functions for nondynamical correlation energy: the valence active space optimized orbital coupled-cluster doubles model, *J. Chem. Phys.*, 1998, **109**,

10669–10678; (b) M. Head-Gordon, T. V. Van Voorhis, S. R. Gwaltney and E. F. C. Byrd, Coupled-cluster methods for bond-breaking. In *Low-Lying Potential Energy Surfaces* (eds M. R. Hoffmann and K. G. Dyall), ACS Symposium Series 828, American Chemical Society, Washington, DC 2002, 93–108.

[20] (a) J. M. Bofill and P. Pulay, The unrestricted natural orbital-complete active space (UNO-CAS) method: an inexpensive alternative to the complete active space-self-consistent-field (CAS-SCF) method, *J. Chem. Phys.*, 1989, **90**, 3637–3646; (b) D. M. Potts, C. M. Taylor, R. K. Chadhuri and K. F. Freed, The improved virtual orbital-complete active space configuration interaction method, a "packageable" efficient ab initio many-body method for describing electronically excited states, *J. Chem. Phys.*, 2001, **114**, 2593–2600; (c) M. L. Abrams and C. D. Sherrill, Natural orbitals as substitutes for optimized orbitals in complete active space wave functions, *Chem. Phys. Lett.*, 2004, **395**, 227–232.

[21] A. I. Krylov, Size-consistent wave functions for bond-breaking: the spin-flip model, *Chem. Phys. Lett.*, 2001, **338**, 375–384.

[22] J. S. Sears, C. D. Sherrill and A. I. Krylov, A spin-complete version of the spin-flip approach to bond breaking: what is the impact of obtaining spin eigenfunctions?, *J. Chem. Phys.*, 2003, **118**, 9084–9094.

[23] (a) A. I. Krylov, Spin-flip configuration interaction: an electronic structure model that is both variational and size-consistent, *Chem. Phys. Lett.*, 2001, **350**, 522–530; (b) A. I. Krylov and C. D. Sherrill, Perturbative corrections to the equation-of-motion spin-flip model: application to bond-breaking equilibrium properties of diradicals, *J. Chem. Phys.*, 2002, **116**, 3194–3203; (c) Y. H. Shao, M. Head-Gordon and A. I. Krylov, The spin-flip approach within time-dependent density functional theory: theory and applications to diradicals, *J. Chem. Phys.*, 2003, **118**, 4807–4818.

[24] P. Piecuch, K. Kowalski, I. S. O. Pimienta and M. J. McGuire, Recent advances in electronic structure theory: method of moments of coupled-cluster equations and renormalized coupled-cluster approaches, *Int. Rev. Phys. Chem.*, 2002, **21**, 527–655.

[25] (a) T. D. Crawford and J. F. Stanton, Investigation of an asymmetric triple-excitation correction for coupled-cluster energies, *Int. J. Quantum Chem.*, 1998, **70**, 601–611; (b) S. R. Gwaltney, E. F. C. Byrd, T. Van Voorhis and M. Head-Gordon, A perturbative correction to the quadratic coupled-cluster method for higher excitations, *Chem. Phys. Lett.*, 2002, **353**, 359–367.

[26] (a) M. Tachikawa and Y. Osamura, Isotope effect of hydrogen and lithium hydride molecules. Application of the dynamic extended molecular orbital method and energy component analysis, *Theor. Chem. Acc.*, 2000, **104**, 29–39; (b) H. Nakai, Simultaneous determination of nuclear and electronic wave functions without Born–Oppenheimer approximation: ab initio NO + MO/HF theory, *Int. J. Quantum Chem.*, 2002, **86**, 511–517; (c) W. P. Webb, T. Iordanov and S. Hammes-Schiffer, Multiconfigurational nuclear-electronic orbital approach: incorporation of nuclear quantum effects in electronic structure calculations, *J. Chem. Phys.*, 2002, **117**, 4106–4118; (d) Y. Ohrn and E. Deumens, Electron nuclear dynamics, *Adv. Chem. Phys.*, 2002, **124**, 323–353; (e) M. Cafiero, S. Bubin and L. Adamowicz, Non-Born–Oppenheimer calculations of atoms and molecules, *Phys. Chem. Chem. Phys.*, 2003, **5**, 1491–1501.

[27] A. D. Bochevarov, E. F. Valeev and C. D. Sherrill, The electron and nuclear orbitals model: current challenges and future prospects, *Mol. Phys.*, 2004, **102**, 111–123.

Section 2
Molecular Modeling Methods

Section Editor: Carlos Simmerling
Center for Structural Biology
Stony Brook University
Stony Brook
NY 11794
USA

CHAPTER 5

A Review of the TIP4P, TIP4P-Ew, TIP5P, and TIP5P-E Water Models

Thomas J. Dick and Jeffry D. Madura

Department of Chemistry and Biochemistry, Center for Computational Sciences, Duquesne University, 600 Forbes Avenue, Pittsburgh, PA 15282, USA

Contents

1. Introduction 59
2. Methods 61
3. 4-site water models 62
   3.1. TIP4P 62
   3.2. TIP4P-Ew 64
4. 5-site water models 65
   4.1. TIP5P 65
   4.2. TIP5P-E 67
5. Conclusions 72
Acknowledgements 72
References 72

Recent advances in simulation techniques have prompted the re-parameterization of earlier water models, which were originally parameterized using older simulation methods. Many of these newly developed water models focused on reproducing specific properties, or a number of properties in a particular phase; other models have been designed to save on computational costs while some others take advantage of faster computational resources by treating various interactions more effectively. In this review, we examine the TIP4P and TIP5P water models and the recent re-parameterizations to include current Ewald summation techniques to treat long-range electrostatics, which has been shown to be superior to that of previous truncation cut-off techniques. The empirical differences between the models along with the change in observed properties will be reviewed, as well as the applications of these models in the literature.

1. INTRODUCTION

Molecular simulations have been used for the past 40 years to observe properties of bio-molecules, organic, and inorganic compounds in their natural environment

ANNUAL REPORTS IN COMPUTATIONAL CHEMISTRY, VOLUME 1
ISSN: 1574-1400 DOI 10.1016/S1574-1400(05)01005-4

on a time scale that is not feasible with current experimental techniques [1,2]. Data from molecular simulations has been used to save huge amount of resources in the scientific community, but still are only as reliable as the work that was put into simulation and its preparation [3]. In order for a molecular model to be as accurate as possible, one must consider all influences in that particular system and the contribution that each makes to the overall stability of the observed molecules. Thus, it is reasonable to assume that the solvent would be one of the most important choices in preparing a molecular simulation, as solvents tend to make up most of the bulk of a bio-molecular simulation. The bulk of bio-molecules, proteins, and ions are found naturally occurring in an aqueous solution, so the natural environment for most molecular simulations is in water or some type of aqueous solution. Prior to computational molecular simulations, there have been many proposed models to try and explain all the unique properties of water [4–7]. The models of water have unique forms and properties that are associated with them and influence molecular simulations accordingly.

As there is no one water model that can provide all the properties of water within experimental uncertainty, there have been numerous water models proposed throughout the literature, each with its advantages and weaknesses [8,9]. In choosing a water model for use in a molecular simulation, the desired properties of interest must be planned, as it will determine which water model would be optimal for the simulation. Efforts have been taken in parameterizing certain empirical water models to make them suitable for simulations under specific conditions [10–12]. An important feature of water molecules is the transferability of the force fields in different simulation programs. In some cases, transferability of the model may not be possible, as some programs use different functional forms for their potential energy. The functional form is also a computational consideration, since more descriptors of the water molecule may lead to better calculated properties, at the cost of the speed and resources needed for the simulation [1,2,13]. These considerations have been dealt with in many of the water molecules, and certain water molecules have emerged from the literature as the more popular water molecules in these types of simulations.

Among some of the most widely used water molecules are the 4-site and 5-site transferable intermolecular potential water molecules, TIP4P and TIP5P, respectively, developed by Jorgensen and coworkers [12,14]. These water molecules were parameterized to be used in a wide range of simulations and reproduce many experimental water properties, with reasonable computational cost. The TIP4P and TIP5P water molecules were originally parameterized using truncated cut-off methods for long-range electrostatics to save on computational costs [12,14]. As computational resources have increased in recent years, the treatment of the long-range electrostatics has become a priority in the simulation community, as the truncated portion can now be treated more precisely and more reasonably from a resource standpoint [15–21]. Re-parameterization of these water molecules has been done in order for Ewald summation techniques to be used in the long-range electrostatic summation, which has been shown to be superior to its predecessor cut-off method. These new models, the TIP4P-Ew and TIP5P-E, have shown improved properties of water over their base models, even without the use of flexibility or polarizability, which has been previously done in the literature [3,22].

With each developed water molecule, there come a number of simulations that show the simulated properties of the water molecule, along with the limits of the molecule for use in bio-molecular and phase transitional simulations. This review will encompass the simulations completed with the previously mentioned water models, the TIP4P, TIP4P-Ew, TIP5P, and TIP5P-E. An examination of the parameters and properties that were used in the parameterization will be presented, along with the literature use of these water models. Comparisons of the water molecules simulated structural, kinetic, and thermodynamic properties will be shown along with their comparisons to water's experimental values.

2. METHODS

All the water models in this review are empirical water models that use a force field equation that involves only non-bonded terms, as the water molecules themselves are rigid. The interaction for any of the water molecules with themselves or external influences can be calculated as:

$$U(r)_{\text{non-bonded}} = \sum(\varepsilon_{ij}[\{(\sigma_{ij}/r_{ij}) \times 12\} - \{(\sigma_{ij}/r_{ij}) \times 6\}] + (q_i q_j)/(\varepsilon_{\text{D}} r_{ij})) \qquad (1)$$

where the $U(r)$ is the potential energy of the system that is summed over all the non-bonded atoms of the system, ε_{ij} the well depth calculated from the individual atoms, σ_{ij} the collision diameter, q the charges on the atoms, ε_{D} the dielectric constant, and r_{ij} the distance between atoms i and j. The well depth and collision diameter for the pairs are calculated from the standard mixing rules that can be shown as [23]:

$$\sigma_{\text{AB}} = [(\sigma_{\text{AA}} + \sigma_{\text{BB}})/2] \qquad (2)$$

and

$$\varepsilon_{\text{AB}} = [(\varepsilon_{\text{AA}}\varepsilon_{\text{BB}}) \times \tfrac{1}{2}] \qquad (3)$$

where the AA and BB terms are the interaction of two like atoms and the AB terms are the mixed terms.

For the potential energy of the system to be calculated, the Lennard-Jones (LJ) and the Coulombic interactions are calculated separately. The LJ interactions are calculated between a pair and are based on a cut-off distance with both a smoothing function and a correction term applied for the energy lost after the cut-off. The electrostatic interaction can be calculated in several ways, including the truncated cut-off method and the Ewald summation method.

The Ewald summation technique has been shown to be superior to that of the truncated cut-off method used in previous simulations. It has been shown that the cut-off technique underestimates the energy of the system due to the interactions that were disposed of after the cut-off boundary. Although the energies after the cut-off might seem negligible at first to the contribution to the overall energy, the conservation of energy in the simulation is affected enough to make an impact on the outcome of the simulation.

Problems arise when treating long-range electrostatics due to the summation of the electrostatic interaction converging quite slowly, which makes the electrostatics quite costly. Methods were used that would save computational costs in the area of the expensive electrostatic contribution summations. Truncated cut-off methods looked at the overall contribution of the electrostatic summation and found a defining radius where indispensable electrostatic contribution was defined, where the interaction beyond this point could be deemed negligible and were excluded from the overall electrostatic contribution. This cut-off distance depended on the type of simulation being done, the types of molecules in the system, the solvent choice, and, as it seems in some cases, the discretion of the user, although explicit spherical cut-offs were set in the original parameterizations.

The Ewald summation technique is effective in treating the long-range electrostatics by splitting the summation into two separate series and then treating the two summations with a self-correction factor. The advantage of splitting the summation into two series is that each of the series converges much more rapidly than the original summation. With the series in the Ewald summation converging much more rapidly than that of the traditional calculation of the long-range electrostatics, it is possible to treat all electrostatic interactions using Ewald summation techniques. For a more thorough discussion on how to calculate or implement the Ewald summation techniques, refer to articles by De Leeuw *et al.* [24–26] and sections in referenced books [23,27].

3. 4-SITE WATER MODELS

3.1. TIP4P

The 4-site transferable intermolecular potential (TIP4P) model was developed by Jorgensen *et al.* to reproduce experimental structural and thermodynamic properties of water at 1 atm pressure [12]. At the time of its development, TIP4P was a balance between computational cost and reliability, with a rigid 4-site model This is computationally less expensive than earlier 5-site models, and reproduces liquid structure and density at 298 K and 1 atm better than some previous 3-site models, including the CF1, TIP3P, and SPC water models [10,12,28].

The orientation of the TIP4P water model can be seen in Fig. 1. It is an adaptation of a 3-site model with the charge of the oxygen moved down to a fictitious site, M. This site M is located in the bisecting plane of the HOH angle and at a distance of 0.15 Å from the oxygen atom. For the geometry of the water model, the experimental gas-phase geometry was chosen, where the distance from the oxygen to the hydrogen, *d*, is set at 0.9572 Å, with an HOH angle of 104.52°. The charge of the midpoint was set to $-1.040e$, with a $0.520e$ charge placed on each of the hydrogen atoms; this yields an overall dipole moment of 2.177 D. This dipole moment is significantly improved over the rigid 3-site models in the literature. For the LJ interactions, the σ was set at 3.15365 Å, with a well depth, ε, of 0.1550 kcal/mol. A summary of the empirical parameters for the TIP4P water model can be seen in Table 1.

Since its development, numerous Monte Carlo (MC) and molecular dynamic (MD) simulation have been performed using the TIP4P water model [29–37]. From these simulations, many property values have been computed with reasonable statistical uncertainty. Densities of liquid water have been calculated over a temperature range for the TIP4P model, and it has been found that the density maximum occurs at $\sim$260 K, with a value of 1.001 ± 0.001 g/cm^3 at STP (298 K and 1 atm) [32,38–41]. The $\Delta_{vap}H$ at STP is shown to be 10.65 ± 0.01 kcal/mol showing $<1\%$ deviation from the experimental value of 10.51 kcal/mol. Other calculated thermodynamic and kinetic values for TIP4P, such as C_p, κ, α, and ε, are within reasonable agreement with the experimental values at STP. TIP4P yields a C_p value of 20.4 ± 0.7 cal/mol K, a $\kappa \times 10^6$ value of 60 ± 5 atm^{-1}, an $\alpha \times 10^5$ value of 44 ± 8 deg^{-1}, and ε values of 53, 61, and 72 [38,42–44]. The corresponding experimental values for C_p, κ, and α are, respectively, 18.0 cal/mol K, 45.8×10^6 atm^{-1}, 25.7×10^5 deg^{-1}, and ε is found to be 78 [12,41,45–47]. While these values are close to values reported from

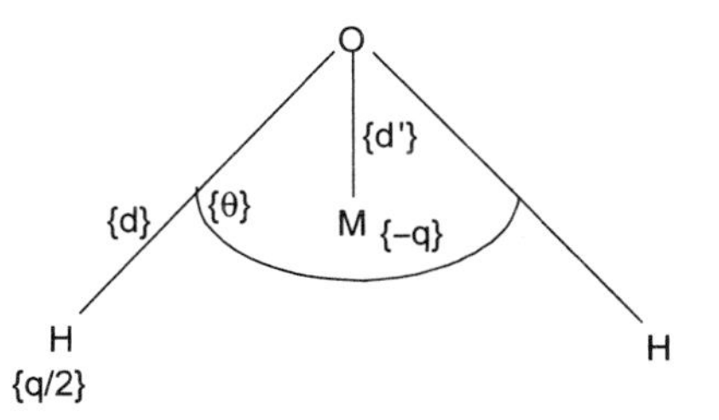

Fig. 1. Structure of TIP4P and TIP4P-Ew water models.

Table 1. Force field parameters for TIP4P and TIP4P-Ew

	TIP4P	TIP4P-Ew
σ (Å)	3.15365	3.16435
ε (kcal/mol)	0.155	0.16275
d (Å)	0.9572	0.9572
θ (°)	104.52	104.52
d' (Å)	0.15	0.125
q (e^-)	1.04	1.04844
Dipole moment, μ (D)	2.177	2.321

TIP3P and SPC water, the radial distribution functions (RDFs) for TIP4P are much closer to experiment, with an OO RDF showing well-defined second and third peaks, without sacrificing the OH and HH RDFs [38].

3.2. TIP4P-Ew

More recently, there has been a move to treat electrostatic interactions more accurately than previous truncated cut-off methods as computational resources have increased and algorithms are more refined. The truncated energy in the original TIP4P water model has been found to play a significant part in the overall energy of the molecular system, and the use of the Ewald summation technique for long-range electrostatics with the original TIP4P water model shows significant deviation from the properties that it was originally parameterized to reproduce [3]. A re-parameterization of the TIP4P water model was undertaken to account for the changes in structural, thermodynamic, and electrostatic properties of liquid water, with the use of the Ewald summation and long-range Lennard-Jones interactions, as it has been shown to lower energies of the order of ~2% [48,49]. This new model has been dubbed the TIP4P-Ew water model, as it is meant for use with electrostatic contributions calculated with Ewald summation techniques and more precise long-range Lennard-Jones interactions.

Precise experimental thermodynamic and structural properties were used to re-parameterize the new water model for use with Ewald. Of these precise experimental quantities, the enthalpy of vaporization, $\Delta_{vap}H(T)$, and the density, $\rho(T)$, were used to parameterize the new model, without sacrificing any of the previous model properties. This was done by minimizing the error of both $\Delta_{vap}H(T)$ and $\rho(T)$, over a desired temperature range, with the new inclusion of the more precise energies. The temperature range selected for the parameter error reduction was from 235.5 to 400 K, the entire liquid range for simulated water; these were all done at 1 atm pressure. The authors of the TIP4P-Ew model also note the precise calculation of the structural and thermodynamic properties of the system with statistical uncertainties weighted appropriately for the analysis.

The TIP4P-Ew water model retains the gas-phase geometry of its predecessor, TIP4P, with changes being done with the other parameters of the water model. The re-parameterization included changing the charge, q, on site M from 1.040e in the original TIP4P model to 1.04844 and subsequently changing the charge on the hydrogens. This change improves the overall dipole moment (μ) of the water molecule to a value of 2.321 D, which is a significant improvement over the TIP4P μ, which is 2.177 D. TIP4P's ε was changed to a value of 0.162750 with a σ value set at 3.16435 Å. Although changes to these values appear only miniscule, they have drastically changed the structural, thermodynamic, and kinetic properties of the simulated bulk water.

The desired properties of the re-parameterization were improved significantly over the original model. Since the TIP4P-Ew model has only recently been released, the calculated properties are reported by Horn *et al.* [3] The bulk density curve moved its maximum to the experimental value of 277 K at 1 atm. This density curve also more closely follows the experimental density curve than TIP4P and even more closely than that of the polarizable TIP4P-pol [50], at lower temperatures, and the 5-site TIP5P, although the shape of the TIP5P models more closely matches the experimental slope. The overall absolute average density error is found to be 0.0056 g/cm^3 for the TIP4P-Ew model. The $\Delta_{vap}H(T)$ over the simulated temperature range for the TIP4P-Ew model closely follows the experimental values and slope, with a maximum error of $+1.7\%$ at 235.5 K; this yields heat capacities, $C_p(T)$, that are slightly high [3].

4. 5-SITE WATER MODELS

4.1. TIP5P

The TIP5P water model was developed by Jorgensen and Mahoney to reproduce known experimental values of water better than its 4-site and 5-site predecessors, including the TIP4P and ST2 models [12,14,51]. The 5-site transferable intermolecular potential (TIP5P) is similar to previous 5-site models, where the sites are located on the atoms of the water molecule and on two lone pair sites, as can be seen in Fig. 2. The geometry of water in the TIP5P model is the same as the TIP4P model, where the experimental gas-phase geometry was adopted. The distance from the oxygen to the lone pairs, r_{OL}, is 0.70 Å, while the θ_{LOL} angle is set to 109.46°. Charges for the water molecule reside on the two hydrogens and the two lone pairs, with LJ terms only residing on the oxygen. The charge for the lone pairs is each 0.241e, with corresponding charge residing on each hydrogen atom; this charge separation yields a dipole moment, μ, of 2.29 D for the molecule. The LJ terms were set as 3.120 Å for σ and 0.160 kcal/mol for ε. All the empirical parameters for the TIP5P model can be seen in Table 2.

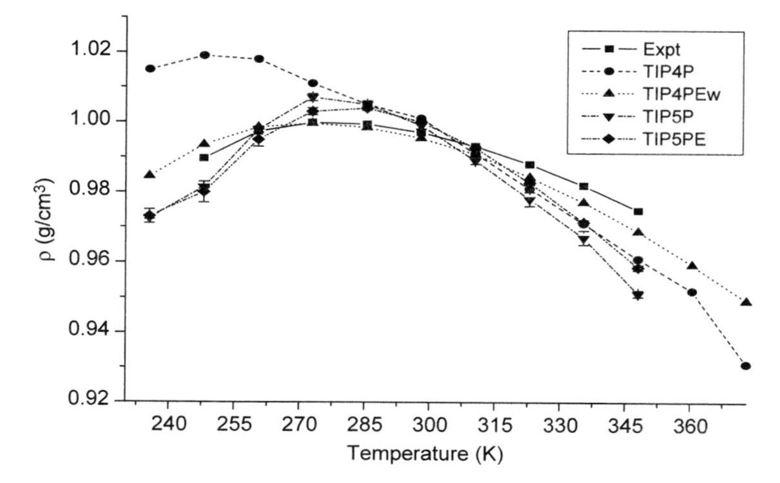

Fig. 2. Temperature dependence of bulk water density for TIP4P, TIP4P-Ew, TIP5P, and TIP5P-E.

TIP5P water reproduces the experimental density curve from 235.5 to 335.5 K remarkably well with an average error in density of only 0.006 g/cm^3 [14]. The maximum density of TIP5P reproduces the experimental maximum at 277 K, with better agreement below 323 K. Trends for C_p, κ, and α are similar to that of TIP4P water, with only slight error from experimental value. However, the error reported from TIP4P is from only STP, while the TIP5P model follows the temperature range from 235.5 to 348 K considerably well, with slight deviation at the extremes [14]. The dielectric constant from 273 to 373 K is reproduced well when compared to experiment, and the slope is similar only translated higher by ~6%, although this would be assumed as the dielectric constant would improve with a better description of the quadrupole moment, as seen in the literature [42,52].

Structural properties of the TIP5P model show similar results to the TIP4P model, which offered significant improvements over 3-site models. The TIP5P

Table 2. Force field parameters for TIP5P and TIP5P-E

	TIP5P	TIP5P-E
σ (Å)	3.12	3.097
ε (kcal/mol)	0.16	0.16275
d_{OH} (Å)	0.9572	0.9572
d'_{OL} (Å)	0.7	0.7
θ_{HOH} (°)	104.52	104.52
θ'_{LOL} (°)	109.47	109.47
q_H (e^-)	0.241	0.241
Dipole moment, μ (D)	2.29	2.29

model reproduces the experimental g_{OO} as well as the TIP4P model, with even better improvements in the shape of the second peak, with some H–H RDF being somewhat high [14]. Hydrogen bonding over the simulated temperature scales seems to be in agreement with the TIP4P model, with more tetrahedral-like geometries being sampled at higher temperatures [14]. Kinetically, TIP5P water reproduces the D_{self} over a wide range of temperatures (235.5–348 K) reasonably well, following the trend with a maximum deviation from experiment occurring at higher temperatures [22,53]. There has been considerable work done on the pressure dependence of the D_{self} of TIP5P at high pressures; it is noted that there is good correlation between experimental values and TIP5P from 1 to ~1700 atm for temperatures between 298 and 363 K [53]. Similar results reported by Rick are with ~2% error [22].

4.2. TIP5P-E

Recently released is the re-parameterization of the TIP5P model to include Ewald summations, dubbed the TIP5P-E model [22]. Similar to that of the 4-site models, the use of Ewald summations lowers the energies and changes the properties of the original water model, to a certain degree, as previously discussed. The TIP5P-E model reproduces many of the experimental properties of water and is as quantitative in reproducing the accuracy that has been associated with the TIP5P model.

The TIP5P re-parameterization, TIP5P-E, has made changes to the LJ terms of the TIP5P model, as can be seen in Table 2. Structural geometries and electrostatic potentials remain the same as the original model, but the LJ potentials were modified for use with the Ewald summations by lowering the short-range repulsion forces by ~2% and raising the long-range attractive forces by ~6% from their original TIP5P values. The well depth was set at 0.162750 kcal/mol and the collision diameter was set to 3.097 Å. Since the charge was not changed in this model, the dipole moment remains the same as the original TIP5P model.

The results reported here for the structural, thermodynamic, and kinetic properties are values reported by Rick [22]. TIP5P-E reproduces the density curve of water over a wide range of temperatures, with an average error in the density of only 0.004 g/cm^3 and a maximum density occurring at 277 K. The enthalpy of vaporization, $\Delta_{vap}H(T)$, maintains a similar slope to that of TIP5P, which is in reasonable agreement over the simulated temperature range of 235.5–348 K. The TIP5P-E $\Delta_{vap}H$ results at 298 K are 10.377 kcal/mol for 512 water molecules, which is $<1\%$ higher than that of TIP5P water at 298 K. $\Delta_{vap}H(T)$, over the temperature range mentioned above, for TIP5P and TIP5P-E shows that the Ewald model reproduces the experimental enthalpy better at lower temperatures, while TIP5P does slightly better at higher temperatures. The temperature

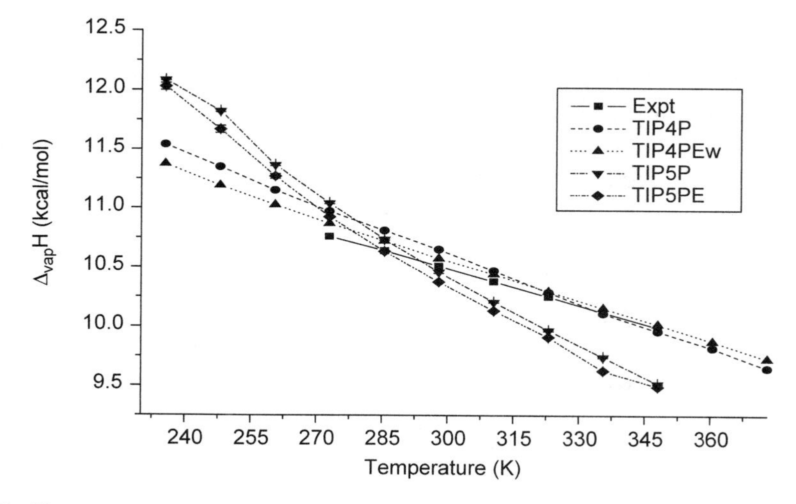

Fig. 3. Temperature dependence of enthalpy of vaporization for TIP4P, TIP4P-Ew, TIP5P, and TIP5P-E.

dependence for TIP5P and TIP5P-E can be seen in Figs 2 and 3, of the density and enthalpy of vaporization, respectively.

Calculated properties of the TIP5P-E model show similar results to that of the TIP5P model. TIP5P reproduces the C_p, κ, and α over the temperature range of 235.5–348 K reasonably well. Results seen in Table 3 show that agreement is seen between the 5-site models presented here and with experimental values. The C_p and α values are closer to the experimental values than TIP5P, for values <260 K; the TIP5P and TIP5P-E results deviate differently here, as the TIP5P-E seems to make a drastic change. TIP5P-E reproduces the κ better than TIP5P for temperatures lower than 298 K, but results are similar between the two models above 298 K. The static dielectric constant is reproduced better with the TIP5P model, but is within the error of the TIP5P-E model over the specified temperature (Fig. 4).

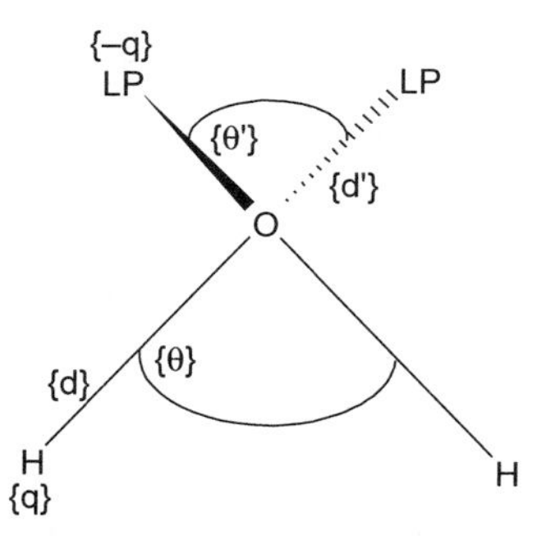

Fig. 4. Structure of TIP5P and TIP5P-E water models.

Table 3. Calculated properties of TIP4P, TIP4P-Ew, TIP5P, and TIP5P-E

Property	Model	235.5 K	248 K	260.5 K	273 K	285.5 K	298 K
ρ (g/cm^3)	TIP4P [38]	1.015	1.019	1.018	1.011	1.005	1.001
	TIP4P-Ew [3]	0.9845 (4)	0.9935 (4)	0.9986 (3)	0.9996 (3)	0.9984 (3)	0.9954 (3)
	TIP5P [53]	0.9725 (3)	0.9814 (4)	0.9979 (8)	1.007 (1)	1.005 (1)	0.999 (1)
	TIP5P-E [22]	0.973 (2)	0.980 (3)	0.995 (2)	1.003 (1)	1.0039 (6)	1.0000 (5)
	Expt [58,59]	0.9688	0.98924	0.99714	0.99981	0.99953	0.99716
$\Delta_{vap}H$ (kcal/mol)	TIP4P [38]	11.54	11.35	11.15	10.97	10.81	10.65
	TIP4P-Ew [3]	11.373 (4)	11.191 (4)	11.025 (4)	10.869 (4)	10.723 (4)	10.575 (4)
	TIP5P [22]	12.084 (3)	11.823 (7)	11.367 (8)	11.041 (8)	10.735 (7)	10.46 (1)
	TIP5P-E [22]	12.03 (3)	11.67 (3)	11.27 (2)	10.924 (7)	10.633 (4)	10.377 (4)
	Expt [58]	11.18	11.0372	10.9029	10.7732	10.6483	10.5176
C_p (cal/molK)	TIP4P [38]	23	23.2	23.3	21.8	20.7	21.4
	TIP4P-Ew [3]	24.2 (3)	21.9 (3)	20.8 (3)	20.1 (3)	19.6 (3)	19.2 (3)
	TIP5P [22]		43.0 (2)	39.4 (3)	33.8 (5)	30.9 (8)	29.1 (8)
	TIP5P-E [22]	19 (2)	22 (1)	32 (3)	31 (2)	29 (1)	27.2 (6)
	Expt [58–60]	23.47	19.34	18.38	18.17	18.048	18.004
$10^5\ \alpha$ (deg^{-1})	TIP4P [38]	−37.2	−11.6	33.8	66.3	39.8	44
	TIP4P-Ew [3]	−8.7 (3)	−5.3 (3)	−2.4 (3)	−0.1 (3)	1.8 (3)	3.4 (3)
	TIP5P [22]		−12.5 (1)	−10.5 (3)	−3.2 (5)	3.3 (7)	6.3 (6)
	TIP5P-E [22]	1.2 (21)	1.1 (14)	−5.0 (30)	−1.8 (22)	1.8 (9)	4.9 (6)
	Expt [59]		−9.674	−3.712	−0.705	1.185	2.558
$10^6\ \kappa$ (atm^{-1})	TIP4P [38]	43.9	51.5	44.1	45.4	52	60.3
	TIP4P-Ew [3]	54.3			48.9		48.1
	TIP5P [22]		17 (1)	24 (1)	31 (1)	36 (1)	41 (2)
	TIP5P-E [22]	18 (3)	29 (3)	48 (4)	52 (4)	53 (3)	52 (3)
	Expt [59]		71.88	58.27	51.56	47.86	45.85

continued on next page

Table 3. *continued*

Property	Model	235.5 K	248 K	260.5 K	273 K	285.5 K	298 K
D_{self} (10^{-9} m^2/s)	TIP4P [61]						3.31 (8)
	TIP4P-Ew [3]	0.172 (1)			1.179 (1)		2.335 (4)
	TIP5P [53]	0.070 (8)	0.14 (2)	0.43 (3)	1.01 (2)	1.87 (8)	2.62 (4)
	TIP5P-E [22]	0.09 (2)	0.17 (2)	0.48 (5)	1.2 (1)	1.9 (1)	2.8 (1)
	Expt [62,63]			0.66	1.1	1.64	2.3
ε	TIP4P						
	TIP4P-Ew [3]	81.9 (5.2)			70.8 (1.4)		63.9 (0.9)
	TIP5P [22]				92 (2)		82 (2)
	TIP5P-E [22]				95 (14)	90 (9)	92 (14)
	Expt [64]				87.74	83.02	78.3
τ_{nmr} (ps)	TIP5P [22]	69 (7)					1.58 (5)
	TIP5P-E [22]	63 (10)	28 (2)	10.3 (9)	4.1 (3)	2.3 (1)	1.55 (4)
	Expt [65]					3.44	2.46
Property	Model	310.5 K	323 K	335.5 K	348 K	360.5 K	373 K
ρ (g/cm^3)	TIP4P [38]	0.991	0.981	0.971	0.961	0.952	0.931
	TIP4P-Ew [3]	0.9908 (3)	0.9843 (3)	0.9771 (3)	0.9688 (3)	0.9594 (3)	0.9492 (3)
	TIP5P [53]	0.989 (1)	0.978 (2)	0.967 (2)	0.9512 (9)		
	TIP5P-E [22]	0.9926 (6)	0.9827 (6)	0.9714 (6)	0.9586 (7)		
	Expt [58,59]	0.99362	0.98838	0.98207	0.97527	0.96737	0.95869
$\Delta_{vap}H$ (kcal/mol)	TIP4P [38]	10.47	10.29	10.11	9.96	9.82	9.65
	TIP4P-Ew [3]	10.444 (4)	10.297 (4)	10.155 (4)	10.018 (4)	9.875 (4)	9.73 (4)
	TIP5P [22]	10.207 (6)	9.967 (3)	9.744 (6)	9.519 (7)		
	TIP5P-E [22]	10.133 (3)	9.910 (3)	9.697 (3)	9.493 (5)		
	Expt [58]	10.3986	10.264	10.1286	9.9993	9.8619	9.7206

C_p (cal/mol K)	TIP4P [38]	22.2	22.4	21.2	19.4	20.5	
	TIP4P-Ew [3]	18.9 (3)	18.7 (3)	18.6 (3)	18.5 (3)	18.5 (3)	18.5 (3)
	TIP5P [22]	27.6 (3)	27 (1)	25.9 (9)	25.9 (8)		
	TIP5P-E [22]	26.6 (5)	25.5 (9)	24.9 (5)	24 (1)		
	Expt [58–60]	17.995	18.004	18.024	18.054	18.096	18.151
$10^5\ \alpha$ (deg^{-1})	TIP4P [38]	77.4	81.3	82.4	76.4	129.2	
	TIP4P-Ew [3]	4.6 (3)	5.5 (3)	6.3 (3)	7.0 (3)	7.6 (3)	8.3 (3)
	TIP5P [22]	8.7 (5)	9.2 (11)	11.0 (10)	12.7 (7)		
	TIP5P-E [22]	6.9 (10)	9.1 (13)	10.6 (9)	11.8 (14)		
	Expt [59]	3.648	4.567	5.379	6.121	6.821	7.498
$10^6\ \kappa$ (atm^{-1})	TIP4P [38]	44.2	50.1	50.1	55.3	63.2	78.6
	TIP4P-Ew [3]		49.4		53.6		59.9
	TIP5P [22]	47 (1)	56 (4)	59 (3)	65 (3)		
	TIP5P-E [22]	58 (3)	60 (3)	64 (4)	67 (4)		
	Expt [59]	44.91	44.76	45.22	46.22		49.65
D_{self} (10^{-9} m^2/s)	TIP4P [61]						
	TIP4P-Ew [3]		3.822 (4)		5.637 (4)		7.709 (4)
	TIP5P [53]	3.70 (9)	4.74 (8)	6.33 (7)	6.78 (10)		
	TIP5P-E [22]	3.88 (6)	5.2 (2)	6.4 (2)	8.0 (2)		
	Expt [62,63]	3.07	3.95	4.96	6.08		
ε	TIP4P						
	TIP4P-Ew [3]		60.0 (0.7)		54.1 (0.7)		48.7 (0.6)
	TIP5P [22]		75 (2)		69 (2)		
	TIP5P-E [22]	80 (2)	77 (9)	80 (2)	72 (1)		
	Expt [64]	74.11	69.91	66.17	62.43		
τ_{nmr} (ps)	TIP5P [22]				0.47 (2)		
	TIP5P-E [22]	1.03 (3)	0.74 (1)	0.58 (1)	0.44 (2)		
	Expt [65]	1.92	1.66	1.41	1.12		

5. CONCLUSIONS

The 4- and 5-site transferable intermolecular potential water models for use with truncated cut-offs and Ewald summation electrostatic treatments are reviewed. The use of Ewald summation techniques is found to lower the energy and change the molecular properties of the model. These changes in the molecular properties are within the error of the original model or have significantly improved upon them, when the models have been parameterized taking into account long-range electrostatics using Ewald techniques.

TIP4P-Ew has been found to reproduce the density curve considerably better than TIP4P. Better agreement between the new parameterization and the original TIP4P model was also found for the enthalpy of vaporization and self-diffusion coefficient, as seen in Table 3. The TIP4P-Ew water model reproduces the density curve and the slope of the enthalpy of vaporization, as seen in Figs 2 and 3, respectively, better than any of the water models reviewed. The TIP4P-Ew model has been tested in BOSS [54], AMBER [55], GROMACS [56], and CHARMM [57]. Table 3 summarizes the calculated properties of the TIP4P, TIP4P-Ew, TIP5P, and TIP5P-E water models along with the experimental values; a review of the presented data shows that the Ewald re-parameterizations of the original 4- and 5-site models have moderate to significant improvements in the calculated properties.

ACKNOWLEDGEMENTS

Special thanks to Dr Hans Horn at IBM Almaden Research Center and Dr Julian Tirado-Rives at Yale University for sharing data and unpublished results.

REFERENCES

[1] G. W. Robinson, S.-B. Zhu, S. Singh and M. W. Evans, *Water in Biology, Chemistry and Physics*, World Scientific, Singapore, 1996, p. 509.
[2] K. B. Lipkowitz and D. B. Boyd, *Reviews in Computational Chemistry*, Vol. 13, Wiley-VCH, New York, 1999, p. 426.
[3] H. W. Horn, W. C. Swope, J. W. Pitera, J. D. Madura, T. J. Dick, G. L. Hura and T. Head-Gordon, *J. Chem. Phys.*, 2004, **120**, 9665–9678.
[4] J. D. Bernal and R. H. Fowler, *J. Chem. Phys.*, 1933, **1**, 515–548.
[5] W. H. Stockmayer, *J. Chem. Phys.*, 1941, **9**, 398–402.
[6] E. J. W. Verwey, *Recl Trav. Chim. Pays-Bas Belg.*, 1941, **60**, 887–896.
[7] N. Bjerrum, *Science (Wash., DC)*, 1952, **115**, 385–390.
[8] K. A. T. Silverstein, A. D. J. Haymet and K. A. Dill, *J. Am. Chem. Soc.*, 1998, **120**, 3166–3175.
[9] K. Dill and S. Bromberg, *Molecular Driving Forces*, Garland Science, New York, 2003, p. 666.
[10] H. J. C. Berendsen, J. P. M. Postma, W. F. Van Gunsteren and J. Hermans, *Jerus. Symp. Quantum Chem. Biochem.*, 1981, **14**, 331–342.

[11] H. J. C. Berendsen, J. R. Grigera and T. P. Straatsma, *J. Phys. Chem.*, 1987, **91**, 6269–6271.
[12] W. L. Jorgensen, J. Chandrasekhar, J. D. Madura, R. W. Impey and M. L. Klein, *J. Chem. Phys.*, 1983, **79**, 926–935.
[13] B. Guillot, *J. Mol. Liq.*, 2002, **101**, 219–260.
[14] M. W. Mahoney and W. L. Jorgensen, *J. Chem. Phys.*, 2000, **112**, 8910–8922.
[15] A. Brodka and P. Sliwinski, *J. Chem. Phys.*, 2004, **120**, 5518–5523.
[16] J. Norberg and L. Nilsson, *Q. Rev. Biophys.*, 2003, **36**, 257–306.
[17] V. Dungsrikaew, J. Limtrakul, K. Hermansson and M. Probst, *Int. J. Quantum Chem.*, 2003, **96**, 17–22.
[18] A. Brodka and A. Grzybowski, *J. Chem. Phys.*, 2002, **117**, 8208–8211.
[19] Z. Wang and C. Holm, *J. Chem. Phys.*, 2001, **115**, 6351–6359.
[20] A. Toukmaji, C. Sagui, J. Board and T. Darden, *J. Chem. Phys.*, 2000, **113**, 10913–10927.
[21] Y. Komeiji, *TheoChem*, 2000, **530**, 237–243.
[22] S. W. Rick, *J. Chem. Phys.*, 2004, **120**, 6085–6093.
[23] A. R. Leech, *Molecular Modeling Principles and Applications*, Prentice Hall, Harlow, England, 2001, p. 726.
[24] S. W. De Leeuw, J. W. Perram and E. R. Smith, *Proc. R. Soc. Lond. Ser. A: Math. Phys. Eng. Sci.*, 1980, **373**, 27–56.
[25] S. W. De Leeuw, J. W. Perram and E. R. Smith, *Proc. R. Soc. Lond. Ser. A: Math. Phys. Eng. Sci.*, 1980, **373**, 56–66.
[26] S. W. De Leeuw, J. W. Perram and E. R. Smith, *Proc. R. Soc. Lond. Ser. A: Math. Phys. Eng. Sci.*, 1980, **373**, 177–193.
[27] D. Frenkel and B. Smit, *Understanding Molecular Simulation: From Algorithms to Applications*, Academic Press, San Diego, 1996, p. 443.
[28] F. H. Stillinger and C. W. David, *J. Chem. Phys.*, 1978, **69**, 1473–1484.
[29] R. B. Ayala and V. Tchijov, *Can. J. Phys.*, 2003, **81**, 11–16.
[30] E. N. Brodskaya and A. I. Rusanov, *Mol. Phys.*, 2003, **101**, 1495–1500.
[31] I. Brovchenko, D. Paschek and A. Geiger, *J. Chem. Phys.*, 2000, **113**, 5026–5036.
[32] W. L. Jorgensen, C. Jenson and M. W. Mahoney, *Book of Abstracts*, 218th ACS National Meeting, New Orleans, August 22–26, 1999, HYS-139.
[33] H. S. Kim, *Chem. Phys. Lett.*, 2000, **321**, 262–268.
[34] T. Kuznetsova and B. Kvamme, *Mol. Phys.*, 1999, **97**, 423–431.
[35] I. Nezbeda and J. Kolafa, *Mol. Phys.*, 1999, **97**, 1105–1116.
[36] Z.-M. Sheng and J.-W. Luo, *Wuli Xuebao*, 2003, **52**, 2342–2346.
[37] E. J. W. Wensink, A. C. Hoffmann, P. J. van Maaren and D. van der Spoel, *J. Chem. Phys.*, 2003, **119**, 7308–7317.
[38] W. L. Jorgensen and C. Jenson, *J. Comput. Chem.*, 1998, **19**, 1179–1186.
[39] P. H. Poole, U. Essmann, F. Sciortino and H. E. Stanley, *Phys. Rev. E: Stat. Phys. Plasmas Fluids Relat. Interdiscip. Top.*, 1993, **48**, 4605–4610.
[40] F. Sciortino, P. H. Poole, U. Essmann and H. E. Stanley, *Phys. Rev. E: Stat. Phys. Plasmas Fluids Relat. Interdiscip. Top.*, 1997, **55**, 727–737.
[41] W. L. Jorgensen and J. D. Madura, *Mol. Phys.*, 1985, **56**, 1381–1392.
[42] K. Watanabe and M. L. Klein, *Chem. Phys.*, 1989, **131**, 157–167.
[43] M. Ferrario and A. Tani, *Chem. Phys. Lett.*, 1985, **121**, 182–186.
[44] M. Neumann, *J. Chem. Phys.*, 1986, **85**, 1567–1580.
[45] N. E. Dorsey, *Properties of Ordinary Water Substance*, Reinhold, New York, 1940, p. 673.
[46] C. A. Angell, M. Oguni and W. J. Sichina, *J. Chem. Phys.*, 1982, **86**, 998.
[47] G. S. Kell, *J. Chem. Eng. Data*, 1967, **12**, 66–69.
[48] D. Paschek, *J. Chem. Phys.*, 2004, **120**, 6674–6690.
[49] M. Lisal, J. Kolafa and I. Nezbeda, *J. Chem. Phys.*, 2002, **117**, 8892–8897.
[50] B. Chen, J. Xing and J. I. Siepmann, *J. Phys. Chem. B*, 2000, **104**, 2391–2401.

[51] F. H. Stillinger and A. Rahman, *J. Chem. Phys.*, 1974, **61**, 4973–4980.
[52] T. Head-Gordon and F. H. Stillinger, *J. Chem. Phys.*, 1993, **98**, 3313–3327.
[53] M. W. Mahoney and W. L. Jorgensen, *J. Chem. Phys.*, 2001, **114**, 363–366.
[54] W. L. Jorgensen, in *BOSS – Biochemical and Organic Simulation System*, Vol. 5 (ed. P. v. R. Schleyer), Wiley, Athens, 1998, **5**, pp. 3281–3285.
[55] S. J. Weiner, P. A. Kollman, D. A. Case, U. C. Singh, G. Ghio, G. Alagona, S. Profeta, Jr. and P. Weiner, *J. Am. Chem. Soc.*, 1984, **106**, 765.
[56] H. J. C. Berendsen, D. van der Spoel and R. van Drunen, *Comput. Phys. Commun.*, 1995, **91**, 43.
[57] B. R. Brooks, R. E. Bruccoleri, B. D. Olafson, D. J. States, S. Swaminathan and M. Karplus, *J. Comp. Chem.*, 1983, **4**, 187–217.
[58] W. Wagner and A. Pruss, *J. Phys. Chem. Ref. Data*, 2003, **31**, 387–535.
[59] G. S. Kell, *J. Chem. Eng. Data*, 1975, **20**, 97–105.
[60] D. G. Archer, *J. Phys. Chem. B*, 2000, **104**, 8563–8584.
[61] M. R. Reddy and M. Berkowitz, *J. Chem. Phys.*, 1987, **87**, 6682–6686.
[62] H. R. Pruppacher, *J. Chem. Phys.*, 1972, **56**, 101.
[63] F. X. Prielmeier, E. W. Lang, R. J. Speedy and H. D. Ludemann, *Phys. Rev. Lett.*, 1987, **59**, 1128–1131.
[64] C. G. Malmberg and A. A. Maryott, *J. Res. Natl Bur. Stand.*, 1956, **56**, 1.
[65] J. Jonas, T. DeFries and D. J. Wilbur, *J. Chem. Phys.*, 1976, **65**, 582–588.

CHAPTER 6

Molecular Modeling and Atomistic Simulation of Nucleic Acids

Thomas E. Cheatham III

Departments of Medicinal Chemistry and of Pharmaceutics and Pharmaceutical Chemistry, College of Pharmacy, University of Utah, 2000 East 30 South Skaggs Hall 201, Salt Lake City, UT 84112, USA

Contents

1. Introduction 75
2. Successes 77
 2.1. Agreement with experiment 77
 2.2. Insight beyond experiment? 78
 2.3. Methodological and force field advances 79
3. Limitations 80
 3.1. Artifacts from the boundary conditions 80
 3.2. Force field issues and sampling limitations 80
4. Conclusions 82
Acknowledgements 82
References 82

1. INTRODUCTION

Over the last decade, molecular modeling and bio-molecular simulation methods have been increasingly applied to give insight into nucleic acid structure, dynamics, and interaction. This includes simulation of RNA and DNA alone, bound to proteins, interacting with drugs, in various damaged and modified forms, and in different environments ranging from the gas phase to varied solvent and ionic environments. This revolution in the application of bio-molecular simulation methods – including molecular dynamics (MD) simulation and various enhanced sampling and free energy methods – has been made possible by a convergence of capabilities. In particular, enabling technologies include the availability of faster computational hardware (and parallelized simulation codes), better force fields for simulating both explicitly and implicitly solvated nucleic acids [1–6], and the emergence of efficient means to accurately treat the long-range electrostatics interactions either through Ewald methods [7,8] or smoothing of the forces at the cutoff [9,10].

The previous decade (from ~1994 to 2004) might aptly be termed the '10-nanosecond era' as it has been characterized by a large series of MD simulations

ANNUAL REPORTS IN COMPUTATIONAL CHEMISTRY, VOLUME 1
ISSN: 1574-1400 DOI 10.1016/S1574-1400(05)01006-6

of small nucleic acid models in explicit solvent almost exclusively on the 1–10 ns time scale. A large international community of researchers has together applied these methods to nucleic acid systems and exposed the promise of these simulation methods by directly complementing experiment and by further providing detailed information not readily accessible by experiment.

In addition to the promise, these rather short time-scale MD simulations have begun to expose both sampling limitations and force field deficiencies. Although these fundamental limitations are not unique to nucleic acids, nucleic acids typify a class of atomistic systems that tend to easily expose such deficiencies. Nucleic acids are highly charged, highly flexible, and subject to motions across large time and size scales. The structures of nucleic acids are strongly influenced by the surrounding solvent and ionic environment and are stabilized *via* a subtle balance between the hydrophobic/dispersion-attraction induced stacking of the bases, the hydrophilic and ionic interactions of the backbone (and grooves), and specific pairing of the bases. Individual motions of the bases, sugars, and backbone are relatively rapid and short ranged, whereas larger fluctuations, such as base pair opening, bending/twisting, compaction and breathing, occur over longer size and time scales. These issues not only complicate theoretical treatments but also plague experiments where it is difficult to probe fast time scale dynamics (such as sugar repuckering or B_I/B_{II} backbone transitions), to avoid artifacts due to the representation (such as crystal packing artifacts or fast time scale disordering of structure), and to otherwise fully understand the nature of nucleic acids at the atomic level. Particular controversy, in both experiment and theory, relates to the interaction and role of ions in nucleic acid structure, distortion, and dynamics [11–21]. The difficulties in understanding the effect of ions on nucleic structure are not limited to monovalent ions, as recent work clearly demonstrates that putative Mg^{2+} ions intricately bound to RNA are not in every case Mg^{2+} cations; reinterpretation of nucleic acid structures in some cases show that bound Cl^- or sulfate anions are often incorrectly assigned in crystal structures as Mg^{2+} ions [22].

We believe that theory and bio-molecular simulation can fill in the gaps in our understanding of nucleic acid structure and dynamics. Proving this, however, turns out to be a tricky proposition at present since we do not always know what the correct answer is (and, therefore, whether what we observe in simulation is correct or significant), nor do we fully understand the limitations in our models, methods, and representations. In this report, we will highlight some of the successes and some of the limitations in the modeling and simulation of the atomic structure of nucleic acid emerging from current work and work in the last few years. Our intent is not to provide a comprehensive view, but to express this author's opinion related to the current state of the art in modeling of nucleic acids. As the capabilities have evolved in this 10-ns era, the larger 'we' of the nucleic acid simulation community has not only revolutionized our study of nucleic acids, but published a large set of detailed reviews [18,21,23–39]. Detailed overviews of the simulation success and progress for the first half of the nanosecond era are available in the reviews by Beveridge *et al.* [24] and by Cheatham and Kollman [25]. An introduction to modeling and simulation methods as applied to nucleic acids is available in a series of papers by Cheatham *et al.* [40–43] and most recently in a wonderful review of the methods by Orozco *et al.* [35]. Rather comprehensive reviews have recently appeared highlighting DNA dynamics [39], association of ions to RNA [21] and DNA curvature and flexibility [38,44] (both from a simulation and experimental perspective). A current review by our group is forthcoming [45].

2. SUCCESSES

2.1. Agreement with experiment

Before anyone will trust MD simulation results and ultimately believe predictions, it is critical to show agreement with experiment. This is a non-trivial exercise for nucleic acids due to their profound flexibility and sensitivity to the environment. In fact, it is not fully clear from experiment what the rate of sugar repuckering is, nor in detail how fast the fast time scale motions (such as crankshaft transitions in the backbone) are. This makes the comparison between the MD simulation and experiment tough since there is a significant gap in the time scales sampled by MD compared to NMR or crystallography. This is demonstrated nicely in recent work that suggests that good agreement between the ^{13}C NMR relaxation measurements and MD can only be obtained by averaging out the high frequency motions [46]. A further issue is that the effects of the environment are subtle, be it a subtle push from a protein side chain to perturb the DNA structure [47] to crystallization artifacts [48,49], to the observed non-linear dependence of melting temperature on Na^+ ion concentration [50]. In reproducing nucleic acid structure, three dominant force fields have emerged. These include the Cornell *et al.* force field [1] and its recent variants [2,3], the CHARMM27 all-atom force field [4,5] and a hybrid force field by Langley [6]. All three of these force fields have been shown to perform rather well in reproducing B-DNA structure [28,51] and in a wide variety of applications as discussed herein and in the cited reviews. The methods have also been shown to give detailed insight into nucleic acid dynamics [39,52–54], including the coupled motions in the DNA backbones [55–59]. Base pair opening has also received considerable attention through the application of umbrella sampling methods that allow estimations of the potential of mean force for opening the pair [60–63]. Similar methods have been applied to probe the distortion of base pair steps by insertion of protein side chains into the minor groove [64]. Such opening, dynamic, and deformability events likely play a key role in DNA repair [65–68] and recognition of nucleic acids by proteins [69,70].

A clear success of the methods has been demonstrated by the Beveridge and Bolton groups who provide a detailed comparison between MD-calculated and NMR-observed properties of the duplex d(CGCGAATTCGCG)$_2$ in solution [71]. Their results show that the 2D NOESY volumes and scalar coupling constants back calculated from a 14-ns MD simulation are in better agreement with the NMR data than back calculations based on canonical A-DNA, B-DNA, or crystal geometries of the same duplex. Another clear set of successes relates to the combination of experiment and theory to understand the interaction of drugs with the minor groove of DNA. Wilson and Neidle have teamed up to investigate the binding of the drug CGP 40215A to AT-rich sequences in DNA using a combination of surface plasmon resonance, footprinting, CD and UV spectroscopy, and MD simulation [72,73]. This group also investigated a series of other bisbenzimidazole derivatives to give insight into 2:1 binding at TTAA *vs.* AATT sequences [74]. Similar work, including NMR, gives insight into nickel–peptide interactions with the minor groove of DNA [75]. We have also demonstrated reasonable estimates of the relative binding affinity of DAPI bound to the minor groove of DNA in various different binding modes (with a variety of different force fields) using the MM-PBSA methodology [76]; to get reasonable binding affinity estimates, we suggest it is necessary to include some of the specifically bound explicit water [77]. The importance of water is also shown by Liedl and co-workers, however, in this case it is due to differential hydration of the free drug [78]. Beyond DNA, simulations have also probed intercalating dye interactions with RNA [79] and the affinity of ligands binding to a RNA aptamer [80]. Calculating absolute binding affinities is difficult, not only due to subtleties in the balance between the molecular mechanical energies and continuum solvation, but also mainly due to the difficulty in accurately estimating the solute entropic components and in particular the loss of rotational and translational freedom. These issues, including a detailed theoretical framework for the MM-PBSA approaches are presented in recent work [81].

2.2. Insight beyond experiment?

The simulations over the past decade suggest that MD simulations applied to nucleic acids have come of age and are ready to give detailed insights that not only complement experiment but also are predictive. Areas where we have glimpsed beyond what is accessible experimentally includes detailed investigation of DNA in the gas phase [82], DNA on surfaces [83], the effect of dynamics on the electronic properties of DNA [84–87], and in the simulation of new models or modified bases and backbones. Examples include investigations into the formation of quadruplex DNA (showing the importance of the integral monovalent ions) [88], antiparallel triplex DNA with modified bases [89], antiparallel Hoogsteen duplexes [90], and damaged DNA alone or interacting with proteins [91–94]. Simulations have also probed the effect of specific base substitution in RNA hairpin loops [95].

2.3. Methodological and force field advances

Although these advances are not limited to applications involving nucleic acids, there have been a number of methodological developments and force field enhancements that directly or indirectly lead to improvements in the analysis, simulation accuracy, or sampling of nucleic acid structures. One common need is methods for fast and accurate estimation of solvation and electrostatics effects. Implicit solvent models can greatly speed up the calculations allowing for significantly longer simulation, post-processing of the MD results with MM-PBSA type methods, or even ligand docking studies. Kang *et al.* optimized a set of parameters for a generalized Born implicit solvent model in the docking program DOCK with reference to a large set of ligand–DNA and ligand–RNA complexes [96]. Using MD simulation and free energy perturbation methods, Banavali and Roux developed a set of continuum radii (for use in Poisson–Boltzmann calculations) that is consistent with the all-atom CHARMM27 force field for nucleic acids [97]. Although, the current set of implicit solvent models are not perfect – and generally not as accurate as simulations in explicit solvent – such methods give ready insight into nucleic acid structure and dynamics. This includes insight into the deformation of DNA induced by protein binding [98], RNA loop dynamics [99] and folding [100,101] mismatches in RNA structure [102], drug binding to DNA [103], and characterization of unusual DNA structures [104].

The best behaved efficient implicit solvent models for use in MD simulation are likely the generalized Born methods [105,106]. The generalized Born model, when compared to simulation in explicit solvent, does a decent job at probing the structural and dynamical implications of replacing a single base with pyrene [107]. However, even with the 'faster' implicit solvent models, sampling is still limited as shown by trapping of the nearby 5′ adenine into both *syn* and *anti* conformations with no inter-conversion. To allow inter-conversion and proper sampling of both states, enhanced sampling methods are required as shown by the application of the locally enhanced sampling (LES) method during simulations in explicit solvent. More recently this group has extended the LES methodology to work correctly within the generalized Born implicit solvent model and applied this to RNA hairpin loop structures [108]. Compared to LES in explicit solvent, more motion is evident (including greater exploration of different transition paths by the copies) in the LES simulations in implicit solvent.

3. LIMITATIONS

3.1. Artifacts from the boundary conditions

In addition to significantly increasing the computational cost, the use of explicit solvent in MD simulations of nucleic acids typically comes at a further price; specifically the potential for artifacts due to the boundary conditions either when periodic boundary conditions are applied (such as in Ewald calculations) or when a finite system (such as a solvent blob) is simulated. In the former case, there is significant worry that the artificial imposition of periodicity could dampen motion [109] and otherwise over-stabilize particular structures [110,111]. Recently, Hunenberger and co-workers [112] have re-investigated this issue in the context of nucleic acid simulation. Contrary to intuition, the induced periodicity tends to *increase* fluctuations in the DNA structure (although the periodicity tends to over-stabilize the native conformation particularly in the absence of ions). Fortunately, the magnitude of these artifacts for a water-solvated system are not too large (on the order of kT) and appear to be less of a problem under conditions of just net-neutralizing salt (as opposed to no salt or excess salt) which represents the most commonly simulated conditions for nucleic acids. Periodicity artifacts are likely to be more of an issue in low dielectric solvents or if sufficient solvation is not present. The alternative to periodic boundaries, for simulations including explicit solvent, is some finite representation. In these cases, there is often a problem due to the discontinuity between the explicit solvent and the external environment (often a vacuum). In the worst case, structures can be completely over-stabilized and dynamics inhibited, such as we have previously shown in the simulation of nucleic acids within a minimal explicit solvent bath applying a distance-dependent dielectric constant [28]. To overcome this, Mazur developed a model that appears to perform well in the representation of the A-DNA $\leftrightarrow$ B-DNA equilibrium in a minimal explicit solvent environment [113]. Also, recently in the context of simulations of lipid bilayers, reaction field methods were applied and compared to Ewald simulation [114]. Although the results with a reaction field differ from the Ewald results, they are significantly better than truncation of the long-range electrostatic interactions and may hold promise for simulation of nucleic acids.

3.2. Force field issues and sampling limitations

Another significant limitation in MD simulation comes not from the nucleic acid force field itself, but from the ionic environment and inconsistency in the parameters. The inconsistencies include the neglect of polarization, the wide variety of parameterizations available, artifacts due to initial ion placement, and poor sampling efficiency. We have touched on issues related to the long-time scales required for equilibration of even monovalent ions and shown that artifacts from initial ion placement can lead to perturbed structure and significant overestimation of the lifetime of bound ions [28]. In some cases, such as the use of the standard Mg^{2+}, K^{+}, and Cl^{-} parameters in the Cornell *et al.* force field [1], we have observed magnesium ions tunnel in between bases to disrupt pairing and even spontaneous formation of salt crystals in simulations of phased A-tracts (Cheatham, unpublished data). That there could be issues in mixing and matching ion parameters is highlighted in a nice systematic study of various parameters for sodium and chloride applied in MD simulations [115]. The simulations suggest incredible differences in the calculated coordination states, transport properties, and radial distribution functions. Sampling limitations are also significant. Very recently, detailed experimental studies on monovalent ion exchange have appeared; these provide estimates of occupancies in the 50% range and lifetimes in the 10 ns to 100 μs for Na^{+} bound into the minor groove of A-tract regions [116]. As our simulations are just beginning to push the 100 ns barrier, and further since there are hints that we are underestimating our ion–nucleic acid interaction strength [117,118], there may be reason for concern. The problems are further compounded by difficulties in interpreting ion densities from experiment. For example, it is tricky to distinguish Na^{+} or K^{+} ions from water. Recently, Auffinger and Westhof performed a systematic analysis of nucleic acid structures, including MD simulation, and convincingly suggest that some of the ions placed in nucleic acid structure may not actually be cations but anions [22]. Despite these issues, MD simulations have given detailed insight into ion influences on DNA structure including B-DNA to A-DNA transitions [113,119–121] and have strongly suggested that the binuclear Mg^{2+} binding motif observed in the crystal structures of the 5S rRNA loop E are better described by partial occupancy by ions [122] and also that clear K^{+} binding sites are evident [123]. Early on, MD simulations also demonstrated clear ion binding to the minor groove of A-tract DNA [12] and also in the major groove of RNA [124], and more recently characterized the interaction of polyamines with model fibers [125,126].

Beyond the hard to reconcile and subtle force field differences seen to date – such as differences in the effective rate of sugar repuckering and backbone transitions to variations in twist and groove widths seen when comparing the nucleic acid force fields – larger deficiencies have started to emerge. This includes significant differences in the effective opening rates for bases in RNA comparing the CHARMM27 all-atom force fields to the Cornell *et al.* force field [124,127] and apparent force field differences emerging in simulations of various RNA loops or bulges [128–131]. At this point in time, there are unresolved issues in that *both* sets of force fields have shown stability in some cases and instability in others. A more dramatic demonstration of issues with loop structures is seen in the recent studies of DNA quadruplex loop geometries, where the converged LES and MD simulation runs in explicit solvent locate a favorable loop geometry that differs considerably from the experimental structures and lacks characteristic ion binding; at this point, it is not clear if the deficiencies relate to the ion parameters, the DNA force field or the MM-PBSA post-processing [132].

4. CONCLUSIONS

Although there are still issues with the force fields and sampling limitations, MD simulation of nucleic acids has proven to be a valuable tool for giving detailed atomistic insight into nucleic acid structure, dynamics, and interactions. With anticipated improvements in the methods and force fields, and continued advances in the computational technology, it is clear that we will break the microsecond barrier in the near future and facilitate investigation of larger nucleic acid assemblies (such as in the recent work on the ribosome [133,134]).

ACKNOWLEDGEMENTS

Cheatham would like to thank his many colleagues in the field for good advice, information sharing, and advancements in the field related to simulating nucleic acids. Particular thanks go to Jiri Sponer, Carlos Simmering, David Beveridge, David Case, Alex MacKerell, Bernie Brooks, Filip Lankas, Pascal Auffinger, David Langley, Piotr Cieplak and Peter Kollman. Financial support (NSF ITR CHE-0326027, NSF ITR CHE-0218739) and computational power (NRAC MCA01S027P, U. of Utah CHPC) are greatly appreciated.

REFERENCES

[1] W. D. Cornell, P. Cieplak, C. I. Bayly, I. R. Gould, K. M. Merz, D. M. Ferguson, D. C. Spellmeyer, T. Fox, J. W. Caldwell and P. A. Kollman, A second generation force field for the simulation of proteins, nucleic acids, and organic molecules, *J. Am. Chem. Soc.*, 1995, **117**, 5179–5197.
[2] T. E. Cheatham, III, P. Cieplak and P. A. Kollman, A modified version of the Cornell *et al.* force field with improved sugar pucker phases and helical repeat, *J. Biomol. Struct. Dyn.*, 1999, **16**, 845–862.
[3] J. Wang, P. Cieplak and P. A. Kollman, How well does a restrained electrostatic potential (RESP) model perform in calculating conformational energies of organic and biological molecules?, *J. Comput. Chem.*, 2000, **21**, 1049–1074.
[4] N. Foloppe and A. D. MacKerell, Jr., All-atom empirical force field for nucleic acids. 1) Parameter optimization based on small molecule and condensed phase macromolecular target data, *J. Comput. Chem.*, 2000, **21**, 86–104.
[5] A. D. MacKerell, Jr. and N. Banavali, All-atom empirical force field for nucleic acids. 2) Application to molecular dynamics simulations of DNA and RNA in solution, *J. Comput. Chem.*, 2000, **21**, 105–120.
[6] D. R. Langley, Molecular dynamics simulations of environment and sequence dependent DNA conformation: the development of the BMS nucleic acid force field

and comparison with experimental results, *J. Biomol. Struct. Dyn.*, 1998, **16**, 487–509.
[7] U. Essmann, L. Perera, M. L. Berkowitz, T. Darden, H. Lee and L. G. Pedersen, A smooth particle mesh Ewald method, *J. Chem. Phys.*, 1995, **103**, 8577–8593.
[8] C. Sagui and T. A. Darden, Molecular dynamics simulations of biomolecules: long-range electrostatic effects, *Ann. Rev. Biophys. Biomol. Struct.*, 1999, **28**, 155–179.
[9] P. J. Steinbach and B. R. Brooks, New spherical-cutoff methods for long-range forces in macromolecular simulation, *J. Comput. Chem.*, 1994, **15**, 667–683.
[10] J. Norberg and L. Nilsson, On the truncation of long-range electrostatic interactions in DNA, *Biophys. J.*, 2000, **79**, 1537–1553.
[11] J. K. Strauss, C. Roberts, M. G. Nelson, C. Switzer and L. J. Maher, III, DNA bending by hexamethylene-tethered ammonium ions, *Proc. Natl Acad. Sci.*, 1996, **93**, 9515–9520.
[12] M. A. Young, B. Jayaram and D. L. Beveridge, Intrusion of counterions into the spine of hydration in the minor groove of B-DNA: fractional occupancy of electronegative pockets, *J. Am. Chem. Soc.*, 1997, **119**, 59–69.
[13] T. K. Chiu, M. Kaczor-Grzeskowiak and R. E. Dickerson, Absence of minor groove monovalent cations in the crosslinked dodecamer CGCGA$\underline{A}$TTCGCG, *J. Mol. Biol.*, 1999, **292**, 589–608.
[14] K. J. McConnell and D. L. Beveridge, DNA structure: what's in charge?, *J. Mol. Biol.*, 2000, **304**, 803–820.
[15] D. Hamelberg, L. McFail-Isom, L. D. Williams and W. D. Wilson, Flexible structure of DNA: ion dependence of minor-groove structure and dynamics, *J. Am. Chem. Soc.*, 2001, **122**, 10513–10520.
[16] N. V. Hud and M. Polak, DNA–cation interactions: the major and minor grooves are flexible ionophores, *Curr. Opin. Struct. Biol.*, 2001, **11**, 293–301.
[17] V. Tereshko, C. J. Wilds, G. Minasov, T. P. Prakash, M. A. Maier, A. Howard, Z. Wawrzak, M. Manoharan and M. Egli, Detection of alkali metal ions in DNA crystals using state-of-the-art X-ray diffraction experiments, *Nucleic Acids Res.*, 2001, **29**, 1208–1215.
[18] M. Egli, DNA–cation interactions: quo vadis?, *Chem. Biol.*, 2002, **9**, 277–286.
[19] F. Mocci and G. Saba, Molecular dynamics simulations of A–T-rich oligomers: sequence-specific binding of Na^+ in the minor groove of B-DNA, *Biopolymers*, 2003, **68**, 471–485.
[20] J. A. Subirana and M. Soler-Lopez, Cations as hydrogen bond donors: a view of electrostatic interactions in DNA, *Annu. Rev. Biophys. Biomol. Struct.*, 2003, **32**, 27–45.
[21] D. E. Draper, A guide to ions and RNA structure, *RNA*, 2004, **10**, 335–343.
[22] P. Auffinger, L. Bielecki and E. Westhof, Anion binding to nucleic acids, *Structure*, 2004, **12**, 379–388.
[23] P. Auffinger and E. Westhof, RNA solvation: a molecular dynamics simulation perspective, *Biopolymers*, 2000, **56**, 266–274.
[24] D. L. Beveridge and K. J. McConnell, Nucleic acids: theory and computer simulation, Y2K, *Curr. Opin. Struct. Biol.*, 2000, **10**, 182–196.
[25] T. E. Cheatham, III and P. A. Kollman, Molecular dynamics simulation of nucleic acids, *Ann. Rev. Phys. Chem.*, 2000, **51**, 435–471.
[26] P. A. Kollman, I. Massova, C. Reyes, B. Kuhn, S. Huo, L. Chong, M. Lee, T. Lee, Y. Duan, W. Wang, O. Donini, P. Cieplak, J. Srinivasan, D. A. Case and T. E. Cheatham, III, Calculating structures and free energies of complex molecules: combining molecular mechanics and continuum models, *Acc. Chem. Res.*, 2000, **33**, 889–897.
[27] A. N. Lane and T. C. Jenkins, Thermodynamics of nucleic acids and their interactions with ligands, *Q. Rev. Biophys.*, 2000, **33**, 255–306.
[28] T. E. Cheatham, III and M. A. Young, Molecular dynamics simulations of nucleic acids: successes, limitations and promise, *Biopolymers*, 2001, **56**, 232–256.

[29] F. Major and R. Griffey, Computational methods for RNA structure determination, *Curr. Opin. Struct. Biol.*, 2001, **11**, 282–286.
[30] W. Wang, O. Donini, C. M. Reyes and P. A. Kollman, Biomolecular simulations: recent developments in force fields, simulations of enzyme catalysis, protein–ligand, protein–protein, and protein–nucleic acid noncovalent interactions, *Ann. Rev. Biophys. Biomol. Struct.*, 2001, **30**, 211–243.
[31] E. Giudice and R. Lavery, Simulations of nucleic acids and their complexes, *Acc. Chem. Res.*, 2002, **35**, 350–357.
[32] R. Lavery, A. LeBrun, J. F. Allemand, D. Bensimon and V. Croquette, Structure and mechanics of single biomolecules: experiment and simulation, *J. Phys. Condens. Matter*, 2002, **14**, R383–R414.
[33] V. Makarov, B. M. Pettitt and M. Feig, Solvation and hydration of proteins and nucleic acids: a theoretical view of simulation and experiment, *Acc. Chem. Res.*, 2002, **35**, 376–384.
[34] J. Norberg and L. Nilsson, Molecular dynamics applied to nucleic acids, *Acc. Chem. Res.*, 2002, **35**, 465–472.
[35] M. Orozco, A. Perez, A. Noy and F. Javier Luque, Theoretical methods for the simulation of nucleic acids, *Chem. Soc. Rev.*, 2003, **32**, 350–364.
[36] J. Sponer and P. Hobza, Molecular interactions of nucleic acid bases. A review of quantum-chemical studies, *Collect. Czech. Chem. Commun.*, 2003, **68**, 2231–2282.
[37] Y. N. Vorobjev, In silico modeling and conformational mobility of DNA duplexes, *Mol. Biol.*, 2003, **37**, 210–222.
[38] D. L. Beveridge, S. B. Dixit, G. Barreiro and K. M. Thayer, Molecular dynamics simulation of DNA curvature and flexibility: helix phasing and premelting, *Biopolymers*, 2004, **73**, 380–403.
[39] F. Lankas, DNA sequence-dependent deformability – insights from computer simulations, *Biopolymers*, 2004, **73**, 327–339.
[40] T. E. Cheatham, III, B. R. Brooks and P. A. Kollman, Molecular modeling of nucleic acid structure. In *Current Protocols in Nucleic Acid Chemistry* (eds S. L. Beaucage, D. E. Bergstrom, G. D. Glick and R. A. Jones), Wiley, New York, 2000, Vol. 1, pp. 7.5.1–7.5.12.
[41] T. E. Cheatham, III, B. R. Brooks and P. A. Kollman, Molecular modeling of nucleic acid structure: setup and analysis. In *Current Protocols in Nucleic Acid Chemistry* (eds S. L. Beaucage, D. E. Bergstrom, G. D. Glick and R. A. Jones), Wiley, New York, 2001, Vol. 1, pp. 7.10.1–7.10.18.
[42] T. E. Cheatham, III, B. R. Brooks and P. A. Kollman, Molecular modeling of nucleic acid structure: electrostatics and solvation. In *Current Protocols in Nucleic Acid Chemistry* (eds S. L. Beaucage, D. E. Bergstrom, G. D. Glick and R. A. Jones), Wiley, New York, 2001, Vol. 1, pp. 7.9.1–7.9.21.
[43] T. E. Cheatham, III, B. R. Brooks and P. A. Kollman, Molecular modeling of nucleic acid structure: energy and sampling. In *Current Protocols in Nucleic Acid Chemistry* (eds S. L. Beaucage, D. E. Bergstrom, G. D. Glick and R. A. Jones), Wiley, New York, 2001, Vol. 1, pp. 7.8.1–7.8.20.
[44] N. V. Hud and J. Plavec, A unified model for the origin of DNA sequence-directed curvature, *Biopolymers*, 2003, **69**, 144–158.
[45] T. E. Cheatham, III, Simulation and modeling of nucleic acid structure, dynamics and interactions, *Curr. Opin. Struct. Biol.*, 2004, **14**, 360–367.
[46] R. J. Isaacs and H. P. Spielmann, Insight into G–T mismatch recognition using molecular dynamics with time-averaged restraints derived from NMR spectroscopy, *J. Am. Chem. Soc.*, 2004, **126**, 583–590.
[47] M. Y. Tolstorukov, R. L. Jernigan and V. B. Zhurkin, Protein–DNA hydrophobic recognition in the minor groove is facilitated by sugar switching, *J. Mol. Biol.*, 2004, **337**, 65–76.

[48] C. A. Bingman, G. Zon and M. Sundaralingam, Crystal and molecular structure of the A-DNA dodecamer d(CCGTACGTACGG). Choice of fragment helical axis, *J. Mol. Biol.*, 1992, **227**, 738–756.
[49] R. E. Dickerson, D. S. Goodsell and S. Neidle, "...the tyranny of the lattice...", *Proc. Natl Acad. Sci. USA*, 1994, **91**, 3579–3583.
[50] R. Owczarzy, Y. You, B. G. Moreira, J. A. Manthey, L. Huang, M. A. Behlke and J. A. Walder, Effects of sodium ions on DNA duplex oligomers: improved predictions of melting temperatures, *Biochemistry*, 2004, **43**, 3537–3554.
[51] S. Y. Reddy, F. Leclerc and M. Karplus, DNA polymorphism: a comparison of force fields for nucleic acids, *Biophys. J.*, 2003, **84**, 1421–1449.
[52] F. Lankas, T. E. Cheatham, III, P. Hobza, J. Langowski, N. Spackova and J. Sponer, Critical effect of the N2 amino group on structure, dynamics and elasticity of DNA polypurine tracts, *Biophys. J.*, 2002, **82**, 2592–2609.
[53] F. Lankas, J. Sponer, J. Langowski and T. E. Cheatham, III, DNA base-pair step deformability inferred from molecular dynamics simulation, *Biophys. J.*, 2003, **85**, 2872–2883.
[54] F. Lankas, J. Sponer, J. Langowski and T. E. Cheatham, III, DNA deformability at the base pair level, *J. Am. Chem. Soc.*, 2004, **126**, 4124–4125.
[55] P. Varnai, D. Djuranovic, R. Lavery and B. Hartmann, alpha/gamma transitions in the B-DNA backbone, *Nucleic Acids Res.*, 2002, **30**, 5398–5406.
[56] C. Rauch, M. Trieb, B. Wellenzohn, M. J. Loferer, A. Voegele, F. R. Wibowo and K. R. Liedl, C5-methylation of cytosine in B-DNA thermodynamically and kinetically stabilizes B_I, *J. Am. Chem. Soc.*, 2003, **125**, 14990–14991.
[57] M. Trieb, C. Rauch, B. Wellenzohn, F. R. Wibowo, T. Loerting and K. R. Liedl, Dynamics of DNA: B-I and B-II phosphate backbone transitions, *J. Phys. Chem. B*, 2004, **108**, 2470–2476.
[58] M. Trieb, C. Rauch, B. Wellenzohn, F. R. Wibowo, T. Loerting, E. Mayer and K. R. Liedl, Daunomycin intercalation stabilizes distinct backbone conformations of DNA, *J. Biomol. Struct. Dyn.*, 2004, **21**, 713–724.
[59] D. Djuranovic and B. Hartmann, DNA fine structure and dynamics in crystals and in solution: the impact of BI/BII backbone conformations, *Biopolymers*, 2004, **73**, 356–368.
[60] P. Varnai and R. Lavery, Base flipping in DNA: pathways and energetics studied with molecular dynamics simulations, *J. Am. Chem. Soc.*, 2002, **124**, 7272–7273.
[61] E. Giudice and R. Lavery, Nucleic acid base pair dynamics: the impact of sequence and structure using free-energy calculations, *J. Am. Chem. Soc.*, 2003, **125**, 4998–4999.
[62] E. Giudice, P. Varnai and R. Lavery, Base pair opening within B-DNA: free energy pathways for GC and AT pairs from umbrella sampling simulations, *Nucleic Acids Res.*, 2003, **31**, 1434–1443.
[63] N. Huang, N. K. Banavali and A. D. MacKerell, Jr., Protein-facilitated base flipping in DNA by cytosine-5-methyltransferase, *Proc. Natl Acad. Sci.*, 2003, **100**, 68–73.
[64] D. Bosch, M. Campillo and L. Pardo, Binding of proteins to the minor groove of DNA: what are the structural and energetic determinants for kinking a basepair step?, *J. Comput. Chem.*, 2003, **24**, 682–691.
[65] E. Seibert, J. B. A. Ross and R. Osman, Role of DNA flexibility in sequence-dependent activity of uracil DNA glycosylase, *Biochemistry*, 2002, **41**, 10976–10984.
[66] E. Seibert, J. B. A. Ross and R. Osman, Contribution of opening and bending dynamics to specific recognition of DNA damage, *J. Mol. Biol.*, 2003, **330**, 687–703.
[67] M. Pinak, 8-Oxoguanine lesioned B-DNA molecule complexed with repair enzyme hOGG1: a molecular dynamics study, *J. Comput. Chem.*, 2003, **24**, 898–907.
[68] H. T. Allawi, M. W. Kaiser, A. V. Onufriev, W. P. Ma, A. E. Brogaard, D. A. Case, B. P. Neri and V. I. Lyamichev, Modeling of flap endonuclease interactions with DNA substrate, *J. Mol. Biol.*, 2003, **328**, 537–554.

[69] B. Wellenzohn, W. Flader, R. H. Winger, A. Hallbrucker, E. Mayer and K. R. Liedl, Indirect readout of the *trp*-repressor–operator complex by B-DNA's backbone conformation transitions, *Biochemistry*, 2002, **41**, 4088–4095.
[70] K. S. Byun and D. L. Beveridge, Molecular dynamics simulations of papilloma virus E2 DNA sequences: dynamical models for oligonucleotide structures in solution, *Biopolymers*, 2004, **73**, 369–379.
[71] H. Arthanari, K. J. McConnell, R. Berger, M. A. Young, D. L. Beveridge and P. H. Bolton, Assessment of the molecular dynamics structure of DNA in solution based on calculated and observed NMR NOESY volumes and dihedral angles from scalar coupling constants, *Biopolymers*, 2003, **68**, 3–15.
[72] B. Nguyen, M. P. H. Lee, D. Hamelberg, A. Joubert, C. Bailly, R. Brun, S. Neidle and W. D. Wilson, Strong binding in the DNA minor groove by an aromatic diamidine with a shape that does not match the curvature of the groove, *J. Am. Chem. Soc.*, 2002, **124**, 13680–13681.
[73] B. Nguyen, D. Hamelberg, C. Bailly, P. Colson, J. Stanek, R. Brun, S. Neidle and W. D. Wilson, Characterization of a novel DNA minor-groove complex, *Biophys. J.*, 2004, **86**, 1028–1041.
[74] F. A. Tanious, D. Hamelberg, C. Bailly, A. Czarny, D. W. Boykin and W. D. Wilson, DNA sequence dependent monomer-dimer binding modulation of asymmetric benzimidazole derivatives, *J. Am. Chem. Soc.*, 2004, **126**, 143–153.
[75] Y.-Y. Fang, B. D. Ray, C. A. Claussen, K. B. Lipkowitz and E. C. Long, Ni(II)-Arg-Gly-His–DNA interactions: investigation into the basis for minor groove binding and recognition, *J. Am. Chem. Soc.*, 2004, **126**, 5403–5412.
[76] J. Srinivasan, T. E. Cheatham, III, P. Cieplak, P. A. Kollman and D. A. Case, Continuum solvent studies of the stability of DNA, RNA and phosphoramidate helices, *J. Am. Chem. Soc.*, 1998, **120**, 9401–9409.
[77] N. Spackova, T. E. Cheatham, III, F. Ryjacek, F. Lankas, L. van Meervelt, P. Hobza and J. Sponer, Molecular dynamics simulations and thermodynamic analysis of DNA–drug complexes. Minor groove binding between 4′,6-diamidino-2-phenylindole (DAPI) and DNA duplexes in solution, *J. Am. Chem. Soc.*, 2003, **125**, 1759–1769.
[78] B. Wellenzohn, M. J. Loferer, M. Trieb, C. Rauch, R. H. Winger, E. Mayer and K. R. Liedl, Hydration of hydroxypyrrole influences binding of ImHpPyPy-beta-Dp polyamide to DNA, *J. Am. Chem. Soc.*, 2003, **125**, 1088–1095.
[79] D. H. Nguyen, T. Dieckmann, M. E. Colvin and W. H. Fink, Dynamics studies of a malachite green–RNA complex revealing the origin of the red-shift and energetic contributions of stacking interactions, *J. Phys. Chem. B*, 2004, **108**, 1279–1286.
[80] H. Gouda, I. D. Kuntz, D. A. Case and P. A. Kollman, Free energy calculations for theophylline binding to an RNA aptamer: comparison of MM-PBSA and thermodynamic integration methods, *Biopolymers*, 2003, **68**, 16–34.
[81] J. M. J. Swanson, R. H. Henchman and J. A. McCammon, Revisiting free energy calculations: a theoretical connection to MM/PBSA and direct calculation of the association free energy, *Biophys. J.*, 2004, **86**, 67–74.
[82] M. Rueda, S. G. Kalko, F. Javier Luque and M. Orozco, The structure and dynamics of DNA in the gas phase, *J. Am. Chem. Soc.*, 2003, **125**, 8007–8014.
[83] K.-Y. Wong and B. M. Pettitt, Orientation of DNA on a surface from simulation, *Biopolymers*, 2004, **73**, 570–578.
[84] R. N. Barnett, C. L. Cleveland, A. Joy, U. Landman and G. B. Schuster, Charge migration in DNA: ion-gated transport, *Science*, 2001, **294**, 567–571.
[85] J. P. Lewis, J. Pikus, T. E. Cheatham, III, E. B. Starikov, H. Wang, J. Tomfohr and O. F. Sankey, A comparison of electronic states in periodic and aperiodic poly(dA)–poly(dT) DNA, *Phys. Status Solidi (b)*, 2002, **233**, 90–100.
[86] J. P. Lewis, T. E. Cheatham, III, H. Wang, E. B. Starikov and O. F. Sankey, Dynamically amorphous character of electronic states in poly(dA)–poly(dT) DNA, *J. Phys. Chem. B*, 2003, **107**, 2581–2587.

[87] B. Bouvier, J. P. Dognon, R. Lavery, D. Markovitsi, P. Millie, D. Onidas and K. Zakrzewska, Influence of conformational dynamics on the exciton states of DNA oligomers, *J. Phys. Chem. B*, 2003, **107**, 13512–13522.
[88] R. Stefl, T. E. Cheatham, III, N. Spackova, E. Fadrna, I. Berger, J. Koca and J. Sponer, Formation pathways of a guanine-quadruplex DNA revealed by molecular dynamics and thermodynamical analysis of the substates, *Biophys. J.*, 2003, **85**, 1787–1804.
[89] A. Avino, E. Cubero, C. Gonzalez, R. Eritja and M. Orozco, Antiparallel triple helices. Structural characteristics and stabilization by 8-amino derivatives, *J. Am. Chem. Soc.*, 2003, **125**, 16127–16138.
[90] E. Cubero, N. G. A. Abrescia, J. A. Subirana, F. Javier Luque and M. Orozco, Theoretical study of a new DNA structure: the antiparallel Hoogsteen duplex, *J. Am. Chem. Soc.*, 2003, **125**, 14603–14612.
[91] M. Wu, S. Yan, D. J. Patel, N. E. Geacintov and S. Broyde, Relating repair susceptibility of carcinogen-damaged DNA with structural distortion and thermodynamic stability, *Nucleic Acids Res.*, 2002, **30**, 3422–3432.
[92] S. X. Yan, M. Wu, D. J. Patel, N. E. Geacintov and S. Broyde, Simulating structural and thermodynamic properties of carcinogen-damaged DNA, *Biophys. J.*, 2003, **84**, 2137–2148.
[93] R. A. Perlow and S. Broyde, Extending the understanding of mutagenicity: structural insights into primer-extension past a benzo[*a*]pyrene diol epoxide–DNA adduct, *J. Mol. Biol.*, 2003, **327**, 797–818.
[94] S. X. Yan, M. Wu, T. Buterin, H. Naegeli, N. E. Geacintov and S. Broyde, Role of base sequence context in conformational equilibria and nucleotide excision repair of benzo[*a*]pyrene diol epoxide–adenine adducts, *Biochemistry*, 2003, **42**, 2339–2354.
[95] J. Sarzynska, L. Nilsson and T. Kulinski, Effects of base substitutions in an RNA hairpin from molecular dynamics and free energy simulations, *Biophys. J.*, 2003, **85**, 3445–3459.
[96] X. Kang, R. H. Schafer and I. D. Kuntz, Calculation of ligand–nucleic acid binding free energies with the generalized-Born model in DOCK, *Biopolymers*, 2004, **73**, 192–204.
[97] N. K. Banavali and B. Roux, Atomic radii for continuum electrostatics calculations on nucleic acids, *J. Phys. Chem. B*, 2002, **106**, 11026–11035.
[98] K. Zakrzewska, DNA deformation energetics and protein binding, *Biopolymers*, 2003, **70**, 414–423.
[99] K. B. Hall and D. J. Williams, Dynamics of the IRE RNA hairpin loop probed by 2-aminopurine fluorescence and stochastic dynamics simulations, *RNA*, 2004, **10**, 34–47.
[100] E. J. Sorin, M. A. Engelhardt, D. Herschlag and V. S. Pande, RNA simulations: probing hairpin unfolding and the dynamics of a GNRA tetraloop, *J. Mol. Biol.*, 2002, **317**, 493–506.
[101] E. J. Sorin, Y. M. Rhee, B. J. Nakatani and V. S. Pande, Insights into nucleic acid conformational dynamics from massively parallel stochastic simulations, *Biophys. J.*, 2003, **85**, 790–803.
[102] G. Villescas-Diaz and M. Zacharias, Sequence context dependence of tandem guanine:adenine mismatch conformations in RNA: a continuum solvent analysis, *Biophys. J.*, 2003, **85**, 416–425.
[103] L. F. De Castro and M. Zacharias, DAPI binding to the DNA minor groove: a continuum solvent analysis, *J. Mol. Recogn.*, 2002, **15**, 209–220.
[104] T. E. Malliavin, J. Gau, K. Snoussi and J. L. Leroy, Stability of the I-motif structure is related to the interactions between phosphodiester backbones, *Biophys. J.*, 2003, **84**, 3838–3847.
[105] V. Tsui and D. A. Case, Molecular dynamics simulations of nucleic acids with a generalized Born solvation model, *J. Am. Chem. Soc.*, 2000, **122**, 2489–2498.

[106] V. Tsui and D. A. Case, Theory and applications of the generalized Born solvation model in macromolecular simulations, *Biopol. Nucleic Acid Sci.*, 2001, **56**, 275–291.
[107] G. Cui and C. Simmerling, Conformational heterogeneity observed in simulations of a pyrene-substituted DNA, *J. Am. Chem. Soc.*, 2002, **124**, 12154–12164.
[108] X. Cheng, V. Hornak and C. Simmerling, Improved conformational sampling through an efficient combination of mean-field simulation approaches, *J. Phys. Chem. B*, 2004, **108**, 426–437.
[109] T. E. Cheatham, III and B. R. Brooks, Recent advances in molecular dynamics simulation towards the reliable representation of biomolecules in solution, *Theor. Chem. Acc.*, 1998, **99**, 279–288.
[110] P. H. Hunenberger and J. A. McCammon, Effect of artificial periodicity in simulations of biomolecules under Ewald boundary conditions: a continuum electrostatics study, *Biophys. Chem.*, 1999, **78**, 69–88.
[111] W. Weber, P. H. Hunenberger and J. A. McCammon, Molecular dynamics simulations of a polyalanine octapeptide under Ewald boundary conditions: influence of artificial periodicity on peptide conformation, *J. Phys. Chem. B*, 2000, **104**, 3668–3675.
[112] M. A. Kastenholz and P. H. Hunenberger, Influence of artificial periodicity and ionic strength in molecular dynamics simulations of charged biomolecules employing lattice sums, *J. Phys. Chem. B*, 2004, **108**, 774–788.
[113] A. K. Mazur, Titration *in silico* of reversible B $\leftrightarrow$ A transitions in DNA, *J. Am. Chem. Soc.*, 2003, **125**, 7849–7859.
[114] M. Patra, M. Karttunen, M. T. Hyvonen, E. Falck and I. Vattulainen, Lipid bilayers driven to a wrong lane in molecular dynamics simulations by subtle changes in long-range electrostatic interactions, *J. Phys. Chem. B*, 2004, **108**, 4485–4494.
[115] M. Patra and M. Karttunen, Systematic comparison of force fields for microscopic simulations of NaCl in aqueous solutions: diffusion, free energy of hydration, and structural properties, *J. Comput. Chem.*, 2004, **25**, 678–689.
[116] F. Cesare Marincola, V. P. Denisov and B. Halle, Competitive Na^+ and Rb^+ binding in the minor groove of DNA, *J. Am. Chem. Soc.*, 2004, **126**, 6739–6750.
[117] T. Darden, D. Pearlman and L. G. Pedersen, Ionic charging free energies: spherical versus periodic boundary conditions, *J. Chem. Phys.*, 1998, **109**, 10921–10935.
[118] A. S. Petrov, G. R. Pack and G. Lamm, Calculations of magnesium-nucleic acid site binding in solution, *J. Phys. Chem. B*, 2004, **108**, 6072–6081.
[119] T. E. Cheatham, III and P. A. Kollman, Insight into the stabilization of A-DNA by specific ion association: spontaneous B-DNA to A-DNA transitions observed in molecular dynamics simulations of d[ACCCGCGGGT]$_2$ in the presence of hexaammine cobalt(III), *Structure*, 1997, **5**, 1297–1311.
[120] T. E. Cheatham, III, M. F. Crowley, T. Fox and P. A. Kollman, A molecular level picture of the stabilization of A-DNA in mixed ethanol–water solutions, *Proc. Natl Acad. Sci.*, 1997, **94**, 9626–9630.
[121] A. N. Real and R. J. Greenall, Influence of spermine on DNA conformation in a molecular dynamics trajectory of d(CGCGAATTCGCG)$_2$: major groove binding by one spermine molecule delays the A $\rightarrow$ B transition, *J. Biomol. Struct. Dyn.*, 2004, **21**, 469–487.
[122] P. Auffinger, L. Bielecki and E. Westhof, The Mg^{2+} binding sites of the 5S rRNA loop E motif as investigated by molecular dynamics simulations, *Chem. Biol.*, 2003, **10**, 551–561.
[123] P. Auffinger, L. Bielecki and E. Westhof, Symmetric K^+ and Mg^{2+} ion-binding sites in the 5S rRNA loop E inferred from molecular dynamics simulations, *J. Mol. Biol.*, 2004, **335**, 555–571.
[124] T. E. Cheatham, III and P. A. Kollman, Molecular dynamics simulations highlight the structural differences in DNA:DNA, RNA:RNA and DNA:RNA hybrid duplexes, *J. Am. Chem. Soc.*, 1997, **119**, 4805–4825.

[125] N. Korolev, A. P. Lyubartsev, A. Laaksonen and L. Nordenskiold, A molecular dynamics simulation study of oriented DNA with polyamine and sodium counterions: diffusion and averaged binding of water and cations, *Nucleic Acids Res.*, 2003, **31**, 5971–5981.
[126] N. Korolev, A. P. Lyubartsev, A. Laaksonen and L. Nordenskiold, Molecular dynamics simulation study of oriented polyamine- and Na-DNA: sequence specific interactions and effects on DNA structure, *Biopolymers*, 2004, **73**, 542–555.
[127] Y. P. Pan and A. D. MacKerell, Jr., Altered structural fluctuations in duplex RNA versus DNA: a conformational switch involving base pair opening, *Nucleic Acids Res.*, 2003, **31**, 7131–7140.
[128] K. Reblova, N. Spackova, J. E. Sponer, J. Koca and J. Sponer, Molecular dynamics simulations of RNA kissing-loop motifs reveal structural dynamics and formation of cation-binding pockets, *Nucleic Acids Res.*, 2003, **31**, 6942–6952.
[129] K. Reblova, N. Spackova, R. Stefl, K. Csaszar, J. Koca, N. B. Leontis and J. Sponer, Non-Watson–Crick basepairing and hydration in RNA motifs: molecular dynamics of 5S rRNA loop E, *Biophys. J.*, 2003, **84**, 3564–3582.
[130] F. Beaurain, C. Di Primo, J. J. Toulme and M. Laguerre, Molecular dynamics reveals the stabilizing role of loop closing residues in kissing interactions: comparison between TAR–TAR* and TAR–aptamer, *Nucleic Acids Res.*, 2003, **31**, 4275–4284.
[131] S. Aci, J. Ramstein and D. Genest, Base pairing at the stem–loop junction in the SL1 kissing complex of HIV-1 RNA: a thermodynamic study probed by molecular dynamics simulation, *J. Biomol. Struct. Dyn.*, 2004, **21**, 833–839.
[132] E. Fadrna, N. Spackova, R. Stefl, J. Koca, T. E. Cheatham, III and J. Sponer, Molecular dynamics simulations of guanine quadruplex loops: advances and force field limitations, *Biophys. J.*, 2004, **87**, 227–242.
[133] C.-S. Tung, S. Joseph and K. Y. Sanbonmatsu, All-atom homology model of the *Escherichia coli* 30S ribosomal subunit, *Nat. Struct. Biol.*, 2002, **9**, 750–755.
[134] K. Y. Sanbonmatsu and S. Joseph, Understanding discrimination by the ribosome: stability testing and goove measurement of codon–anticodon pairs, *J. Mol. Biol.*, 2003, **328**, 33–47.

CHAPTER 7

Empirical Force Fields for Proteins: Current Status and Future Directions

Alexander D. MacKerell Jr.

Department of Pharmaceutical Sciences, School of Pharmacy, University of Maryland, 20 Penn Street, Baltimore, MD 21201, USA

Contents

1. Introduction 91
2. Protein force fields 92
   2.1. Gas-phase *versus* condensed-phase target data 94
   2.2. Free energies of aqueous solvation 96
   2.3. Comments on enhancements to protein force fields 96
   2.4. United-atom protein force fields 97
   2.5. Future directions 97
3. Summary 99
Acknowledgements 99
References 99

1. INTRODUCTION

Understanding structure–activity relationships (SAR) of biological macromolecules has been greatly facilitated by theoretical methods based on empirical force fields [1]. Empirical force-field-based methods allow for atomic detail interpretation of a variety of experimental data as well as impart the ability to access short-lived conformations occurring in structural transitions and chemical catalysis [2] that are not readily accessible to experimental approaches. While a variety of algorithmic improvements have greatly facilitated the application of force-field-based techniques to biological systems, the quality of the empirical force fields themselves makes a huge contribution to the accuracy of the method.

Force fields have undergone significant improvements in the last 30+ years. Most early force fields primarily focused on geometries and conformational energies of small molecules, while the consideration of both intramolecular and intermolecular terms dominates current force-field development. These developments have been fueled by increased computational resources allowing more rigorous evaluation of force-field accuracy in condensed phase simulations as well as allowing for higher level quantum mechanical (QM) calculations on model compounds representative of biomolecules. In this review an overview of force

ANNUAL REPORTS IN COMPUTATIONAL CHEMISTRY, VOLUME 1
ISSN: 1574-1400 DOI 10.1016/S1574-1400(05)01007-8

fields commonly used for protein simulations will be presented, including discussion of approaches used in the development of the force fields and the relevance of those methods to their applicability. This will be followed by a short section on future force-field developments and a closing summary section.

2. PROTEIN FORCE FIELDS

Protein simulations typically involve thousands to hundreds of thousands of atoms for durations of nanoseconds or more. To access systems of these sizes and time scales, it is necessary to use a simple equation to calculate the energy as a function of the structure or conformation of the system of interest. The typical potential energy function used in biomolecular simulations includes terms for the intramolecular or internal portion of the potential energy function along with intermolecular (*aka* external or nonbonded) terms. The form of the potential energy function common to protein force fields has been presented elsewhere [3,4]. Such energy functions contain terms describing the structure and the parameters that allow for the simple potential energy functions to treat complex systems such as proteins. It is the combination of the form of the potential energy function and the parameters used in that function that comprise a force field. As discussed below, the parameters are optimized to reproduce a variety of target data from both QM and experimental studies. The ultimate quality of a force field lies in its ability to accurately reproduce a wide variety of experimental target data, thereby insuring that the results obtained from empirical force-field-based studies are representative of the experimental regimen.

The first biomolecular MD simulation was performed on BPTI in the gas phase using a force field based primarily on small molecule parameters [5]. Since then a number of force fields have been developed and applied to simulation studies of proteins. Readers are referred to a recent review on protein force fields by Ponder and Case [6] for additional information as well as an alternative point of view on protein force fields. Currently, the three most commonly used all-atom force fields are the OPLS/AA [7,8], CHARMM22 [4] and AMBER (PARM99) [3] models. Parameters for all three force fields were extensively optimized based on small molecular weight compounds with the resultant parameter set then extended to proteins. In all three force fields, nonbonded parameters were carefully optimized at the small molecule level to reproduce a variety of condensed-phase properties. With OPLS and CHARMM22 the partial atomic charges were based on HF/6-31G* supramolecular data while the AMBER charges are based on the restrained electrostatic potential (RESP) method. In the supramolecular approach the charges are optimized to reproduce QM-determined interaction energies and geometries of the model compound with, typically, individual water molecules although model compound dimers are often used. Such charges are generally developed for functional groups, so that they may be transferred between molecules allowing for charge assignment to novel molecules to be performed readily. RESP fitting involves optimization of charges to reproduce a QM-determined electrostatic potential mapped onto a grid surrounding the model compound. Such methods are convenient and a number of charge fitting methods based on this approach have been developed. In both methods the HF/6-31G* level of theory was used for the QM calculations. This level typically overestimates dipole moments, thereby approximating the influence of the condensed phase on the obtained charge distribution. This, ideally, yields a charge distribution that is implicitly polarized, allowing for satisfactory condensed-phase properties to be obtained.

LJ parameters for all three force fields have primarily been based on the reproduction of condensed-phase properties, typically neat liquids, based on the pioneering work of Jorgensen [9]. Following assignment of partial atomic charges, the LJ parameters for a model compound are adjusted to reproduce experimentally determined heats of vaporization and density as well as isocompressibilities and heat capacities when available. Alternatively, heats or free energies of aqueous solvation or heats of sublimation and lattice geometries can be used as the target data for the LJ optimization [10]. Targeting experimental data for LJ parameter optimization insures that satisfactory condensed-phase properties for proteins will be obtained, including packing of the protein interior. However, the parameter correlation problem (i.e., the fact that force-field parameters are typically underdetermined, allowing for multiple combinations of parameters to yield similar properties) allows for LJ parameters for different atoms in a molecule (e.g., H and C in ethane) to compensate for each other such that it is difficult to accurately determine the 'correct' LJ parameters of a molecule based on the reproduction of condensed-phase properties alone [11]. To overcome this problem, a method was developed that determines the relative value of the LJ parameters based on high level QM data and the absolute values based on the reproduction of experimental data [12,13], thereby decreasing, though not eliminating, the parameter correlation problem.

Careful optimization of the nonbond parameters is essential to maximize the quality of a force field in accurately reproducing experimental observables, including the treatment of atomic interactions. The latter is important because the atomic details of SAR in proteins from simulation studies are of extreme interest and often difficult to obtain experimentally, especially for short-lived intermediates. The partial atomic charges and LJ parameters in a force field largely dictate the atomic nature of intermolecular interactions. With the partial atomic charges, the overall trends are similar for the three force fields, although differences are present, even in the backbone charges [6]. Notably, even though parameter optimization methodologies used in the CHARMM and OPLS/AA force fields are similar, significant differences in the atomic details of interactions have been observed [13]. Such differences emphasize the importance of evaluating the results from force-field-based studies with respect to the strategy used in the optimization of the parameters. In some cases, it may be considered desirable to reproduce 'interesting' results from an MD simulation with a second force field, if feasible. In addition, it should be emphasized that the LJ parameters and partial atomic charges are highly correlated, such that LJ parameters determined for a given set of charges are typically not appropriate for charges determined *via* another methodology.

Accurate treatment of the internal portion of the biomolecular force fields insures that the intramolecular distortions the proteins undergo during MD simulations will be representative of the experimental regimen. Essential to the quality of protein force fields is their treatment of conformational energies associated with ϕ,ψ (i.e., the Ramachandran map [14]), as they dictate, to a large extent, the sampling of conformational space in MD simulations. The alanine dipeptide is the quintessential model compound for the optimization of dihedral parameters associated with ϕ,ψ and, accordingly, has been the subject of a variety of QM studies [15,16]. In the case of AMBER and OPLS/AA the final adjustment of the dihedral parameters that control the ϕ,ψ conformational energies was dominated by QM data, while with CHARMM22 a combination of fitting to QM data and empirical adjustments based on simulations of carboxymyoglobin was applied (see below). Recently, results from MD simulations showing the oversampling of π helices [17] and protein-folding studies indicating overstabilization of different secondary structures by the different force fields [18,19] have motivated additional optimization of the protein backbone parameters in CHARMM and AMBER, as discussed below.

To date, all three force fields have each been used in hundreds of simulation studies of proteins, attesting to their general utility and lack of any catastrophic problems in all three cases. Thus, all three force fields may be considered of similar quality, as validated by a recent comparison, showing all three to reproduce experimental structures in a similar manner in MD simulations of proteins [20]. However, differences in the optimization strategies for the force fields emphasize the need for users to interpret results from simulations with care, accounting for biases in the force fields that may impact the obtained results.

2.1. Gas-phase *versus* condensed-phase target data

A general question in protein, as well as other biomolecular force fields, is the validity of directly applying gas-phase QM data to produce a force field that will be used in condensed-phase simulations. An example of the importance of this consideration is the geometry of the peptide bond. Upon going from the gas to the condensed phase there is a significant decrease in the length of the peptide bond, while the carbonyl bond length increases [21]. Alterations also occur in selected bond angles and it is known that in the gas phase the peptide bond is nonplanar [22,23]. These alterations are associated with changes in the delocalization of the amide nitrogen lone pair due to hydrogen bonding in the condensed phase. Such effects are also present in the conformational energies of the alanine dipeptide and related compounds [24,25]. This phenomenon was initially shown to be important for protein force fields, where it was necessary to deviate from gas-phase energetic data based on the alanine dipeptide in order to better reproduce conformational distributions of ϕ,ψ in simulations of carboxymyoglobin [4]. More recent studies have verified this observation (see below). Thus, it is often preferable to optimize parameters to reproduce condensed-phase target data *versus* gas-phase data, due to the inability of current force fields to adequately model changes in electronic structure, such as lone pair delocalization, as the environment is altered [26].

Recently, we have tested, in the context of the CHARMM22 force field, the ability of the current form of potential energy function and extensions of the potential energy function to accurately treat the entire ϕ,ψ energy surface of the alanine dipeptide and how that impacts results from MD simulations [27,28]. This work involved extension of the force field to include a 2D dihedral energy grid correction map (CMAP) that accounts for the energy difference between a target energy surface (e.g., a QM energy surface) and the empirical force-field surface. This approach allows the force field to reproduce the target surface to near-quantitative accuracy. Based on this approach, application of a ϕ,ψ gas-phase QM energy surface for the alanine dipeptide to MD simulations of proteins in their crystal environment resulted in systematic differences in ϕ between calculated and experimental crystal structures. This motivated empirical adjustments to the alanine dipeptide ϕ,ψ energy surface leading to improved ϕ,ψ sampling in MD simulations as judged by the reproduction of survey data from the PDB [29]. It is anticipated that this approach will have general applicability in empirical force fields for proteins.

Recent adjustments of the AMBER all-atom force field have been performed. The current AMBER force field (PARM99) was optimized to reproduce conformational energies for both the alanine di- and tetra-peptides [30]. Other modification of the ϕ,ψ dihedral parameters has focused on improving the backbone conformational properties in peptide simulations, typically performed using continuum solvation models [18,31]. In another study the AMBER ϕ,ψ dihedral parameters were optimized to reproduce full surfaces of the alanine and glycine dipeptide [32]. Interestingly, the QM data were obtained using a reaction field solvation model with a dielectric constant of 4 to model the condensed-phase environment present in protein simulations. The use of reaction field models to include condensed-phase effects in QM data is an interesting alternative for target data for force-field optimization, although extensive tests of such data have yet to be performed.

2.2. Free energies of aqueous solvation

Another area in which improvements are required is in the thermodynamics of solvation of the amino acid side chains. In a recent study, free energies of solvation of model compounds representative of protein side chains were calculated for the AMBER, CHARMM and OPLS-AA force fields and compared with experiment [33]; similar studies have been reported elsewhere [34–36]. Overall, all three force fields performed well, with OPLS-AA being the best of the three. However, with all three force fields poor results were obtained for selected compounds. These results indicate that improvements in the nonbonded aspects of the force fields can be made.

2.3. Comments on enhancements to protein force fields

While the need for improvements in the current protein force fields is evident, such improvements must be performed with care. The CMAP correction implemented in CHARMM22 is based on the assumption that all other aspects of the force field have been properly optimized. This is necessary to insure that all the terms in the potential energy function, excluding the ϕ,ψ related dihedral parameters, are making an appropriate contribution to the backbone conformational energies, thereby assuring that the atomic contributions to the conformational properties are representative of the experimental regimen. It is then appropriate to include an energy correction such as CMAP, which is, in essence, accounting for a variety of limitations in the potential energy function. It should be noted that this is not a new approach, as in the majority of force fields final optimization of selected dihedral parameters has been performed to obtain target conformational energies, as discussed above for the alanine dipeptide. It should be emphasized that since the CMAP type of correction can, in principle, be used to make any collection of atoms reproduce a selected target surface (i.e., CMAP could be applied to pentane to yield an energy surface identical to that of the alanine dipeptide), such an approach must applied with care to avoid 'hiding' other problems in a force field.

Changes in the partial atomic charges for AMBER based on high level QM data have been performed in conjunction with adjustments to the backbone dihedral parameters [32]. Alterations to the charge distribution of a protein force field should be done with great care. New charges should be accompanied by reevaluation of all aspects of the internal portion of the force field as changes in geometries, vibrational spectra and conformational energies will occur due to electrostatic contributions to those properties. For example, the electrostatic contribution to the energy difference between the C5 and $C7_{eq}$ conformations of the alanine dipeptide is 4.0 kcal/mol, *versus* a total energy difference of 0.9 kcal/mol in CHARMM22.

While the motivation for additional optimization of any force field is to make improvements, care must be taken to avoid the creation of a collection of divergent force fields, which may lead to problems in comparing results from different studies as well as make it difficult to perform future enhancements in a coherent fashion. Furthermore, with all adjustments it is essential that tests be performed on a wide variety of proteins to insure that the changes in the force field are not biased by a limited set of target data.

2.4. United-atom protein force fields

United- or extended-atom force fields are models where only polar hydrogens are included explicitly, while nonpolar hydrogens (e.g., aliphatic and aromatic hydrogens) are treated as part of their parent carbon. These force fields dominated early theoretical studies of proteins due to the savings in CPU associated with the decreased number of atoms and are still widely used, especially in the area of protein folding. United-atom force fields include OPLS/UA [9], the early AMBER force fields [37], GROMOS87 and 96 [38] and CHARMM PARAM19 [39]. The GROMOS united-atom force fields [38] are still widely used in MD simulations that include explicit solvent representations. Enhancements in GROMOS96 have included condensed-phase tests [40] and additional optimization of LJ parameters to reproduce experimental condensed-phase properties [41]. For protein-folding studies the PARAM19 force field currently dominates. This is associated with, in part, the development of a variety of implicit solvent models consistent with this model. These include EEF1 [42], ACE [43], several GB models [44–46] and a model by Caflisch and coworkers [47]. It should be noted that these and other implicit solvent models are often used with the all-atom force fields discussed above [48].

2.5. Future directions

As emphasized above, there is room for improvements in the current force fields for proteins. Such improvements can be made within the context of the current form of the potential energy function or based on extended forms. Within the context of the current form, improvements can be made by simply improved optimization of the parameters. Such enhanced optimization is often based on additional experimental or QM target data that allow for more rigorous testing of force fields as well as the ability to perform condensed-phase simulations at a rate that allows for comparisons with target data in an iterative fashion as required for parameter optimization. Protein-folding calculations are having a significant impact in this area. Comprehensive free energies of solvation of model compounds representative of the protein backbone and side chains, in combination with pure solvent thermodynamic data, offer the potential to implement improvements in the nonbonded portion of the force fields. Again, changes in the nonbonded portion of the force field should include reevaluation of the internal portion of the model to assure that consistency between the nonbonded and internal aspects of the force field is maintained. With respect to the nonbonded parameters, improvements could be made *via* the explicit inclusion of lone pairs. Lone pairs have been shown to yield improvements for specific interactions, such as in-plane *versus* out-of-plane interactions between pyridine and water [49] and the recently developed TIP5P water model contains lone pairs [50]. In general, the addition of lone pairs to a force field will improve the accuracy due to the increased number of parameters available. However, the addition of lone pairs could lead to complications in parameter optimization due to the parameter correlation problem [11], such as the determination of partial atomic charges *via* fitting to QM electrostatic potentials, which is already problematic as emphasized by the need to include restraints (i.e., RESP method) when using this approach [51].

Concerning extended forms of the potential energy function, the 2D grid correction map (CMAP) is a good example of how improvements can be made by adding new terms to the potential energy function. Another extension that is anticipated to have a significant impact on protein force fields is the inclusion of electronic polarizability [6,52]. While current additive force fields are optimized to yield charge distributions that include implicit polarization, thereby allowing them to satisfactorily treat an aqueous environment, they cannot accurately model electrostatics over a wide range of environments. A good example is the need to overestimate the interaction energy of the gas-phase water dimer in order to accurately treat the pure solvent [53]. A recent review [54] covers the approaches used to treat polarizability as well as applications of those methods. Notably, to date, published work on the application of polarizable models to full proteins has been limited. Kaminski *et al.* presented results on polarizable MD simulations of a collection of proteins, although the simulations were only of 2 ps duration and performed in the gas phase [55]. More recently, Patel *et al.* presented results for several proteins in solution using a full polarizable model (i.e., both solvent and protein were polarizable) based on the CHARMM22 force field that yielded stable structures of the proteins in MD simulations of up to several nanoseconds in duration [56]. Other efforts towards polarizable protein force fields are ongoing [57–59]. Thus, slow progress is being made towards the development of polarizable force fields for proteins as well as other biomolecules. It should be emphasized that the usefulness of such force fields, as judged by improved accuracy in a variety of scenarios, will not be possible until well-optimized polarizable models have been developed.

3. SUMMARY

As is evident a variety of force fields for empirical-based studies of proteins have been produced. Of these the AMBER (PARM99), CHARMM22, OPLS-AA and GROMOS 87/96 force fields are the most widely used. As each of these force fields has been employed in hundreds of studies, it is clear that they are all satisfactory for simulation studies of proteins. The recent focus on protein-folding studies, the π-helical phenomena discussed above and several comprehensive studies on the free energies of solvation of model compounds representative of amino acid side chains indicate that additional improvements in protein force fields in the context of the present potential energy function as well as in extended functions can be made. Motivated by these studies, variants of both the CHARMM and AMBER force fields have been presented. While in all cases improvements are evident based on the particular target data used in the study, the true value of these modifications in leading to improved accuracy for a wide variety of peptides and proteins in different environments has yet to be shown. Finally, it should be emphasized that results from empirical force-field-based calculations, as with all scientific investigations, have to be evaluated within the context of the assumptions and models used in those studies. Accordingly, it is important for users of force fields to be aware of the assumptions made in their development and gauge the robustness of the obtained results in that context.

ACKNOWLEDGEMENTS

Financial support from the NIH (GM51501) is acknowledged.

REFERENCES

[1] O. M. Becker, A. D. MacKerell, Jr., B. Roux and M. Watanabe (eds), *Computational Biochemistry and Biophysics*, Marcel Dekker, New York, 2001, p. 512.

[2] M. Garcia-Viloca, J. Gao, M. Karplus and D. G. Truhlar, How enzymes work: analysis by modern rate theory and computer simulations, *Science*, 2004, **303**, 186–195.

[3] W. D. Cornell, P. Cieplak, C. I. Bayly, I. R. Gould, K. M. Merz, D. M. Ferguson, D. C. Spellmeyer, T. Fox, J. W. Caldwell and P. A. Kollman, A second generation force field for the simulation of proteins, nucleic acids, and organic molecules, *J. Am. Chem. Soc.*, 1995, **117**, 5179–5197.

[4] A. D. MacKerell, Jr., D. Bashford, M. Bellott, R. L. Dunbrack, Jr., J. Evanseck, M. J. Field, S. Fischer, J. Gao, H. Guo, S. Ha, D. Joseph, L. Kuchnir, K. Kuczera, F. T. K. Lau,

C. Mattos, S. Michnick, T. Ngo, D. T. Nguyen, B. Prodhom, I. Reiher, W. E. B. Roux, M. Schlenkrich, J. Smith, R. Stote, J. Straub, M. Watanabe, J. Wiorkiewicz-Kuczera, D. Yin and M. Karplus, All-atom empirical potential for molecular modeling and dynamics studies of proteins, *J. Phys. Chem. B*, 1998, **102** (18), 3586–3616.
[5] J. A. McCammon, B. R. Gelin and M. Karplus, Dynamics of folded proteins, *Nature*, 1977, **267** (5612), 585–590.
[6] J. W. Ponder and D. A. Case, Force fields for protein simulations, *Adv. Protein Chem.*, 2003, **66**, 27–85.
[7] W. L. Jorgensen, D. S. Maxwell and J. Tirado-Rives, Development and testing of the OPLS all-atom force field on conformational energetics and properties of organic liquids, *J. Am. Chem. Soc.*, 1996, **118**, 11225–11236.
[8] G. Kaminski, R. A. Friesner, J. Tirado-Rives and W. L. Jorgensen, Evaluation and reparametrization of the OPLS-AA force field for proteins via comparison with accurate quantum chemical calculations on peptides, *J. Phys. Chem. B*, 2001, **105**, 6474–6487.
[9] W. L. Jorgensen and J. Tirado-Rives, The OPLS potential functions for proteins. Energy minimizations for crystals of cyclic peptides and crambin, *J. Am. Chem. Soc.*, 1988, **110**, 1657–1666.
[10] N. Foloppe and A. D. MacKerell, Jr., All-atom empirical force field for nucleic acids: 1) Parameter optimization based on small molecule and condensed phase macromolecular target data, *J. Comput. Chem.*, 2000, **21**, 86–104.
[11] A. D. MacKerell, Jr., Atomistic models and force fields. In *Computational Biochemistry and Biophysics* (eds O. M. Becker, *et al.*), Marcel Dekker, New York, 2001, pp. 7–38.
[12] D. Yin and A. D. MacKerell, Jr., Combined ab initio/empirical approach for the optimization of Lennard–Jones parameters, *J. Comput. Chem.*, 1998, **19**, 334–348.
[13] I.-J. Chen, D. Yin and A. D. MacKerell, Jr., Combined ab initio/empirical optimization of Lennard–Jones parameters for polar neutral compounds, *J. Comput. Chem.*, 2002, **23**, 199–213.
[14] G. N. Ramachandran, C. Ramakrishnan and V. Sasisekharan, Stereochemistry of polypeptide chain configurations, *J. Mol. Biol.*, 1963, **7**, 95–99.
[15] S. Ono, M. Kuroda, J. Higo, N. Nakajima and H. Nakamura, Calibration of force-field dependency in free energy landscapes of peptide conformations by quantum chemical calculations, *J. Comput. Chem.*, 2002, **23**, 470–476.
[16] R. Vargas, J. Garza, B. P. Hay and D. A. Dixon, Conformational study of the alanine dipeptide at the MP2 and DFT levels, *J. Phys. Chem. A*, 2002, **106**, 3213–3218.
[17] M. Feig, A. D. MacKerell, Jr. and C. L. Brooks, III, Force field influence on the observation of π-helical protein structures in molecular dynamics simulations, *J. Phys. Chem. B*, 2002, **107**, 2831–2836.
[18] A. Okur, B. Strockbine, V. Hornak and C. Simmerling, Using PC clusters to evaluate the transferability of molecular mechanics force fields for proteins, *J. Comput. Chem.*, 2003, **24**, 21–31.
[19] S. Gnanakaran and A. E. Garcia, Validation of an all-atom protein force field: from dipeptides to larger peptides, *J. Phys. Chem. B*, 2003, **107**, 12555–12557.
[20] D. J. Price and C. L. Brooks, III, Modern protein force fields behave comparably in molecular dynamics simulations, *J. Comput. Chem.*, 2002, **23**, 1045–1057.
[21] H. Guo and M. Karplus, Ab initio studies of hydrogen bonding of *N*-methylacetamide: structure, cooperativity, and internal rotational barriers, *J. Phys. Chem.*, 1992, **96**, 7273–7287.
[22] J. D. Evanseck and M. Karplus, Theoretical investigation of peptide nonplanar distortion, amide nitrogen and carbonyl carbon pyramidalization and prediction of secondary structure from amide I and II infrared bands, Unpublished work, 1994.
[23] M. Ramek, C.-H. Yu, J. Sakon and L. Schäfer, Ab initio study of the conformational dependence of the nonplanarity of the peptide bond group, *J. Phys. Chem. A*, 2000, **104**, 9636–9645.

[24] H. S. Shang and T. Head-Gorden, Stabilization of helices in glycine and alanine dipeptides in a reaction field model of solvent, *J. Am. Chem. Soc.*, 1994, **116**, 1528–1532.

[25] C. Park and W. A. Goddard, III, Solvent effects on the secondary structures of proteins, *J. Phys. Chem. A*, 2000, **104**, 2498–2503.

[26] B. E. Mannfors, N. G. Mirkin, K. Palmo and S. Krimm, Analysis of the pyramidalization of the peptide group nitrogen: implications for molecular mechanics energy functions, *J. Phys. Chem. A*, 2003, **107**, 1825–1832.

[27] A. D. MacKerell, Jr., M. Feig and C. L. Brooks, III, Accurate treatment of protein backbone conformational energetics in empirical force fields, *J. Am. Chem. Soc.*, 2004, **126**, 698–699.

[28] A. D. MacKerell, Jr., M. Feig and C. L. Brooks, III, Extending the treatment of backbone energetics in protein force fields: limitations of gas-phase quantum mechanics in reproducing protein conformational distributions in molecular dynamics simulations, *J. Comput. Chem.*, 2004, **25**, 1400–1415.

[29] R. L. Dunbrack, Jr. and F. E. Cohen, Bayesian statistical analysis of protein sidechain rotamer preferences, *Prot. Sci.*, 1997, **6**, 1661–1681.

[30] M. D. Beachy, D. Chasman, R. B. Murphy, T. A. Halgren and R. A. Friesner, Accurate ab initio quantum chemical determination of the relative energetics of peptide conformations and assessment of empirical force fields, *J. Am. Chem. Soc.*, 1997, **119**, 5908–5920.

[31] A. E. Garcia and K. Y. Sanbonmatsu, α-Helical stabilization by side chain shielding of backbone hydrogen bonds, *Proc. Natl Acad. Sci. USA*, 2002, **99**, 2782–2787.

[32] Y. Duan, C. Wu, S. Chowdhury, M. C. Lee, G. Xiong, W. Zhang, R. Yang, P. Ceiplak, R. Luo, T. Lee, J. Caldwell, J. Wang and P. Kollman, A point-charge force field for molecular mechanics simulations of proteins based on condensed-phase quantum mechanical calculations, *J. Comput. Chem.*, 2003, **24**, 1999–2012.

[33] M. R. Shirts, J. W. Pitera, W. C. Swope and V. S. Pande, Extremely precise free energy calculations of amino acid side chain analogs: comparison of common molecular mechanics force fields for proteins, *J. Chem. Phys.*, 2003, **119**, 5740–5761.

[34] A. Villa and A. E. Mark, Calculation of the free energy of solvation for neutral analogs of amino acid side chains, *J. Comput. Chem.*, 2002, **23**, 548–553.

[35] J. L. MacCallum and P. Tieleman, Calculation of the water–cyclohexane transfer free energies of amino acid side-chain analogs using the OPLS all-atom force field, *J. Comput. Chem.*, 2003, **24**, 1930–1935.

[36] Y. Deng and B. Roux, Hydration of amino acid side chains: non-polar and electrostatic contributions calculated from staged molecular dynamics free energy simulations with explicit water molecules, *J. Phys. Chem. B*, 2004, **108**, 16567–16576.

[37] S. J. Weiner, P. A. Kollman, D. A. Case, U. C. Singh, C. Ghio, G. Alagona, S. Profeta and P. Weiner, A new force field for molecular mechanical simulation of nucleic acids and proteins, *J. Am. Chem. Soc.*, 1984, **106**, 765–784.

[38] W. F. van Gunsteren, S. R. Billeter, A. A. Eising, P. H. Hünenberger, P. Krüger, A. E. Mark, W. R. P. Scott and I. G. Tironi, *Biomolecular Simulation: The GROMOS96 Manual and User Guide*, BIOMOS b.v., Zürich, 1996.

[39] E. Neria, S. Fischer and M. Karplus, Simulation of activation free energies in molecular systems, *J. Chem. Phys.*, 1996, **105**, 1902–1919.

[40] X. Daura, P. H. Hünenberger, A. E. Mark, E. Querol, F. X. Avilés and W. F. van Gunsteren, Free energies of transfer of Trp analogs from chloroform to water: comparison of theory and experiment and the importance of adequate treatment of electrostatics and internal interactions, *J. Am. Chem. Soc.*, 1996, **118**, 6285–6294.

[41] L. D. Schuler, X. Daura and W. F. van Gunsteren, An improved GROMOS96 force field for aliphatic hydrocarbons in the condensed phase, *J. Comput. Chem.*, 2001, **22**, 1205–1218.

[42] T. Lazaridis and M. Karplus, Effective energy function for proteins in solution, *Proteins*, 1999, **35**, 133–152.
[43] M. Schaefer, C. Bartels, F. LeClerc and M. Karplus, Effective atoms volumes of implicit solvent models: comparison between Voronoi volumes and minimum fluctuation volumes, *J. Comput. Chem.*, 2001, **22**, 1857–1879.
[44] M. S. Lee, F. R. Salsbury, Jr. and C. L. Brooks, III, Novel generalized Born methods, *J. Chem. Phys.*, 2002, **116**, 10606–10614.
[45] M. S. Lee, M. Feig, F. R. Salsbury, Jr. and C. L. Brooks, III, New analytical approximation to the standard molecular volume definition and its application to generalized Born calculations, *J. Comput. Chem.*, 2003, **24**, 1348–1356.
[46] W. Im, M. S. Lee and C. L. Brooks, III, Generalized Born model with a simple smoothing function, *J. Comput. Chem.*, 2003, **24** (14), 1691–1702.
[47] P. Ferrara, J. Apostolakis and A. Caflisch, Evaluation of a fast implicit solvent model for molecular dynamics simulations, *Proteins*, 2002, **46**, 24–33.
[48] M. Feig and C. L. Brooks, III, Recent advances in the development and application of implicit solvent models in biomolecular simulations, *Curr. Opin. Struct. Biol.*, 2004, **14**, 217–224.
[49] R. W. Dixon and P. A. Kollman, Advancing beyond the atom-centered model in additive and non-additive molecular mechanics, *J. Comput. Chem.*, 1997, **18**, 1632–1646.
[50] M. W. Mahoney and W. L. Jorgensen, A five-site model for liquid water and the reproduction of the density anomaly by rigid, nonpolarizable potential functions, *J. Chem. Phys.*, 2000, **112**, 8910–8922.
[51] C. I. Bayly, P. Cieplak, W. D. Cornell and P. A. Kollman, A well-behaved electrostatic potential based method using charge restraints for deriving atomic charges: the RESP model, *J. Phys. Chem.*, 1993, **97**, 10269–10280.
[52] T. A. Halgren and W. Damm, Polarizable force fields, *Curr. Opin. Struct. Biol.*, 2001, **11**, 236–242.
[53] G. Lamoureux, A. D. MacKerell, Jr. and B. Roux, A simple polarizable model of water based on classical Drude oscillators, *J. Chem. Phys.*, 2003, **119**, 5185–5197.
[54] S. W. Rick and S. J. Stuart, Potentials and algorithms for incorporating polarizability in computer simulations, *Rev. Comput. Chem.*, 2002, **18**, 89–146.
[55] G. A. Kaminski, H. A. Stern, B. J. Berne, R. A. Friesner, Y. X. Cao, R. B. Murphy, R. Zhou and T. A. Halgren, Development of a polarizable force field for proteins via ab initio quantum chemistry: first generation model and gas phase tests, *J. Comput. Chem.*, 2002, **23**, 1515–1531.
[56] S. Patel, A. D. MacKerell, Jr. and C. L. Brooks, III, CHARMM fluctuating charge force field for proteins: II. Protein/solvent properties from molecular dynamics simulations using a non-additive electrostatic model, *J. Comput. Chem.*, 2004, **25**, 1504–1514.
[57] P. Cieplak, J. W. Caldwell and P. A. Kollman, Molecular mechanical models for organic and biological systems going beyond the atom centered two body additive approximations: aqueous solution free energies of methanol and *N*-methyl acetamide, nucleic acid base, and amide hydrogen bonding and chloroform/water partition coefficients of the nucleic acid bases, *J. Comput. Chem.*, 2001, **22**, 1048–1057.
[58] P. Ren and J. W. Ponder, Consistent treatment of inter- and intramolecular polarization in molecular mechanics calculations, *J. Comput. Chem.*, 2002, **23**, 1497–1506.
[59] V. M. Anisimov, I. V. Vorobyov, G. Lamoureux, S. Noskov, B. Roux and A. D. MacKerell, Jr., CHARMM all-atom polarizable force field parameter development for nucleic acids, *Biophys. J.*, 2004, **86**, 415a.

CHAPTER 8

Nonequilibrium Approaches to Free Energy Calculations

Adrian E. Roitberg

Quantum Theory Project, Department of Chemistry, University of Florida, P.O. Box 118435, Gainesville, FL 32611-8435, USA

Contents

1. Introduction 103
2. The original Jarzynski method 104
3. Experimental applications 108
4. Theoretical developments 108
5. Computational uses 109
6. Conclusions 110
Acknowledgements 110
References 110

1. INTRODUCTION

Free energies are central quantities to both thermodynamics and kinetics, relating to experimentally determined properties such as equilibrium constants and reaction rates. Even though proper computation of enthalpies is relatively simple at particular molecular conformations, estimates of the entropic factors require sampling over large numbers of conformations. Modern applications of free energy calculations in computational chemistry are used for ligand binding [1], free energy profiles in mixed quantum–classical enzymatic calculation [2] and hydration free energies [3]. These calculations are done under (if possible) equilibrium conditions, or with as full sampling as possible. We will review recent work done using nonequilibrium calculations of free energies, based on the so-called Jarzynski relationship [4–8] which has been extended and shown to be part of a subset of classical thermodynamics dealing with very small systems, as well as with fluctuations in macroscopic properties. This review is assembled as follows: first, we will introduce the Jarzynski relationship and discuss its connection to more classical methods for the computation of free energies. Then we will move into recent work in theoretical applications and experimental advances associated with nonequilibrium free energy calculations.

ANNUAL REPORTS IN COMPUTATIONAL CHEMISTRY, VOLUME 1
ISSN: 1574-1400 DOI 10.1016/S1574-1400(05)01008-X

2. THE ORIGINAL JARZYNSKI METHOD

The free energy difference between two states A and B (described by a single variable λ bounded between 0 and 1) is formally described by

$$\Delta G_{0\to 1} = G_1 - G_0 = -\frac{1}{\beta}\ln\left(\frac{Z_1}{Z_0}\right) = -\frac{1}{\beta}\ln\left(\frac{\int dr \exp(-\beta H_1)}{\int dr \exp(-\beta H_0)}\right) \tag{1}$$

where ΔG represents the Gibbs free energy, and 0 and 1 correspond to the end points of the system (A and B, respectively). Z stands for the canonical partition functions, which are explicitly written in terms of Boltzmann weights in the right-hand side of equation (1).

Methods for the computation of such quantities have a long history. All of them have to surmount an important hurdle: in order to compute the partition function, very extensive (one might say complete) sampling of the phase space at *both* A and B must be done. This is of course mostly unattainable, and one must rely on a number of approximations. One widely used idea can be traced to Zwanzig [9] and is widely known as free energy perturbation (FEP). It requires the calculation of the energy difference between states A and B, ensemble averaged over the initial ensemble A and is used as in

$$\Delta G_{0\to 1} = -\frac{1}{\beta}\ln\langle \exp(-\beta(H_1 - H_0))\rangle_A \tag{2}$$

There are alternative formulations, requiring a Hamiltonian that varies smoothly between A and B, for instance.

$$H(\lambda) = H(0) + \lambda[(H(1) - H(0))] \tag{3}$$

Other forms of the interpolation scheme can be used as well as Hamiltonian decomposition into energy terms. The definition of the λ-dependent Hamiltonian allows for the computation of free energy derivatives with respect to λ, which in turn enables the calculation of the free energy difference between A and B using the so-called thermodynamics integration (TI) method.

$$\Delta G_{0\to 1} = \int_0^1 d\lambda \left\langle \frac{\partial H(\lambda)}{\partial \lambda} \right\rangle_\lambda \tag{4}$$

If one thinks about the system as evolving from A to B with a time-dependent Hamiltonian, then the above equations can be rewritten simply by assuming a perturbation parameter, $\lambda = \lambda(t)$ (which obviously, according to equation (3), immediately means a Hamiltonian $H(t)$).

The second law of thermodynamics requires that the ensemble average of the work done onto the system by the external perturbation be larger or equal to the free energy difference, with the difference being the dissipative (nonuseful) work. Under a quasi-static (QS) transformation from A to B (in infinite time), the perturbation is continuously very close to equilibrium conditions, and only then the work is exactly equal to the free energy difference, namely

$$\langle W_{A \to B} \rangle_A \geq \Delta G_{A \to B} \tag{5}$$

In the case of simulations, the average is taken over members of the equilibrium ensemble of state A. For QS changes, all realizations of the experiment will give the same value of W, and a well-defined value for ΔG. This is equivalent to the statement that the distribution of work values under a QS transformation is a delta function around the exact value of ΔG.

It is then clear that under non-QS changes from A to B, a number of statements must be true:

1. The average work will be larger than ΔG.
2. The distribution of work values will have a width > 0.
3. Any individual realization could give rise to work values lower than ΔG.

Even though these are interesting points, they were of no use at all until a seminal article by Jarzynski in 1997 [4]. In that paper, he proved the so-called Jarzynski relationship (JR) that states:

$$\Delta G_{A \to B} = -\frac{1}{\beta} \ln \langle \exp(-\beta W_{A \to B}) \rangle_A \tag{6}$$

where W is the out of equilibrium work done onto the system when going from point A to point B in phase space, and the exponential average is done over an equilibrium ensemble of state A only. This formula seems very counterintuitive at first, since it makes a clear connection between nonequilibrium work values (which are, by definition, path dependent), with the equilibrium free energy, a state function (and hence, not path dependent). Moreover, the only two requirements for this equality to work are that the initial ensemble over state A be equilibrated, and that the exponential average be converged, which in turn requires large numbers of realizations of the transformation. There is no requirement as to how the switch from state A to state B should be done (in computational implementation, how fast can one switch the system Hamiltonian from A to B), which seems counterintuitive. First, let us see that this setup reduces to known expressions under certain limits. Clearly, if the switch is done close to infinitely slowly, then the transformation is QS. In that case, the work W is equal to the free energy (there is no dissipative work) and the JR holds true. In the other extreme, one could switch from state A to state B instantaneously. In that regime, the work done on the system is simply the enthalpy change between initial and final points as in

$$\Delta G_{A \rightarrow B} = -\frac{1}{\beta} \ln \langle \exp(-\beta \Delta H_{A \rightarrow B}) \rangle_A \tag{7}$$

Note that this is an already well-known formulation known as FEP and that equation (7) is the same as the previously described equation (2).

In a general situation, the transformation of the system from A to B and the application of the Jarzynski relation can be seen in Fig. 1. At an initial state A ($\lambda = 0$), the system is equilibrated. This is represented by the vertical line at left. This initial ensemble could be equilibrated by long molecular dynamics or Monte Carlo runs, or by advanced sampling techniques such as replica exchange [10,11]. Once this is done, a number N of initial frames is taken from the initial ensemble. They are then transformed into state B (and all states between), at $\lambda = 1$ at a finite rate. The work for each realization is then computed, and the overall free energy is extracted by using the JR as shown in equation (6).

The demonstration of the validity of the JR is beyond the scope of this review, but the interested reader is encouraged to read the original Jarzynski's article [4]. However, it is important to provide a simple description of why this seemingly strange equality might work. Figure 2 presents a hint as to the behavior of the system. Under near-equilibrium conditions, the distribution of work values could be expected to be roughly Gaussian (this requirement is not needed, but makes explanations clearer). The vertical line at 1 unit of work (arbitrary units) represents

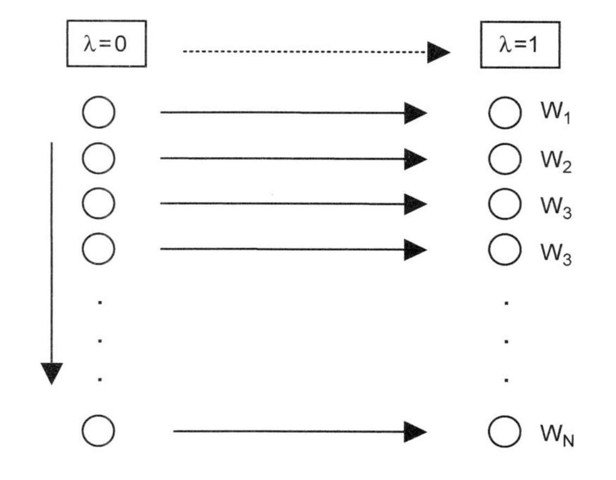

Fig. 1. General scheme for the use of the JR. The system is equilibrated at $\lambda = 0$. N Snapshots are taken from that equilibrated ensemble and transformed into the final state ($\lambda = 1$) in a finite time. Work is computed for each realization (W_i).

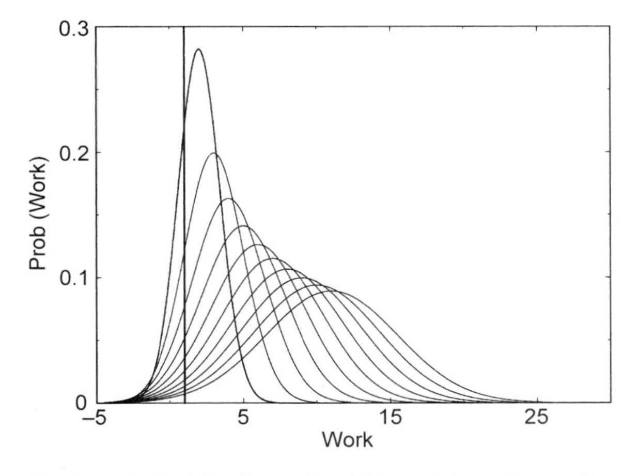

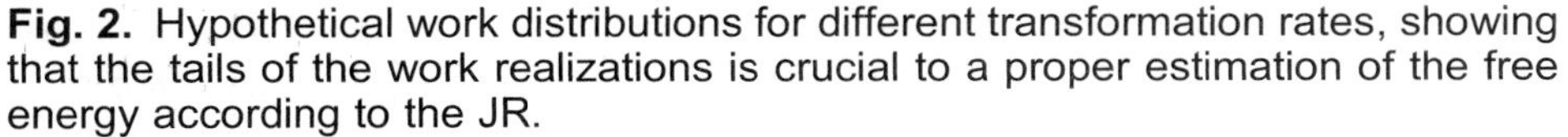
Fig. 2. Hypothetical work distributions for different transformation rates, showing that the tails of the work realizations is crucial to a proper estimation of the free energy according to the JR.

the free energy difference in going from a state with $\lambda = 0$ to $\lambda = 1$. Under QS conditions, the distribution of work values is very nearly a delta function (a very narrow Gaussian). However, under any transformation rate larger than zero, two things happen at the same time. The average work gets larger as the rate increases, while the width of the work distribution also increases. The JR takes the exponentially weighted average of this distribution. The net effect of this nonlinear averaging is to pick, from the work distribution, trajectories that are low in work values. The 'number' of these trajectories decreases drastically with an increase of the transformation rate, and hence the effort required to converge the nonexponential average increases quickly. There is also a result that seems to contradict the second law of thermodynamics – the probability of an individual realization of the transformation from A to B having work lower than the free energy difference between A and B is nonnegligible. This would seem to indicate a negative dissipative work, which is of course not possible. The reader is invited to see that the second law applies only to macroscopic systems. Under those conditions, the proper quantity to compare to the free energy is the average work, which indeed is always larger than ΔG.

At this point, a remarkable early work of Hermans [12] comes into focus. He described that one could improve on the simple use of many free energy calculations starting from different initial conditions, by not simply averaging linearly, but making use of the standard deviation of the work calculations (or measured).

$$\Delta G \approx \langle W \rangle - \frac{\beta}{2}\sigma_{\mathrm{W}}^{2} \tag{8}$$

This relationship, which has been substantially used in the literature, turns out to be a simple linear response expansion of equation (6). This may be shown in two interrelated ways. First, if one assumes a Gaussian distribution of work values (a reasonable zeroth-order approximation), then equation (8) is exact. Second, if one describes the exponential average as a cumulant expansion, then equation (8) becomes simply the first- and second-order cumulants. There are also third and higher order cumulants (which are all exactly zero in the case of linear response), which alternate in sign and are very slowly converging.

3. EXPERIMENTAL APPLICATIONS

The year 2003 saw the JR used by a number of groups, as an interpretational aid to the nonequilibrium measurements these groups performed. These measurements were mainly of the force-induced unfolding type, where an external potential (an optical trap, an AFM tip for instance) is used to change end-to-end distances in attached biomolecules such as RNA [13] and DNA [14]. 'Before' Jarzynski such measurements could not make use of free-energy-related data, since the average work could not be easily interpreted. Currently, many of these experiments are simply repeated enough times until the exponential average of equation (6) or the approximation of equation (8) can be safely used.

Protein–protein interactions have also been studied using force-induced unbinding, and interpreted by means of JR-type relations [15]. This technique has been applied to regulation of integrin activity, cell adhesion and leukocyte rolling.

A more basic experimental study was used to show local, single particle violations of the second law of thermodynamics, by following entropy production and consumption over short trajectory durations [16]. Of course, this violation disappears when averaging is taken over longer time scales and many molecules at once.

4. THEORETICAL DEVELOPMENTS

New derivations of the JR are appearing, in what seems to be a trend to put the original work into a more general framework. A recent article by Evans highlights the close relationship between the nonequilibrium free energy theorems and the fluctuation theorem [17]. In a different derivation, Mittag [18] generalized the fluctuation theorem to thermostated dissipative systems which respond to time-dependent dissipative fields. This chapter highlights the deep connection between the second law of thermodynamics, time-reversal symmetry and the fluctuation theorem.

Despite the many applications of the Jarzynski estimator, its behavior is not yet fully understood. Among the still unresolved issues, one outstanding is the computation of errors, both in term of bias and sampling. The convergence of the estimator and the error bars with the number of realizations has been described by Gore *et al.* [19].

Hummer discussed recently a simple procedure to extract kinetic information from pulling experiments and illustrated it for the I27 subunits of the protein titin [20].

A curious extension of the JR is the work of Mukamel [21] where the equality is extended to quantum systems. There is an analogy between the nonequilibrium trajectories and the phase fluctuations over phase space, described by the use of the stochastic Liouville equation.

One of the potential drawbacks of the JR is the fact that many trajectories are generated as realization of the transformations, and are averaged in post-processing. This opens the possibility of using better choices of trajectories. Sun [22] proposes the use of Monte Carlo sampling of very short nonequilibrium trajectories which can yield good estimates of the equilibrium free energy change. This work has very close relation with the path sampling techniques of Chandler and coworkers [23].

5. COMPUTATIONAL USES

As an example of the current state of use of the JR, we still see a larger number of theoretical developments, compared to a much smaller number of pure application articles. As the usefulness of this technique extends, we expect the actual uses to increase, to encompass a large number of possible applications.

Li [24] used steered nonequilibrium simulations in which the unfolding dynamics of the I27 domain of the muscle protein titin is studied by a series of nonequilibrium, steered molecular dynamics simulations. They find that the distribution of the unfolding force as well as its dependence on the pulling rate predicted by their simulations is found to be in agreement with atomic force microscopy experiments.

There is a recent study by Park *et al.* [25] where the Jarzynski equality is applied to the helix–coil transition of deca-alanine in vacuum as an example. With about 10 trajectories sampled, the second-order cumulant expansion of equation (8), among the various averaging schemes examined, yields the most accurate estimates. They also show a curious but important analytical result; if the distribution of work values at short times is Gaussian, then this distribution will remain Gaussian at all times after. The usefulness of this result resides in the fact that a second-order expansion of the JR is exactly valid, and rapidly converging.

6. CONCLUSIONS

The calculation of free energies of many processes has a long history within computational chemistry. With the introduction of the JR, and the recent formal and computational work associated with it, the field is ripe to see a widespread use of this technique. As the ideas find their way into widely accessible software, we are bound to see new, yet unpredicted applications.

ACKNOWLEDGEMENTS

I would like to thank Christopher Jarzynski and Tom Woolf for discussing many of the finer issues in nonequilibrium calculations with me. Hui Xiong has been invaluable for our own implementation of nonequilibrium free energy calculations. Part of this work was funded by the Department of Energy through grant DE-FG02-02ER45995.

REFERENCES

[1] W. L. Jorgensen, The many roles of computation in drug discovery, *Science*, 2004, **303** (5665), 1813–1818.
[2] L. Ridder, Imcm Rietjens, J. Vervoort and A. J. Mulholland, Quantum mechanical/molecular mechanical free energy simulations of the glutathione *S*-transferase (M1-1) reaction with phenanthrene 9,10-oxide, *J. Am. Chem. Soc.*, 2002, **124** (33), 9926–9936.
[3] M. R. Shirts, J. W. Pitera, W. C. Swope and V. S. Pande, Extremely precise free energy calculations of amino acid side chain analogs: comparison of common molecular mechanics force fields for proteins, *J. Chem. Phys.*, 2003, **119** (11), 5740–5761.
[4] C. Jarzynski, Equilibrium free-energy differences from nonequilibrium measurements: a master-equation approach, *Phys. Rev. E*, 1997, **56** (5), 5018–5035.
[5] C. Jarzynski, Equilibrium free energies from nonequilibrium processes, *Acta Phys. Pol. B*, 1998, **29** (6), 1609–1622.
[6] C. Jarzynski, Microscopic analysis of Clausius–Duhem processes, *J. Stat. Phys.*, 1999, **96** (1–2), 415–427.
[7] C. Jarzynski, Hamiltonian derivation of a detailed fluctuation theorem, *J. Stat. Phys.*, 2000, **98** (1–2), 77–102.
[8] C. Jarzynski, How does a system respond when driven away from thermal equilibrium?, *Proc. Natl Acad. Sci. USA*, 2001, **98** (7), 3636–3638.
[9] R. W. Zwanzig, High-temperature equation of state by a perturbation method. 1. Nonpolar gases, *J. Chem. Phys.*, 1954, **22** (8), 1420–1426.
[10] Y. Okamoto, Generalized-ensemble algorithms: enhanced sampling techniques for Monte Carlo and molecular dynamics simulations, *J. Mol. Graph. Model.*, 2004, **22** (5), 425–439.
[11] B. S. Kinnear, M. F. Jarrold and U. H. E. Hansmann, All-atom generalized-ensemble simulations of small proteins, *J. Mol. Graph. Model.*, 2004, **22** (5), 397–403.
[12] J. Hermans, Simple analysis of noise and hysteresis in (slow-growth) free-energy simulations, *J. Phys. Chem.*, 1991, **95** (23), 9029–9032.

[13] B. Onoa, S. Dumont, J. Liphardt, S. B. Smith, I. Tinoco and C. Bustamante, Identifying kinetic barriers to mechanical unfolding of the *T. thermophila* ribozyme, *Science*, 2003, **299** (5614), 1892–1895.

[14] C. Danilowicz, V. W. Coljee, C. Bouzigues, D. K. Lubensky, D. R. Nelson and M. Prentiss, DNA unzipped under a constant force exhibits multiple metastable intermediates, *Proc. Natl Acad. Sci. USA*, 2003, **100** (4), 1694–1699.

[15] J. W. Weisel, H. Shuman and R. I. Litvinov, Protein–protein unbinding induced by force: single-molecule studies, *Curr. Opin. Struct. Biol.*, 2003, **13** (2), 227–235.

[16] G. M. Wang, E. M. Sevick, E. Mittag, D. J. Searles and D. J. Evans, Experimental demonstration of violations of the second law of thermodynamics for small systems and short time scales, *Phys. Rev. Lett.*, 2002, **89** (5).

[17] D. J. Evans, A non-equilibrium free energy theorem for deterministic systems, *Mol. Phys.*, 2003, **101** (10), 1551–1554.

[18] E. Mittag and D. J. Evans, Time-dependent fluctuation theorem, *Phys. Rev. E*, 2003, **67** (2).

[19] J. Gore, F. Ritort and C. Bustamante, Bias and error in estimates of equilibrium free-energy differences from nonequilibrium measurements, *Proc. Natl Acad. Sci. USA*, 2003, **100** (22), 12564–12569.

[20] G. Hummer and A. Szabo, Kinetics from nonequilibrium single-molecule pulling experiments, *Biophys. J.*, 2003, **85** (1), 5–15.

[21] S. Mukamel, Quantum extension of the Jarzynski relation: analogy with stochastic dephasing, *Phys. Rev. Lett.*, 2003, **90** (17).

[22] S. X. Sun, Equilibrium free energies from path sampling of nonequilibrium trajectories, *J. Chem. Phys.*, 2003, **118** (13), 5769–5775.

[23] C. Dellago, P. G. Bolhuis and D. Chandler, *J. Chem. Phys.*, 1998, **108**, 9236.

[24] P. C. Li and D. E. Makarov, Theoretical studies of the mechanical unfolding of the muscle protein titin: bridging the time-scale gap between simulation and experiment, *J. Chem. Phys.*, 2003, **119** (17), 9260–9268.

[25] S. Park, F. Khalili–Araghi, E. Tajkhorshid and K. Schulten, Free energy calculation from steered molecular dynamics simulations using Jarzynski's equality, *J. Chem. Phys.*, 2003, **119** (6), 3559–3566.

CHAPTER 9

Calculating Binding Free Energy in Protein–Ligand Interaction

Kaushik Raha and Kenneth M. Merz Jr.

Department of Chemistry, The Pennsylvania State University, University Park, PA 16802, USA

Contents

1. Introduction 113
2. Calculating binding free energy 114
3. Scoring functions 119
   3.1. Physical chemical 120
   3.2. Empirical scoring functions 122
   3.3. Knowledge-based scoring functions 123
4. Conclusions 125
Acknowledgements 126
References 126

1. INTRODUCTION

A deeper understanding of how a protein recognizes its biologically relevant ligand or a small molecule inhibitor will have a profound effect on our understanding of biological recognition processes and on our ability to design small molecule therapeutics. Because of their importance, these interactions have been the subject of intense research and significant advances have been made in understanding the overall process. A worldwide effort in structural biology has led to the elucidation of the structures of a large number of protein–ligand complexes at atomic resolution which are now available in the Protein Data Bank (PDB) [1]. A complementary experimental effort by the academic and biopharmaceutical industry has resulted in the characterization of numerous ligands bound to a diverse range of protein targets. As a result, a unique opportunity has been presented to the theoretical/computational biology community where hypotheses to describe protein–ligand interactions can be formulated and validated against the wealth of available structural and experimental data. Moreover, the past decade has also seen staggering advances in computer technology with the price/performance ratio consistently falling and the rise of new computing paradigms such as distributed computing and cluster-based computing [2]. Many computing problems that were

ANNUAL REPORTS IN COMPUTATIONAL CHEMISTRY, VOLUME 1
ISSN: 1574-1400 DOI 10.1016/S1574-1400(05)01009-1

deemed intractable just a few years ago are now well within reach, and more problems are going to enter the realm of possibility in the coming years.

As a result, computational methods that characterize protein–ligand interactions from 3D structures based on the laws of physics and chemistry have been a subject of much recent research. The success or failure of these methods is measured not only by their ability to qualitatively describe protein–ligand interaction but also by their ability to quantify the strength of interaction. The strength of interaction is determined by the free energy of binding and can be measured experimentally. Computational methods strive to calculate the free energy of binding from 3D structures and evaluate their performance by comparing with experimentally observed free energies of binding. In spite of all the recent developments in this area the accurate prediction of the free energy of binding using computational methods based on a description of the energetic components of binding has proved to be a major challenge. A physically satisfying and accurate computational method will have widespread practical application in structure-based drug design and virtual screening protocols [3–5].

There is relentless pressure on the pharmaceutical industry to reduce costs, because of the extreme difficulty in bringing a compound to the market as a drug. *In silico* structure-based screening has been a very attractive and potentially cost-saving alternative because of its ability to screen a large number and broad range of compounds [6]. Virtual screening experiments screen a database of compounds against a protein target and identify those compounds that are thought to bind to the protein target. This process entails docking of compounds into a protein binding site and then scoring the docked 'poses' to determine their activity [7–9]. Scoring is related to the strength of the interaction between the ligand and the protein, which experimentally is expressed as the free energy of binding. Hence computational protocols for explicitly calculating the binding free energy, in addition to being of interest as a fundamental problem in molecular recognition, also have practical implications.

In this chapter we will discuss the basic principles that govern protein–ligand interactions and how they can be quantified in terms of a binding free energy. We will also discuss some computational methodologies that take a physically based approach towards the calculation of the free energy of binding. We also review scoring functions that are available in the literature to calculate protein–ligand binding affinities. Finally, we interpret the results from these studies and assess our understanding of binding in light of these observations.

2. CALCULATING BINDING FREE ENERGY

Binding affinity can be estimated experimentally by kinetic experiments that measure the inhibition of the protein or enzyme in the presence of both the inhibitor and the substrate, and is reported as an inhibition constant K_i. Under equilibrium conditions, the free energy of binding is given as the dissociation constant K_d which is the ratio of the concentration of the reactants (protein and ligand) to products (complex):

$$K_d = \frac{[P][L]}{[PL]} \quad (1)$$

Finally, the free energy of binding is related to the dissociation of [PL] as

$$\Delta G = \Delta G^0 - RT \ln K_d \quad (2)$$

where ΔG is the free energy change for the reaction, R the gas constant, and T the temperature. ΔG^0 is the free energy change associated with the reaction at standard conditions where all concentrations are at 1 M, temperature is 298 K, and pressure is 1 atm. Theoretical calculations determine the free energy of binding in a more direct fashion by calculating the properties of individual structures of the protein, ligand, and the complex, or of their ensembles. Binding free energy is a state function and is treated as such in these calculations, which means that it is independent of the path taken from the reactants to the product. Hence, the free energy of the reaction calculated from the dissociation constant K_d can be compared with theoretically calculated free energies of binding. The free energy of binding can be calculated as a difference of the free energy of the reactants (protein and ligand) and the free energy of product as

$$\Delta G_{bind} = \Delta G_{complex} - (\Delta G_{protein} + \Delta G_{ligand}) \quad (3)$$

where ΔG_{bind} is the free energy of binding, $\Delta G_{complex}$ the free energy of the protein–ligand complex, $\Delta G_{protein}$ and ΔG_{ligand} the free energies of the protein and ligand, respectively. The free energy of binding is usually decomposed into different free energy components that are additive and are represented by a single equation or a 'master equation'. We will use the master equation proposed by Ajay and Murcko [10], and discuss this master equation in light of recent protocols used for the calculation of the individual components that make up this equation. The master equation has the following form

$$\Delta G_{bind} = \Delta G_{int} + \Delta G_{solv} + \Delta G_{motion} + \Delta G_{conf} \quad (4)$$

where the first term ΔG_{int} is the free energy due to the interaction of the reactants that form the complex, and it is dominated by enthalpic contributions from steric and electrostatic interactions upon complex formation. These interactions are generally strong and short range due to the close proximity of the ligand and the protein.

These interactions can be calculated at the classical or quantum level of theory. The steric interactions are usually modeled by a pairwise Lennard-Jones (LJ) potential [7,11–14] and the electrostatic interaction energy is usually calculated *via* Coulomb's law using atom-centered point charges. The effect of the protein environment is mimicked by scaling the Columbic interaction by a distance-dependent dielectric constant that is usually set to 4 [15]. Both steric and electrostatic interactions are calculated for non-bonded atoms in molecular mechanics force fields such as AMBER [16], CHARMM [13], MMFF [17], OPLS [18], and MM3 [12]. However, classical point charge based models of electrostatic interaction represent a significant approximation since higher order effects like polarization and charge transfer are ignored. Polarization and charge transfer play a significant role in molecular interaction as shown by us and others [19–21]. Such effects can be captured using quantum mechanics. Rigorous and exhaustive calculation of electrostatic interaction energies can be performed at high levels of quantum theory. Such calculations are feasible only for small model chemistries and are severely restricted by the computational cost associated with the study of larger molecules. *Ab initio* and density functional theories (DFT) are still not being applied to macromolecular biological systems routinely. However, linear scaling technologies in quantum mechanics are trying to bridge this gap and have made significant advances in the past decade [22–24]. These advances are applicable to density functional, Hartree–Fock, and semiempirical quantum theories and can now solve the Schrödinger equation for large molecular systems containing thousands of atoms [25,26]. Semiempirical approaches have been shown to be very useful in studying electrostatic interactions in protein-folding and protein–ligand interaction [20,27,28]. Recently we have used our linear scaling divide and conquer computer program DivCon [29], to calculate the electrostatic interaction energy using the semiempirical AM1 Hamiltonian [21]. DivCon utilizes the divide and conquer algorithm (D&C), which scales linearly with the size of the system. The D&C method divides a molecular system into overlapping subsystems and the localized Roothan-Hall equation

$$(F^{a}C^{a} = S^{a}C^{a}E^{a}) \tag{5}$$

is solved for each of the subsystems a. Here C^{a} is the subsystem coefficient matrix, F^{a} the subsystem Fock matrix, and E^{a} the diagonal matrix of the orbital energies for subsystem a. S^{a} is the overlap matrix and is the identity matrix for standard Semiempirical Hamiltonians. The diagonalization of the global Fock matrix is the most expensive part of the calculation and the D&C method replaces global diagonalization with numerous 'subsystem' diagonalizations that results in linear scaling. Local subsystem density matrices are used to assemble the global density matrix and the total energy is calculated from the global density matrix. Finally, the heat of formation of a molecular system is calculated as the sum of the electronic energy, core–core repulsion, and individual heats of formation of the constituent atoms:

$$\Delta H_{\mathrm{f}} = E_{\mathrm{electronic}} + E_{\mathrm{core-core}} + \Sigma \Delta H_{f\ atoms} \tag{6}$$

Both classical and semiempirical quantum mechanical methods can calculate pairwise interaction energy terms. This allows the calculation of the gas-phase interaction energy between the ligand and the protein in the complex. The interaction energy is enthalpic in nature and hence the free energy ΔG_{int} can be thought of as being dominated by enthalpic contributions. In a recent study, Nikitina *et al.* [30] calculated the binding enthalpy of eight protein–ligand complexes from the PDB using semiempirical QM at the PM3 level of theory. The calculated binding enthalpies are within 2 kcal/mol of the available experimental data. Semiempirical methods are thus capable of calculating the enthalpic contribution to binding.

The free energy of solvation, ΔG_{sol}, has a very important role to play in binding. The role of solvent (usually water) in binding is well known [31–36]. Water has bulk or macroscopic properties that account for long-range effects such as the hydrophobic effect and dielectric relaxation which are not pairwise additive. Water also has microscopic properties in the active site of a protein where it mediates key hydrogen bond interactions between the ligand and the protein. The entropic part of the solvation free energy ΔS_{solv} results from an increase in the entropy of water when it is released from the active site upon complexation. ΔS_{solv} has been described computationally by surface area burial in protein–ligand and protein–protein interactions to within experimental error [10,37–39]. This is a fast and approximate way of accounting for this effect and most protocols for the calculation of binding free energy use surface area burial to estimate solvent entropy. The enthalpic part of the solvation free energy arises from the electrostatic and van der Waals interaction of water with itself and with the protein–ligand complex.

Free energy of solvation can also be calculated using molecular dynamics (MD) simulation of the protein–ligand complex in explicit water under periodic boundary conditions. The linear interaction energy (LIE) method estimates the effect of solvation on protein–ligand complexation explicitly *via* MD simulations [40–42]. However, this method is time consuming because of the nature of the solvation free energy. Macroscopic properties of water like the dielectric constant and hydrophobic effect are average effects that can be accurately calculated by MD simulation in the limit of infinite time. Therefore, long MD simulations are critical for calculating meaningful free energies of binding which make this approach less practical in structure-based design efforts.

The last decade has seen tremendous advances in the development of implicit models for calculating solvation free energies [31,35,36,43–50]. The Poisson–Boltzmann (PB) equation can be used to calculate solvation free energies where the solvent is represented implicitly by a dielectric constant ε (for water $\varepsilon = 78.3$) and the dielectric boundary is defined by the van der Waals surface or the solvent-accessible surface of the molecule. The Poisson–Boltzmann equation relates the electrostatic potential $\Phi(r)$ to the charge density $\rho(r)$ as

$$\nabla\varepsilon(r)\nabla\Phi(r) - \kappa'\Phi(r) = -4\pi\rho(r) \quad (7)$$

where κ' is the Debye–Huckel screening constant. $\Phi(r)$ is calculated using a finite difference scheme on a grid or using the boundary element method [50–53]. Computer programs such as DELPHI [50,54], UHBD [44], and APBS [55] solve the PB equation using the finite difference scheme. We have also developed a PB/SCRF methodology in our group [46]. The electrostatic part of the solvation free energy is largely due to the polarization of the solute electron density because of the presence of an external reaction field potential. These effects, while hard to capture by classical atom-centered fixed point charge models, are perfectly suited for QM methods. We capture the perturbation of the gas-phase Hamiltonian due to the presence of a solvent reaction field using our Poisson–Boltzman based Self-Consistent Reaction Field (SCRF) method implemented in DivCon [29]. The perturbed Hamiltonian contains the potential energy operator of the interaction between the solvent-polarized surface charges with the solute electrons and the nuclei and has been described in detail elsewhere [46]. Using these approaches the solvation free energy G_{solv} is described in terms of electrostatic and non-polar effects:

$$G_{\text{solv}} = G_{\text{elec}} + G_{\text{non-polar}} \quad (8)$$

The electrostatic part of solvation comes from the PB equation, while the non-polar part comes from parameterized surface area terms [46]. Related to these PB approaches is the Generalized Born/SA approaches that are also capable of giving solvation free energies and are used in ligand-binding free energy calculations [56].

The free energy change associated with changes in the 'motion' of the protein, ligand, and the complex is ΔG_{motion}. When a protein and ligand form a complex, three degrees of rotational and three degrees of translational motion are lost. The entropy and internal thermal energy of a system is calculated from partition functions related to the translational, rotational, and vibrational components of binding. Standard statistical mechanical formulations are used to calculate these contributions [57]. For protein–ligand interactions, the free energy associated with translation and rotation towards binding has been estimated by Schwarzl *et al.* [33] to be of the order of 16–18 kcal/mol for benzaminidine-based trypsin inhibitors. The vibrational component is significantly lower at about -4.0 kcal/mol for the same set of inhibitors. Based on this data, it appears that the loss of translation and rotational motions during complexation is unfavorable and is partially compensated by the vibrational component.

The free energy due to conformational changes during complexation is ΔG_{conf}. This depends on the conformation of the ligand and the protein before and after binding. ΔG_{conf} has enthalpic and entropic components. Both protein and ligand lose conformational entropy during binding. This enthalpic part arises from the internal energy of the protein and the ligand. Calculating ΔG_{conf} accurately is challenging because many times the atomic resolution structure of the free protein and ligand are unavailable. In such a scenario, assumptions are made about the initial state of the reactants, by generally assuming that the conformation of the protein and ligand are not very different from their conformations in the complex. This assumption may or may not be true depending on the flexibility of the ligand and the active site. Both ligand and receptor flexibility can be explicitly or implicitly modeled. The conformational entropy of the ligand can be implicitly estimated from the number of rotatable bonds that are 'frozen' during binding. Estimates for the energy penalty associated with this process are 0.4–0.9 kcal/mol [58]. However, many times the number of rotatable bonds are fit in order to optimize the correlation with experiment [59,60]. For example, Ishchenko and Shakhnovich found the optimized value for an sp^3–sp^3 bond as 1.2 kcal/mol and an sp^3–sp^2 bond as -0.5 kcal/mol. For the protein, the loss of conformational flexibility of the amino acid side-chains that are involved in binding can be estimated from the entropic scale derived by Creamer [61] for all 20 amino acids. The internal energy change of the ligand and the protein during complexation can be calculated using any potential function. Molecular mechanics potentials derived from AMBER, CHARMM, MMFF, etc., are routinely used for such calculations. In a recent study, Lin *et al.* account for receptor and ligand conformational flexibilities using the 'relaxed receptor scheme'. They used MD simulation to generate an ensemble of receptor conformations [62]. The ligand was then docked into the active site of the receptor ensemble using a docking program. The binding free energy was calculated using the Molecular Mechanics/Poisson–Boltzmann Surface Area (MM/PBSA) [15,63–65] protocol. The interesting finding of this study was that the binding free energy of the ligand calculated using this method varies over 3 kcal/mol (which is a 100 to 1000-fold difference in the dissociation constants). Thus, microscopic simulations to estimate the ΔG_{conf} can be a source of uncertainty in the computation of the binding free energy.

3. SCORING FUNCTIONS

Master equations are used to guide the development of scoring functions. Scoring functions are also designed to predict binding modes of inhibitors and discriminate between true ('native') and false ('decoy') binding modes. Scoring functions have been reasonably successful in virtual High Throughput Screening (vHTS) experiments in drug discovery, where consensus schemes have been used to dock a database of small molecules into protein-derived pharmacophore [69–71]. We will review scoring functions only from the point of view of affinity ranking or binding affinity prediction. However, there are numerous reviews on scoring functions and their ability to predict binding modes [72–74]. Scoring functions include either all or some of the important energetic contributions discussed above. Most of the present generation of scoring functions discard a more rigorous representation of the binding free energy and opt for an empirical expression that takes into account only a few dominant contributions. This is often due to the requirement of these functions to be fast and reasonably accurate in predicting binding modes or rank potential leads in virtual screening experiments often from a rigid receptor structure. Since different scoring functions emphasize different contributions to the free energy of binding, consensus scoring schemes with multivariate statistical analysis methods are used to improve structure-based virtual screening [75]. Present generation scoring functions can be categorized into: physical chemical, empirical, and knowledge based. This section will review some of the main scoring functions that have been reported.

3.1. Physical chemical

The MM/PBSA method for calculating the free energy of binding is physically based. The main equation for MM/PBSA calculation can be summarized as

$$\Delta G_{\text{bind}} = \Delta G_{\text{MM}} + \Delta G_{\text{sol}}^{\text{C}} - \Delta G_{\text{sol}}^{\text{P}} - \Delta G_{\text{sol}}^{\text{L}} - T\Delta S \tag{9}$$

where

$$\Delta G_{\text{MM}} = \Delta G_{\text{int}}^{\text{ele}} + \Delta G_{\text{int}}^{\text{vdw}} \tag{10}$$

The electrostatic and van der Waals interaction energies between the ligand and the protein are calculated using a molecular mechanics potential like AMBER [14] or CHARMM [13]. The electrostatic part of the solvation energy is calculated using the Poisson–Boltzmann method, and the non-polar part is calculated from the surface area burial. The $T\Delta S$ term is calculated using the usual statistical mechanical partition functions [57]. All the parts that contribute to binding in the above equation are ensemble averages. Molecular dynamics simulations of the complex are carried out to calculate the MM/PBSA energy of a finite number of snapshots from the ensemble. Details of the simulation such as treatment of long-range electrostatics and cutoff for non-bonded interactions are important factors that impact the quality of the prediction in such calculations. MM/PBSA-based methods have also been used with docking simulations to calculate the binding affinities of ligands docked into a hydrophobic cavity created in lysozyme [66]. Kuhn and Kollman studied nine streptavidin and avidin ligands and have calculated the free energy of binding using the MM/PBSA protocol. They obtained very good agreement with the experiment (correlation coefficient R^2 of 0.92) [63]. This study sheds light on the nature of the compensatory effects in binding. In particular, the role of electrostatics is quite revealing. MM/PBSA studies in the literature consistently find the total electrostatic interaction to be unfavorable towards binding [15,63–65]. While the gas-phase interaction between the protein and the ligand is favorable, the solvent-screened electrostatic interaction is unfavorable. This is due to the large desolvation penalty that the ligand and the protein pay during complex formation. Hence, it appears that the strength of binding is driven by short-range (van der Waals) and long-range non-polar (hydrophobic) forces. This observation has also been reported elsewhere in the literature [67]. Indeed the hydrophobicity-based computational model HINT that used parameters derived from experimental Log P values for small organic molecules was shown to correlate with the binding free energy for 53 protein–ligand complexes with a standard error of 2.6 kcal/mol [68]. For the case of electrostatic interactions in solution, the desolvation penalty paid by the ligand and the protein offsets the favorable electrostatic interactions between them in the complex. Moreover, the ionization state(s) of the ligand and binding pocket are often assigned at physiological pH, or as observed in the apo-crystal structure of the protein. Almost all examples of binding affinity calculation do not consider the effect of the changing environment as a result of protein–ligand interactions in terms of shifted pK_as of the ionizable groups or the effect of polarization on the ionization state of the functional groups in the protein and ligand. The errors due to these approximations have not been quantified in the literature since it is non-trivial to calculate the perturbed pK_as of the ligand and the protein side-chains in the active site. Therefore perturbed pK_as and ionization states can only be assigned by empirical observations which often require detailed analysis of the active site and its electrostatic properties. In spite of all the advances in this field, the treatment of electrostatic interaction energies is still a significant challenge and has scope for further development.

Recently, we have developed a quantum mechanics based scoring function that takes into account the role of metal ions in binding. We use semiempirical QM to calculate the gas phase electrostatic interaction energy between the protein and the ligand at the AM1 level of theory using DivCon [29]. The solvation correction to the gas phase electrostatic energy was calculated using our PB/SCRF method that accounts for polarization of the solute charges in the presence of a solvent reaction field [46]. The attractive/dispersive part of the non-polar interaction was calculated using a Lennard-Jones potential based on AMBER force field parameter set. The solvent entropy was calculated from surface area burial due to complex formation. The conformational entropy was evaluated from the number of degrees of freedom that were lost in the small molecule and side chains in the active site during binding. Finally, the active site of the uncomplexed protein was modeled with a zinc-bound water molecule. Using this scoring function, we successfully predicted the binding affinity for 23 ligands bound to the zinc metalloenzymes carbonic anhydrase and carboxypeptidase [21]. We achieved good agreement with available experimental binding free energies (R^2 of 0.69 ($R = 0.83$) and a mean standard error of 1.5 kcal/mol). Another important observation from this study was the observed charge transfer between the protein and the ligand of the order of 1*e* for carbonic anhydrase, and 0.5*e* for carboxypeptidase.

3.2. Empirical scoring functions

These scoring functions relate the binding affinity to weighted contributions from different energy terms that are thought to play a role in binding. Often these terms are simple potentials such as the van der Waals bump potential or a hydrogen bond potential based on geometric measures such as distance and angle [7,76–86]. In some scoring functions, individual terms are calculated using force field based physical potentials [11,77]. However, the coefficients for these terms are derived by regression methods that are used to fit the observations to experimental binding affinities, either for a set of ligands bound to one target receptor or for ligands that are bound to a diverse set of protein targets. The coefficients are optimized to maximize the correlation between computed and experimental binding affinities in the training set. Linear and non-linear regression methods can be used to derive the coefficients for the different terms in the scoring function. The model for prediction, which is a combination of the weighted terms, is then applied to a prediction set [78,80,81,85]. Empirical scoring functions trained on a particular protein target work best for calculating the relative binding affinity of ligands that bind to the same target.

AUTODOCK is an example of a scoring function that makes use of force-field equations and parameters to calculate the binding energy. The binding free energy is described as a sum of the intermolecular interactions between the ligand and the protein and the internal steric energy of the ligand [11,76]. It can be represented by the equation:

$$E_{\text{AUTODOCK}} = E_{\text{vdw}} + E_{\text{H-bond}} + E_{\text{electrostatic}} + E_{\text{internal}} \tag{11}$$

The van der Waals interaction is calculated using a LJ 6-12 potential between the protein and the ligand atoms. The steric part of the H-bond term is calculated using a LJ 10-12 potential. The intermolecular electrostatic interaction is calculated using Coulomb's law. The internal energy of the ligand is a sum of steric and electrostatic interactions calculated for non-bonded ligand atoms. LUDI is an empirical scoring function that was developed by Bohm [78]. It calculates binding free energy as a sum of polar and non-polar interaction terms. The polar interactions are represented as a sum of the H-bond interactions while non-polar interactions are estimated from hydrophobic burial. LUDI differentiates between neutral and ionic (salt-bridge) interactions in its hydrogen bond term. It also has an entropic term that accounts for the loss of rotatable bonds in the ligand, which are presumed 'frozen' in the active site of the protein. ChemScore is similar to LUDI but has an additional term that accounts for metal ion coordination in the active site [76]. F-Score is similar to LUDI and was developed to discriminate between native and decoy binding modes [60].

D-Score is implemented in DOCK [7,77]. It has a very simple form based on a LJ 6-12 interaction term and a Coulombic interaction term between the protein and the ligand atoms. It only accounts for the ΔG_{int} component in the master equation. G-Score is a scoring function implemented in GOLD [87]. It was parameterized for binding mode prediction rather than ranking binding affinities of ligands. It is also force field based and has the following form:

$$E_{\text{total}} = E_{\text{complex}} + E_{\text{H-bond}} + E_{\text{internal}} \tag{12}$$

E_{complex} is calculated using a LJ 8-4 potential. The H-bond interaction energy is calculated using a function that depends on the type and geometry of the donor and acceptor involved in the interaction. The internal energy is calculated for the ligand and is a sum of steric interactions calculated using a LJ 6-12 potential and a torsional energy term [87]. XScore is an empirical scoring function that combines three individual scoring schemes HSScore, HPScore and HMScore that are differentiated by their interpretation of the hydrophobic effect [85]. The common terms in these three schemes are the van der Waals term calculated using a LJ 8-4 potential, a hydrogen bonding term, and a conformational entropy term represented by the number of rotatable bonds. These different terms were parameterized to reproduce experimental binding affinities.

3.3. Knowledge-based scoring functions

This class of scoring functions is based on statistical potentials derived from a database of protein–ligand complexes. The PDB has been used to design statistical potentials for both protein structure prediction and protein–ligand interactions [88–90]. Statistical potentials rely on the occurrences or counts of interacting atom pairs from the PDB. It is assumed that the structures of protein–ligand complexes in the PDB are in a state of thermodynamic equilibrium that represents the global free energy minimum of the complex and the distribution of the atoms in the complex obey Boltzmann's law. Hence, probabilities of atom pair occurrences in the PDB can be related to the potential of mean force (PMF) of interaction between the ligand and the protein. Knowledge-based potentials (KBP) are first built from a structural database by calculating distance dependent probability distributions of atom-pairs. This can then be used to calculate the Helmholtz free energy (A_{ij}) or PMF of atom-pair interaction by

$$A_{ij}(r) = -k_{\mathrm{B}}T \ln\left[\rho_{ij}(r)/\rho_{ij}(\mathrm{bulk})\right] \quad (13)$$

where $\rho_{ij}(r)$ is the number density, or the pair correlation function of an atom-pair of type ij at a distance between r and $r + \delta r$, while ρ_{ij}(bulk) is a normalization factor that is the bulk density for the atom-pair when they are not interacting in the range between r and $r + \delta r$. Atom-pairs are non-interacting when for a particular distance 'bin' atom type i and atom type j are not interacting with each other but are interacting with some other protein or ligand atom. Pair correlation functions for atom types can thus be calculated at different distance bins from the database. The pairwise nature of such a potential makes it easy to use because the pre-computed pair potentials are simply summed up for all protein–ligand atom pairs to form a score that relates to the binding affinity. Various groups have designed KBPs but there are differences in the approach in terms of the normalization factors and implicit inclusion of effects such as desolvation. In this section we briefly review the main KBP and their predictive capability.

Wallqvist *et al.* proposed a KBP based on receptor–ligand contact preferences for different atom types [91]. This scoring function used a surface area based packing score and atom–atom preference for each atom pair. Verkhivker *et al.* designed the KBP piecewise linear potential (PLP) and then used it to compute binding free energies of HIV-Protease inhibitors [84,92]. PLP divides the distance dependent potential function into linear pieces for calculating the interaction between two atom types. The interaction types considered are steric interactions that depend on the radii of the interacting atoms and hydrogen bond interactions that include an angle dependent scaling function [84]. DeWitte and Shakhnovich developed the Small Molecule Growth (SMoG) algorithm that used KBPs to estimate the free energy of binding [90]. Ishchenko and Shakhnovich [59] redefined the reference state and the normalization of contact probabilities based on composition of the atom types in the database. This was implemented in SMoG2001 that was successfully used in a computational combinatorial experiment to predict two enantiomeric ligands that bound to Human Carbonic Anhydrase with high potency [71]. PMFScore is a KBP developed by Muegge and Martin, which has been validated to predict the binding affinity across a wide range of protein targets [93]. The main feature of this KBP is similar to SmoG, where the number densities of atom-pair interactions over radial distance bins were used to calculate the PMF of interaction between the interacting atom-types. However, PMFScore also has an important ligand volume correction term that implicitly captures the desolvation effect in the pair-potentials. Binding affinities calculated using PMFScore showed a good correlation with experiment with an R^2 of 0.61 and standard deviation of 1.8 log K_i units for a set of 77 protein–ligand complexes representing a diverse set of protein targets and ligands. Endothiapepsin inhibitors ($R^2 = 0.22$; SD = 1.89 log K_i units) and sugar binding proteins ($R^2 = 0.48$; SD = 0.86 log K_i units) fared poorly with this KBP. Gohlke *et al.* designed DrugScore for recognizing binding modes, predicting binding affinities, and to identify hotspots in protein–ligand complexes [79]. In addition to distance dependent pair preferences, they also included a solvent accessible surface area dependent term for protein and ligand atoms. They demonstrated the physically intuitive behavior of pair potentials with respect to minima of atom pair interactions. The success of DrugScore is good compared to others. It succeeds in predicting the binding affinities in log K_i units for serine proteases ($R^2 = 0.86$) and metalloproteases ($R^2 = 0.7$). Interestingly this KBP, like PMFScore, also fares poorly in predicting binding affinities for sugar binding proteins ($R^2 = 0.22$) and endothiapepsin ($R^2 = 0.30$). This suggests that KBPs are superior to empirical scoring functions because they use fewer parameters and they show good prediction for a diverse range of protein–ligand targets. However, KBPs capture the bulk or macroscopic features of the binding phenomenon. They account for important features of binding such as the hydrophobic effect and solvation from the shape of the active site, penetration of the ligand and the statistical nature of atom-pair interactions. Also, protein–ligand complexes used for validating KBPs have strong molecular weight dependence. Scoring with KBPs perform well for the serine protease class, which is a set that has a strong molecular weight dependence. The R^2 for a fit of molecular weight of the inhibitor *versus* the binding free energy for the 16 serine protease inhibitors used in the validation of PMFScore is 0.86. On the other hand the binding free energy of endothiapepsin inhibitors have a very weak molecular weight dependence ($R^2 = 0.29$). Another drawback of the KBPs is that if an interaction is not statistically significant in the databank, the scoring function will fail to take it into account. Hence while the role of metal ions in binding is important, KBPs completely ignore such interactions since their occurrences in the database are not statistically significant [93].

4. CONCLUSIONS

The major conclusions that can be drawn from this review are as follows: theoretical treatment of the phenomenon of protein–ligand interaction has scope for further improvement. While our understanding of the basics is satisfactory, computational methods that can accurately calculate the free energy of binding based on theoretical understanding are still lacking. Specifically, electrostatic interactions are far from being well understood or well characterized and their role in binding based on review of the current literature is unclear. Electrostatic calculations involve significant approximations or simplistic treatments that ignore many features of these interactions between the solvent and the participating species. Moreover, understanding the role electrostatics play in altering the protonation state of the ligand and/or protein is in its infancy. Validation is still a major issue, with no widely accepted standard 'test' set being available to rigorously test scoring functions. This, combined with many structural problems with available PDB structures, is a major bottleneck for systematic improvement. The role molecular weight plays in many scoring functions is also a source of concern. Concerted, well defined, and organized efforts are clearly needed to explore the complex role of electrostatics and other interactions in the phenomenon of binding.

ACKNOWLEDGEMENTS

We thank the NCSA for supercomputer time, the NSF (MCB-0211639) and the NIH (GM44974) for support.

REFERENCES

[1] H. M. Berman, J. Westbrook, Z. Feng, G. Gilliland, T. N. Bhat, H Weissing, I. N. Shindyalov and P. E. Bourne, The protein data bank, *Nucleic Acids Res.*, 2000, **28**, 235–242.
[2] M. Baker and K. Buyya, *Cluster Computing at a Glance*, Prentice Hall, Englewood Cliffs, NJ, 1999.
[3] A. Cheng, D. Diller, S. Dixon, W. Egan, G. Lauri and K. M. Merz, Jr., Computation of the physiochemical properties and data mining of large molecular collections, *J. Comput. Chem.*, 2002, **23**, 172–183.
[4] W. L. Jorgensen, The many roles of computation in drug discovery, *Science*, 2004, **303**, 1813–1818.
[5] A. Good, Structure-based virtual screening protocols, *Curr. Opin. Drug Discov. Dev.*, 2001, **4**, 301–307.
[6] J. Drews, Drug discovery: a historical perspective, *Science*, 2000, **287**, 1960–1964.
[7] E. C. Meng, B. K. Shoichet and I. D. Kuntz, Automated docking with grid-based energy evaluation, *J. Comput. Chem.*, 1992, **13**, 380–397.
[8] D. J. Diller and K. M. Merz, Jr., High throughput docking for library design and library prioritization, *Proteins Struct. Funct. Genet.*, 2001, **43**, 113–124.
[9] M. L. Lamb, K. W. Burdick, S. Toba, M. M. Young, A. G. Skillman, X. Zou, J. R. Arnold and I. D. Kuntz, Design, docking, and evaluation of multiple libraries against multiple targets, *Proteins Struct. Funct. Genet.*, 2001, **42**, 296–318.
[10] Ajay and M. A. Murcko, Computational methods to predict binding free energy in ligand–receptor complexes, *J. Med. Chem.*, 1995, **38**, 4953–4967.
[11] D. S. Goodsell and A. J. Olson, Automated docking of substrates to proteins by simulated annealing, *Proteins Struct. Funct. Genet.*, 1990, **8**, 195–202.
[12] N. L. Allinger, Y. H. Yuh and J.-H. Lii, Molecular mechanics – the MM3 force field for hydrocarbons, *J. Am. Chem. Soc.*, 1989, **111**, 8551–8566.
[13] B. R. Brooks, R. E. Bruccoleri, B. D. Olafson, D. J. States, S. Swaminathan and M. Karplus, CHARMM: a program for macromolecular energy minimization and dynamical calculations, *J. Comput. Chem.*, 1983, **4**, 182–217.
[14] D. A. Case, J. W. Caldwell, T. E. Cheatham II, W. S. Ross, C. L. Simmerling, T. A. Darden, K. M. Merz, R. V. Stanton, A. L. Cheng, J. J. Vincent, M. Crowley,

D. M. Ferguson, R. J. Radmer, G. L. Seibel, U. C. Singh, P. K. Weiner and P. A. Kollman, *AMBER 5.0*, University of California, San Francisco.
[15] W. Wang, W. A. Lim, A. Jakalian, J. Wang, J. Wang, R. L. C. Bayly and P. A. Kollman, An analysis of the interactions between the Sem-5 SH3 domain and its ligands using molecular dynamics, free energy calculations, and sequence analysis, *J. Am. Chem. Soc.*, 2001, **123**, 3986–3994.
[16] W. D. Cornell, P. Cieplak, C. I. Baylay, I. R. Gould, K. M. Merz, D. M. Ferguson, D. C. Spellmeyer, T. Fox, J. W. Caldwell and P. A. Kollman, A second generation force field for the simulation of proteins, nucleic acids, and organic molecules, *J. Am. Chem. Soc.*, 1995, **117**, 5179–5197.
[17] T. A. Halgren, Merck molecular force field: 1. Basis, form, scope, parameterization, and performance of MMFF94, *J. Comput. Chem.*, 1996, **17**, 490–519.
[18] W. L. Jorgensen, D. S. Maxwell and J. Tirado-Rives, Development and testing of the OPLS all-atom force field on conformational energetics and properties of organic liquids, *J. Am. Chem. Soc.*, 1996, **118**, 11225–11236.
[19] M. Garcia-Viloca, D. G. Truhlar and J. Gao, Importance of substrate and cofactor polarization in the active site of dihydrofolate reductase, *J. Mol. Biol.*, 2003, **372**, 549–560.
[20] A. van der Vaart and K. M. Merz, Jr., The role of polarization and charge transfer in the solvation of biomolecules, *J. Am. Chem. Soc.*, 1999, **121**, 9182–9190.
[21] K. Raha and K. M. Merz, Jr., A quantum mechanics based scoring function: study of zinc-ion mediated ligand binding, *J. Am. Chem. Soc.*, 2004, **126**, 1020–1021.
[22] A. van der Vaart, V. Gogonea, S. L. Dixon and K. M. Merz, Jr., Linear scaling molecular orbital calculations of biological systems using the semiempirical divide and conquer method, *J. Comput. Chem.*, 2000, **21**, 1494–1504.
[23] S. L. Dixon and K. M. Merz, Jr., Semiempirical molecular orbital calculations with linear system size scaling, *J. Chem. Phys.*, 1996, **104**, 6643–6649.
[24] D. Suarez, V. Gogonea, A. van der Vaart and K. M. Merz, Jr., New developments in applying quantum mechanics to proteins, *Curr. Opin. Struct. Biol.*, 2001, **11**, 217–223.
[25] W. Yang and T.-S. Lee, A density-matrix divide-and-conquer approach for electronic structure calculations of large molecules, *J. Chem. Phys.*, 1995, **103**, 5674–5678.
[26] S. L. Dixon and K. M. Merz, Jr., Fast, accurate semiempirical molecular orbital calculations for macromolecules, *J. Chem. Phys.*, 1997, **107**, 879–893.
[27] A. van der Vaart, B. D. Bursulaya, C. L. Brooks, III and K. M. Merz, Jr., Are many-body effects important in protein folding?, *J. Phys. Chem. B*, 2000, **104**, 9554–9563.
[28] S. Antonczak, G. Monard, M. F. Ruiz-Lopez and J.-L. Rivail, Modeling of peptide hydrolysis by thermolysin. A semiempirical and QM/MM study, *J. Am. Chem. Soc.*, 1998, **120**, 8825–8833.
[29] S. L. Dixon, A. van der Vaart, V. Gogonea, J. J. Vincent, E. N. Brothers, *et al.*, *DIVCON99*, The Pennsylvania State University.
[30] E. Nikitina, D. Sulimov, V. Zayets and N. Zaitseva, Semiempirical calculations of binding enthalpy for protein–ligand complexes, *Int. J. Quantum Chem.*, 2004, **97**, 747–763.
[31] M. Orzco and F. J. Luque, Theoretical methods for the description of solvent effects in biomolecular systems, *Chem. Rev.*, 2000, **100**, 4187–4225.
[32] C. N. Schutz and A. Warshel, What are the dielectric 'constants' of proteins and how to validate electrostatic models, *Proteins Struct. Funct. Genet.*, 2001, **44**, 400–417.
[33] S. M. Schwarzl, T. B. Tschopp, J. C. Smith and S. Fischer, Can the calculation of ligand binding free energies be improved with continuum solvation electrostatics and an ideal-gas entropy correction, *J. Comput. Chem.*, 2002, 23.
[34] T. Simonson, G. Archontis and M. Karplus, Continuum treatment of long-range interactions in free energy calculations: application to protein–ligand binding, *J. Phys. Chem. B*, 1997, **101**, 8349–8362.
[35] J. Tomasi and M. Persico, Molecular interactions in solution: an overview of methods based on continuous distributions of the solvent, *Chem. Rev.*, 1994, **94**, 2027–2094.

[36] N. Arora and D. Bashford, Solvation energy density occlusion approximation for evaluation of desolvation penalties in biomolecular interactions, *Proteins*, 2001, **43**, 12–27.
[37] D. Eisenberg and A. D. McLachlan, Solvation energy in protein folding and binding, *Nature*, 1986, **319**, 199–203.
[38] J. S. Bardi, I. Luque and E. Freire, Structure-based thermodynamic analysis of HIV-1 protease inhibitors, *Biochemistry*, 1997, **36**, 6588–6596.
[39] J. Novotny, R. E. Bruccoleri and F. A. Saul, On the attribution of binding energy in antigen–antibody complexes MCPC 603, D1.3 and Hyhel-5, *Biochemistry*, 1989, **28**, 9700–9711.
[40] R. Zhou, R. A. Friesner, A. Ghosh, R. C. Rizzo, W. L. Jorgensen and R. M. Levy, New linear interaction method for binding affinity calculations using continuum solvation model, *J. Phys. Chem. B*, 2001, **105**, 10388–10397.
[41] J. Aqvist, C. Medina and J. E. Samuelsson, A new method for predicting binding affinity in computer-aided drug design, *Protein Eng.*, 1994, **7**, 385–391.
[42] J. Aqvist, V. B. Luzhkov and B. O. Brandsdal, Ligand binding affinities from MD simulations, *Acc. Chem. Res.*, 2002, **35**, 358–365.
[43] C. J. Cramer and D. G. Truhlar, PM3-SM3: a general parameterization for including aqueous solvation effects in the PM3 molecular orbital method, *J. Comput. Chem.*, 1992, **12**, 1089–1097.
[44] M. E. Davis, J. D. Madura, B. A. Luty and J. A. McCammon, Electrostatics and diffusion of molecules in solution: simulation with the University of Houston Brownian Dynamics program, *Comput. Phys. Commun.*, 1991a, **62**, 187–197.
[45] M. K. Gilson and B. H. Honig, Calculation of electrostatic potentials in an enzyme active site, *Nature*, 1987, **330**, 84–86.
[46] V. Gogonea and K. M. Merz, Jr., Fully quantum mechanical description of proteins in solution. Combining linear scaling quantum mechanical methodologies with the Poisson–Boltzmann equation, *J. Phys. Chem. A*, 1999, **103** (26), 5171–5188.
[47] V. Gogonea and K. M. Merz, Jr., Charge flow between ions and a dielectric continuum. Variational method for distributing charge into the dielectric, *J. Phys. Chem. B*, 2000, **104**, 2117–2122.
[48] W. C. Still, A. Tempczyk, R. C. Hawley and T. Hendrickson, Semianalytical treatment of solvation for molecular mechanics and dynamics, *J. Am. Chem. Soc.*, 1990, **112**, 6127–6129.
[49] A. Warshel and A. Papazyan, Electrostatic effects in macromolecules: fundamental concepts and practical modeling, *Curr. Opin. Struct. Biol.*, 1998, **8**, 211–217.
[50] B. Honig and A. Nicholls, Classical electrostatics in biology and chemistry, *Science*, 1995, **268**, 1144–1149.
[51] Y. N. Vorobjev and H. A. Scheraga, A fast adaptive multigrid boundary element method for macromolecular electrostatic computations in a solvent, *J. Comput. Chem.*, 1997, **18**, 569–583.
[52] R. J. Zauhar and A. Varnek, A fast and space-efficient boundary element method for computing electrostatic and hydration effects in large molecules, *J. Comput. Chem.*, 1996, **17**, 864–877.
[53] M. Gilson, K. Sharp and B. H. Honig, Calculating the electrostatic potential of molecules in solution, *J. Comput. Chem.*, 1988, **9**, 327–335.
[54] W. Rocchia, E. Alexov and B. Honig, Extending the applicability of the nonlinear Poisson–Boltzmann equation: multiple dielectric constants and multivalent ions, *J. Phys. Chem. B*, 2001, **105**, 6507–6514.
[55] N. A. Baker, D. Sept, S. Joseph, M. J. Holst and J. A. McCammon, Electrostatics of nanosystems: application to microtubules and the ribosome, *Proc. Natl Acad. Sci. USA*, 2001, **98**, 10037–10041.

[56] X. Zou, Y. Sun and I. D. Kuntz, Inclusion of solvation in ligand binding free energy calculations using the generalized-born model, *J. Am. Chem. Soc.*, 1999, **121**, 8033–8043.
[57] D. A. McQuarrie and J. D. Simon, *Molecular Thermodynamics*, University Science Books, New York, 1999.
[58] M. S. Searle and D. H. Williams, The cost of conformational order – entropy changes in molecular association, *J. Am. Chem. Soc.*, 1992, **114**, 10690–10697.
[59] A. V. Ishchenko and E. I. Shakhnovich, Small molecule growth 2001 SMoG2001: an improved knowledge-based scoring function for protein–ligand interactions, *J. Med. Chem.*, 2002, **43**, 2770–2780.
[60] M. Rarey, B. Kramer, T. Lengauer and G. Klebe, A fast flexible docking method using an incremental construction algorithm, *J. Mol. Biol.*, 1996, **261**, 470–489.
[61] T. P. Creamer, Side-chain conformational entropy in protein unfolded states, *Proteins Struct. Funct. Genet.*, 2000, **40**, 443–450.
[62] J. H. Lin, A. L. Perryman, J. R. Schames and J. A. McCammon, Computational drug design accommodating receptor flexibility: the relaxed receptor scheme, *J. Am. Chem. Soc.*, 2002, **124**, 5632–5633.
[63] B. Kuhn and P. A. Kollman, Binding of a diverse set of ligands to avidin and streptavidin: an accurate quantitative prediction of their relative affinities by combination of molecular mechanics and continuum solvation models, *J. Med. Chem.*, 2000, **43**, 3786–3791.
[64] T. Hou, S. Guo and X. Xu, Predictions of binding of a diverse set of ligands to gelatinase-A by a combination of molecular dynamics and continuum solvation models, *J. Phys. Chem.*, 2002, **106**, 5527–5535.
[65] K. M. Masukawa, P. A. Kollman and I. D. Kuntz, Investigaion of neuraminidase-substrate recognition using molecular dynamics and free energy calculations, *J. Med. Chem.*, 2003, **46**, 5628–5637.
[66] B. Q. Wei, W. A. Baase, L. H. Weaver, B. W. Matthews and B. K. Shoichet, A model binding site for testing scoring functions in molecular docking, *J. Mol. Biol.*, 2002, **322**, 339–355.
[67] P. H. Hunenberger, V. Helms, N. Narayana, S. S. Taylor and J. A. McCammon, Determinants of ligand binding to cAMP-dependent protein kinase, *Biochemistry*, 1999, **38**, 2358–2366.
[68] P. Cozzini, M. Fornabaio, A. Marabotti, D. J. Abraham, G. E. Kellogg, *et al.*, Simple intuitive calculations of free energy of binding for protein–ligand complexes. 1. Models without explicit constrained water, *J. Med. Chem.*, 2002, **45**, 2469–2483.
[69] S. Gruneberg, M. T. Stubbs and G. Klebe, Successful virtual screening for novel inhibitors of human carbonic anhydrase: strategy and experimental confirmation, *J. Med. Chem.*, 2002, **45**, 3588–3602.
[70] P. S. Charifson, J. J. Corkery, M. A. Murcko and W. P. J. Walters, Consensus scoring: a method for obtaining improved hit rates from docking databases of three-dimensional structures into proteins, *J. Med. Chem.*, 1999, **42**, 5100–5109.
[71] B. A. Grzybowski, A. V. Ishchenko, C. Y. Kim, G. Topalov, R. Chapman, *et al.*, Combinatorial computational method gives new picomolar ligands for a known enzyme, *Proc. Natl Acad. Sci. USA*, 2002, **99**, 1270–1273.
[72] R. Taylor, P. J. Jewsbury and J. W. Essex, A review of protein-small molecule docking methods, *J. Comput.-Aided Mol. Des.*, 2002, **16**, 151–166.
[73] I. Halperin, M. Buyong, H. Wolfson and R. Nussinov, Principles of docking: an overview of search algorithms and a guide to scoring functions, *Proteins Struct. Funct. Genet.*, 2002, **47**, 409–443.
[74] G. M. Verkhivker, D. Bouzida, D. K. Gehlhaar, P. A. Rejto and S. Arthurs, Deciphering common failure in molecular docking of ligand–protein complexes, *J. Comput.-Aided Mol. Des.*, 2000, **14**, 731–751.

[75] M. Jacobsson, P. Liden, E. Strenschantz, H. Bostrom and U. Norinder, Improving structure-based virtual screening by multivariate analysis of scoring data, *J. Med. Chem.*, 2003, **46**, 5781–5789.
[76] M. D. Eldridge, C. W. Murray, T. R. Auton, G. V. Paolini and R. P. Mee, Empirical scoring functions: I. The development of a fast empirical scoring function to estimate the binding affinity of ligands in receptor complexes, *J. Comput.-Aided Mol. Des.*, 1997, **11**, 425–445.
[77] T. J. Ewing, S. Makino, A. G. Skillman and I. D. Kuntz, DOCK 4.0: search strategies for automated molecular docking, *J. Comput.-Aided Mol. Des.*, 2001, **15**, 411–428.
[78] H. J. Bohm, The development of a simple empirical scoring function to estimate the binding constant for a protein–ligand complex of known three-dimensional structure, *J. Comput.-Aided Mol. Des.*, 1994, **8**, 243–256.
[79] H. Gohlke, M. Hendlich and G. Klebe, Predicting binding modes, binding affinities, and 'hot spots' for protein–ligand complexes using a knowledge based scoring function, *Perspect. Drug Discov. Des.*, 2000, **20**, 115–144.
[80] A. Jain, Scoring non-covalent protein–ligand interactions: a continuous differentiable function tuned to compute binding affinities, *J. Comput.-Aided. Mol. Des.*, 1996, **10**, 427–440.
[81] R. D. Head, M. L. Smyte, T. I. Opera, C. L. Waller, S. M. Green and G. R. Marshall, VALIDATE: a new method for the receptor-based prediction of binding affinities of novel ligands, *J. Am. Chem. Soc.*, 1996, **118**, 3959–3969.
[82] B. Kramer, M. Rarey and T. Lengauer, Evaluation of the FlexX incremental construction algorithm for protein–ligand docking, *Proteins Struct. Funct. Genet.*, 1999, **37**, 228–241.
[83] T. I. Opera and G. R. Marshall, Receptor bases prediction of binding affinities, *Perspect. Drug Discov. Des.*, 1998, **9**, 35–61.
[84] G. M. Verkhiver, K. Appelt, S. T. Freer and J. E. Villafranca, Empirical free energy calculations of ligand–protein crystallographic complexes: I. Knowledge-based ligand–protein interaction potentials applies to the prediction of HIV-1 protease binding affinity, *Protein Eng.*, 1995, **8**, 677–691.
[85] R. Wang, Y. Gao and L. Lai, Further development and validation of empirical scoring functions for structure-based binding affinity prediction, *J. Comput.-Aided Mol. Des.*, 2002, **16**, 11–26.
[86] W. Welch, J. Ruppert and A. N. Jain, Hammerhead: fast, fully automated docking of flexible ligands to protein binding sites, *Chem. Biol.*, 1996, **3**, 449–462.
[87] G. Jones, P. Willett, R. C. Glen, A. R. Leach and R. Taylor, Development and validation of a genetic algorithm for flexible docking, *J. Mol. Biol.*, 1997, **267**, 727–748.
[88] I. Muegge, A knowledge-based scoring function for protein–ligand interaction: probing the reference state, *Perspect. Drug Discov. Des.*, 2000, **20**, 99–114.
[89] W. A. Koppensteiner and M. J. Sippl, Knowledge-based potentials – back to the roots, *Biochemistry (Mosc.)*, 1998, **63**, 247–252.
[90] R. S. DeWitte and E. I. Shakhnovich, De novo design method based on simple, fast and accurate free energy estimates. 1. Methodology and supporting evidence, *J. Am. Chem. Soc.*, 1996, **118**, 11733–11744.
[91] A. Wallqvist, R. L. Jernigan and D. G. Covell, A preference-based free energy parameterization of enzyme-inhibitor binding: applications to HIV-1 protease design, *Protein Sci.*, 1995, **4**, 1881–1903.
[92] G. M. Verkhivker, Empirical free energy calculations of human immunodeficiency virus type 1 protease crystallographic complexes. II. Knowledge-based ligand–protein interaction potentials applied to thermodynamic analysis of hydrophobic mutations, *Pac. Symp. Biocomput.*, 1996, 638–652.
[93] I. Muegge and Y. C. Martin, A general and fast scoring function for protein–ligand interactions: a simplified potential approach, *J. Med. Chem.*, 1999, **42**, 791–804.

Section 3
Advances in QSAR/QSPR

Section Editor: Yvonne Martin
Abbott Laboratories
Abbott Park
IL 60064
USA

CHAPTER 10

Computational Prediction of ADMET Properties: Recent Developments and Future Challenges

David E. Clark

Argenta Discovery Ltd, 8/9 Spire Green Centre, Flex Meadow, Harlow, Essex CM19 5TR, UK

Contents

1. Introduction 133
2. Intestinal permeability 134
3. Aqueous solubility 135
4. Human intestinal absorption 137
5. Human oral bioavailability 138
6. Active transport 139
7. Efflux by P-glycoprotein 140
8. Blood–brain barrier permeation 140
9. Plasma protein binding 142
10. Metabolic stability 142
11. Interaction with cytochrome P450s 143
12. Toxicity 144
13. Conclusions 144
References 146

1. INTRODUCTION

Over the last 5 years, the integration of ADMET (absorption, distribution, metabolism, excretion and toxicity) studies into the early phases of drug discovery has been universally implemented within the pharmaceutical industry. The motivation for this paradigm shift has been the requirement to reduce what is termed the 'attrition rate', i.e., the number of drug candidates that fail during late-stage development or clinical trials due to poor ADMET properties. The proportion of failures ascribed to this cause is often quoted as being in the region of 40–50%. If as many as possible of these expensive failures can be identified and eliminated early in the drug discovery process, there is considerable scope for improving the efficiency and cost-effectiveness of the industry. The maxim 'fail fast, fail cheap' is now firmly embedded in the minds of all drug discovery research managers.

ANNUAL REPORTS IN COMPUTATIONAL CHEMISTRY, VOLUME 1
ISSN: 1574-1400 DOI 10.1016/S1574-1400(05)01010-8

To this end, a battery of *in vitro* ADMET assays is now routinely applied during the hit-to-lead and lead optimization phases of drug discovery [1]. To allow timely testing of the numbers of compounds typically of interest at these stages of a project, many assays have been developed in automated, high-throughput formats permitting the testing of hundreds or thousands of compounds per week. In spite of these technical and methodological advances, *in vitro* assays still require that time, effort and money be invested in the synthesis or acquisition of the compounds in question.

The ultimate goal of computational research into ADMET prediction is to be able to identify compounds liable to later stage failure before they are even synthesized, bringing even greater efficiency benefits [2]. While this 'Holy Grail' is still beyond the grasp of the current generation of *in silico* approaches, it is nonetheless attractive enough for much research effort to have been poured into its attainment in recent years. In what follows, we attempt to survey the current state of the art in the computational prediction of some key ADMET properties and to identify some of the current obstacles on the road to what some reviewers have named 'prediction paradise' [3]. Given the breadth of the field and the limited space, the treatment here will necessarily be brief. However, some excellent reviews have been published in the not-too-distant past and the reader is referred to these for further details [3,4]. Reviews specific to particular aspects of ADMET will be cited in the relevant sections.

2. INTESTINAL PERMEABILITY

For reasons of ease of administration and patient compliance, there is an overwhelming preference for drugs to be orally bioavailable. One of the key requirements for oral bioavailability is that a compound be capable of permeating the intestinal epithelium, crossing from the gut into the systemic circulation. For obvious reasons, human data pertaining to permeability are fairly sparse, but those that are available have been modeled by Winiwarter *et al.* using Partial Least Squares (PLS). In their initial work [5], they discovered that the permeability data could be well modeled by polar surface area, a count of hydrogen bond donors and a lipophilicity descriptor ($C\log P$). Similar results were obtained in more recent work [6], which examined a broader range of hydrogen-bonding descriptors. One of the resulting models is shown in equation (1)

$$\log P_{\mathrm{eff}} = -3.128 - 0.0088\mathrm{PSA} - 0.215\mathrm{HBD} + 0.172\log P_{\mathrm{Cr}}$$
$$n = 13,\ r^2 = 0.945,\ q^2 = 0.932 \qquad (1)$$

where $\log P_{eff}$ is the logarithm of the human effective intestinal permeability measured by a regional jejunal perfusion system [7], PSA the polar surface area, HBD a count of the number of hydrogen bond donors and $\log P_{Cr}$ the calculated logarithm of the octanol/water partition coefficient calculated using the Ghose–Crippen method [8]. This model was able to predict the $\log P_{eff}$ values of a small ($n = 4$) test set with an error of less than 0.5 log units. While 0.5 log units is quite a large error on the scale of the $\log P_{eff}$ training set data, which range from -5.5 to -3.0, given the high variability in the experimental data, it might be unrealistic to expect a model to perform much better than this.

Given the difficulty in obtaining human permeability data, the pharmaceutical industry has resorted to a surrogate model for permeability – the Caco-2 cell monolayer [9]. While it is now possible to measure Caco-2 permeability in high-throughput assays, only relatively small collections of data are currently available in the public domain. Efforts to generate predictive models from these data have been recently reviewed by the author [10] and so only a single illustration will be given here. One of the difficulties facing modelers is the notorious inter-laboratory variation in Caco-2 permeability values, which makes the compilation of data from different experimental sources a risky endeavor. For this reason, Artursson and co-workers [11,12] have generated a relatively small set of high-quality Caco-2 permeability data ($\log P_{app}$) in their own laboratories and used it as the basis for predictive modeling. In the most recent work [12], a combination of partitioned molecular surface area descriptors with PLS analysis led to the development of a model showing good training set statistics ($n = 13$, $r^2 = 0.93$, $q^2 = 0.83$) and reasonable performance on an external test set ($n = 26$, RMSE $= 0.85$). An analysis of the model showed that hydrogen-bonding features (e.g., PSA and the surface area specifically attributable to doubly bonded oxygens and N–H groups) were the primary determinants of Caco-2 cell permeability.

As a final remark in this section, it is worth stressing in passing that the Caco-2 model of intestinal permeability is just that – a model – and so data from Caco-2 cell assays (and predictions based upon them) may need careful interpretation, especially when Caco-2 cell permeability is (predicted to be) low. In such instances, it is still possible that compounds may exhibit reasonable intestinal permeability due to the morphological differences between the Caco-2 cell monolayer and the intestinal epithelium [9].

3. AQUEOUS SOLUBILITY

Although not always thought of as an ADMET property, the aqueous solubility of a compound is a key factor in determining its oral absorption – if a compound is poorly soluble in the gut, then only a small fraction of it will be available to permeate the intestinal epithelium. Consequently, the interest in predicting aqueous solubility has intensified over the last few years, not least due to the work of Lipinski [13,14], who has brought to the fore the plethora of issues that can result from poor compound solubility. It is often forgotten that one of the intentions of the now ubiquitously adopted 'rule-of-5' is to identify compounds with poor solubility, not just permeability [15]. Although numerous attempts have been made to generate predictive models for aqueous solubility (Lobell and Sivarajah [16] recently counted at least 17 different approaches), it is proving a difficult quantity to predict accurately [17–20]. The publicly available data are biased towards non-drug-like compounds of lower molecular weight and lipophilicity than many compounds of interest in drug discovery. Additionally, the solubility values have often been determined using different protocols. It has been estimated that the average experimental error in log S (log of molar solubility) measurements is of the order of 0.6 log units, so computational models should not be expected to better this in their performance [17]. In what follows, recent work in this area is summarized.

In an attempt to move away from non-drug-like training sets, Lobell and Sivarajah [16] assembled a training set of 202 drug-like compounds and a test set of 442 compounds. These compounds were determined to be predominantly uncharged at the pH at which the solubility measurement was carried out. From the training set, a simple linear model with a single lipophilicity descriptor ($A \log P98$) was derived ($r^2 = 0.64$, MAE $= 0.54$). When applied to the test set, the simple $A \log P$-based equation outperformed a set of nine other solubility models, many of which had been trained on non-drug-like compound sets.

Cheng and Merz [21] have provided a thoughtful discussion of the issues surrounding solubility prediction and presented a fast QSPR model developed using the Cerius^2 software [22]. The model was trained on 775 compounds with an r^2 value of 0.84. Among the eight descriptors in the regression equation were lipophilicity, the product of the number of hydrogen bond donors and acceptors (believed to be related to the crystal-packing energy of a solute), and various topological indices coding for molecular size. When applied to a test set of 1665 compounds, the model was able to predict log S values with an unsigned error of 0.77 log units, which appears respectable in the light of the comments above regarding the likely magnitude of experimental error.

Using data for 1293 organic compounds, Yan and Gasteiger [23,24] have developed solubility prediction models using two different modeling techniques (multiple linear regression (MLR) and a back-propagation neural network (BPNN)) and two different sets of descriptors. In the first paper, each compound was characterized by a radial distribution function code derived from its 3-D structure [25], together with eight additional descriptors encoding features such as the relative aromatic/aliphatic balance in a compound and its hydrogen-bonding capacity. Using these descriptors, the BPNN model was able to predict the log S values of a 496-compound test set with a standard deviation of 0.59 log units [23].

More recently, the same set of compounds has been represented by 18 topological descriptors. In this case, on a 552-compound test set, the BPNN model was able to predict log *S* with a standard deviation of 0.52 log units [24].

Group contribution approaches have been used successfully in solubility prediction in the past (e.g., [26]). More recently, using a dataset of similar size to that of Yan and Gasteiger, Hou *et al.* [27] used an *atom*-contribution approach to generate a solubility prediction model. A set of 76 atom types was employed and the resulting model was shown to perform well on a small, but widely used, test set of 21 compounds ($r = 0.94$, $s = 0.84$, $\mathrm{MAE} = 0.52$). The authors suggest that using atoms, rather than molecular fragments, as descriptors may help to eliminate the 'missing fragment' problem that bedevils group contribution approaches.

Support Vector Machines (SVMs) constitute an emerging technique for regression and classification across the spectrum of ADME properties. Their use for solubility prediction has been reported by Lind and Maltseva [28]. In this work, data from a training set of 883 compounds characterized by molecular fingerprints were used to generate a model ($r^2 = 0.88$, $\mathrm{RMSE} = 0.62$). The model was then applied to predict the log *S* values of a test set of 412 compounds ($r^2_{\mathrm{pred}} = 0.89$, $\mathrm{RMSE} = 0.68$).

4. HUMAN INTESTINAL ABSORPTION

The prediction of the fraction of a drug absorbed in humans (usually denoted HIA or FA) has been aided in recent years by the Herculean efforts of Abraham and co-workers who carefully compiled and analyzed a set of FA data for 241 drugs from the literature [29]. (Notably, this article was awarded the Ebert Prize, given by the American Pharmaceutical Association for the best paper published in *J. Pharm. Sci.* in 2001.) Of the 241 data, 169 were deemed to be reliable and these were used to generate a QSPR model using Abraham's solute descriptors

$$\mathrm{FA} = 90 + 2.11E + 1.70S - 20.7A - 22.3B + 15.0V$$

$$n = 38,\ r^2 = 0.83,\ r^2(\mathrm{CV}) = 0.75,\ s = 16,\ F = 31 \qquad (2)$$

where *E* is an excess molar refraction, *S* the dipolarity/polarizability, *A* the hydrogen bond acidity of the compound, *B* is the hydrogen-bond basicity of the compound and *V* its characteristic (McGowan) volume. The model suggests, in keeping with other work, that increasing the hydrogen-bonding capacity is deleterious to facile intestinal absorption. The performance of this model on the test set of 131 compounds was quite impressive: an RMSE of 14% and an average absolute error (AAE) of 11%. Given the inherent variability in experimental absorption data, this level of prediction error would seem to be quite respectable. One point to note here, which has been well made by Burton *et al.* [30] in their excellent perspective on absorption prediction, is the skew in the data set towards well-absorbed compounds (of the 241 compounds, 108 have FA $> 90\%$).

This data set has been seized on by other workers and used to help generate a number of other models. Deretey *et al.* [31] used two descriptors generated by the MOE program [32] to create a non-linear absorption model based on 124 of the Abraham data. The descriptors in question were a total hydrogen bond descriptor, taken as the sum of donor and acceptor atomic centers computed in MOE, and $S \log P$, which is the logarithm of the octanol–water partition coefficient, computed using the method of Wildman and Crippen [33]. Additionally, Sun [34] developed a Partial Least Squares Discriminant Analysis (PLS-DA) model based on a set of atom-type descriptors that was able to classify the 169 compounds from Abraham's work into three classes with a good level of accuracy ($r^2 = 0.92$).

Recursive partitioning tools within the Algorithm Builder package [35] were applied by Zmuidinavcius and co-workers to a large set of FA data [36]. The decision tree built for a set of molecules in the molecular weight range 255–580 featured three descriptors: topological polar surface area, Abraham's hydrogen bond acidity (a measure of hydrogen bond donating ability) and calculated lipophilicity. Such descriptors appear frequently in absorption/permeation models and it is particularly interesting to note the presence of a descriptor relating to hydrogen bond donation (cf. equation (1)). There is a growing body of evidence that hydrogen bond donors are more deleterious to absorption than acceptors. Oprea [37] has hypothesized that this is due to the favorable interactions that can occur between hydrogen bond donors and the ester moieties located within lipid headgroups.

Hydrogen bond donors also figure in the model reported by Klopman *et al.* [38]. The count of OH + NH groups was used along with 36 structural fragments generated by the CASE program [39] to model FA data from 417 compounds. On a 50 compound test set, the prediction statistics were $r^2 = 0.79$, $s = 12.32$. Again, this level of accuracy seems in keeping with the probable level of experimental error in the data.

More detailed reviews of absorption prediction are available in some recent publications [10,40].

5. HUMAN ORAL BIOAVAILABILITY

Fewer attempts have been made to predict human oral bioavailability than oral absorption, probably because of the greater complexity inherent in the former, which includes the effects of first-pass metabolism in the gut and liver. For this reason, it is necessary to delve a little deeper into the past to obtain a sense of progress.

Andrews *et al.* [41] collated a set of 591 structures with human oral bioavailability data from public and proprietary sources and used stepwise regression to create a model in which compounds were described by 85 substructural fragments. The r^2 value for the model was 0.71 and the cross-validated r^2 (leave-one-out) was 0.63, indicating a reasonable level of internal predictivity. The RMSE value for the model was 18%, which in the context of an experimental RMSE of 12% is encouraging. For more complete validation, application to an external test set is required.

A classification, rather than regression, approach was adopted by Yoshida and Topliss [42] to model the bioavailability of 232 drugs. This test set was split into four classes. A novel parameter $\Delta \log D$ ($\log D_{6.5} - \log D_{7.4}$) was developed that, together with 15 substructural fragment descriptors, was able to classify 71% of the drugs correctly and 97% to within one class. Using a separate test set of 40 compounds, a classification accuracy of 60% (95% to within one class) was obtained. The difficulty in applying this model in a purely *in silico* manner is the requirement for accurate prediction of log D, which in turn requires the accurate prediction not only of log P but also for ionizable species, pK_a.

More recently, Pintore *et al.* [43] have expanded the data set used by Yoshida and Topliss and split it into four classes in the same manner. The compounds were characterized by a set of 164 molecular descriptors. Using a hybrid genetic algorithm/stepwise method for descriptor selection and adaptive fuzzy partitioning for classification, a superior performance was obtained on the Yoshida and Topliss set (82% correct classification on the training set and 75% on the test set). A good level of performance was also observed on the expanded set of 432 compounds (70% correct classification).

Other work pertaining to bioavailability prediction has been reported by Turner *et al.* [44,45] and Bains *et al.* [46]. Mandagere and Jones [47] have recently reviewed the prediction of bioavailability both with *in vitro* and *in silico* techniques. A previously cited review is also of relevance [40].

6. ACTIVE TRANSPORT

To date, predictive approaches to permeability/absorption prediction have largely confined themselves to compounds that are transported across the intestinal mucosa by predominantly passive (usually transcellular) absorption mechanisms. However, there are well-known classes of drug (e.g., many ACE inhibitors and beta-lactam antibiotics) that rely on active transport systems to convey them from the gut to the bloodstream. Our knowledge of these transporters and understanding of their implications for drug delivery is growing all the time.

Of the active transporters identified to date, most attention has been focused on Peptide Transporter 1 (PepT1), a member of the proton-coupled, oligopeptide transporter (POT) family [48]. Initial modeling work on PepT1 substrates was carried out by Swaan *et al.* [49,50] and Bailey *et al.* [51]. More recently, Gebauer *et al.* [52] developed CoMFA and CoMSIA models from a set of 79 dipeptide PepT1 substrates for which affinity data had been generated under consistent conditions. Both approaches yielded models with good statistics (CoMFA: $r^2 = 0.901$, $q^2 = 0.642$; CoMSIA: $r^2 = 0.913$, $q^2 = 0.776$) and when applied to a test set of 19 compounds, all but one of the predicted affinity values were within 1 log unit of the experimental result. By analyzing the CoMSIA maps, six recognition elements for binding to PepT1 were proposed.

A fairly recent review of other modeling efforts in this area has been provided by Zhang *et al.* [53].

7. EFFLUX BY P-GLYCOPROTEIN

Efflux by P-glycoprotein (P-gp), the most extensively studied of the ATP-binding cassette (ABC) transporters, is a serious liability for potential drug compounds, particularly those seeking to cross the blood–brain barrier (BBB) [54]. The considerable challenges facing modelers attempting to create predictive models for interaction with P-gp have been delineated by Stouch and Gudmundsson [55] and Seelig *et al.* [56]. A brief review of recent work has been presented by Ekins [57] and so only subsequent developments will be covered here.

Gombar *et al.* [58] used a training set of 95 compounds (63 substrates, 32 non-substrates) with efflux data derived from MDR1-MDCK transport assay to derive a linear discriminant function capable of accurately classifying the compounds as substrates/non-substrates. The function comprised 27 molecular descriptors including E-state indices, computed molar refractivity and lipophilicity. When applied to a test set of 58 compounds (35 substrates, 23 non-substrates), the model proved able to identify the substrates very accurately (33/35 correct) and the non-substrates with reasonable accuracy (17/23 correct). As a simple rule-of-thumb, Gombar *et al.* report that compounds with a molecular E-state (MolES, representing the molecular bulk of a compound) value of >110 are likely to be P-gp substrates, while those with a MolES value <49 are likely to be non-substrates.

8. BLOOD–BRAIN BARRIER PERMEATION

Many QSAR approaches to the prediction of BBB permeation have been developed in recent years [59]. The field has been reviewed fairly recently by the author [60] and so only newer work will be discussed here. Generally speaking, approaches to date can be divided into two classes: regression models seeking to predict log BB (=log([brain]/[blood])) and classification models seeking to classify compounds correctly as BBB^+ (i.e., brain permeating) or BBB^- (i.e., non-brain-permeating). Activity in both these directions continues.

Hou and Xu [61] have reported a log BB model based on a training set of 78 compounds that uses only three descriptors: calculated lipophilicity, highly charged polar surface area and excessive molecular weight greater than 360 ($r^2 = 0.77$, $s = 0.36$). The combination of polar surface area and lipophilicity has been shown to be useful for predicting BBB permeation by a number of workers in the past [59]. Applied to two test sets of 14 and 23 compounds, the model predicted the log BB values of the compounds with RMSEs of 0.26 and 0.48, respectively. Hutter [62] has developed a somewhat more computationally expensive approach for log BB prediction using descriptors derived from semi-empirical molecular orbital (AM1) calculations. Key variables in the final regression equation ($r^2 = 0.87$, $s = 0.31$) included properties derived from the electrostatic potential, hydrogen bond donor and acceptor capacities, shape and molecular flexibility. Other log BB prediction models have been reported recently by Sun [34] and Subramanian and Kitchen [63].

In recognition of the important role that P-gp plays at the BBB, a classification approach has been followed by Adenot and Lahana [64] who assembled a data set comprising 1336 BBB^+ compounds, 259 BBB^- compounds and 91 P-gp substrates (which were either BBB^+ or BBB^-). Discriminant functions for BBB permeation and P-gp substrate/non-substrate classification were developed allowing the simultaneous evaluation of the efflux liability and brain permeation characteristics of a compound. When considering just brain permeation, a simple rule-of-thumb emerged: most BBB^- compounds have more than eight heteroatoms (where a heteroatom may be N, O, S, P or halogen) while most BBB^+ compounds have fewer than nine heteroatoms. A similar observation has been made by Norinder and Haeberlein [59].

An alternative, but less commonly used to date, measure of brain permeation is log PS (where PS denotes a permeability–surface area product). Unlike log BB, which represents an equilibrium distribution between brain and blood, log PS is a direct measure of brain permeability and, in principle, is not confounded by plasma and brain tissue binding. It has been suggested that log PS may therefore be a more relevant parameter in brain permeation studies [65]. A recent paper has reported a QSPR model based on log PS data obtained for 23 passively transported, drug-like compounds [65]

$$\log \mathrm{PS} = -2.19 + 0.262 \log D + 0.0683\mathrm{vsa\_base} - 0.009\mathrm{TPSA}$$
$$n = 23,\ r^2 = 0.74,\ s = 0.5,\ F = 18.2 \tag{3}$$

where log D (at pH 7.4) was calculated by ACDlabs software [66], vsa_base is the van der Waals' surface area due to basic atoms and TPSA is the topological polar surface area – the latter descriptors being computed by MOE [32]. Here again, the combination of polar surface area and lipophilicity descriptors is noteworthy.

9. PLASMA PROTEIN BINDING

There have been relatively few attempts at modeling plasma protein binding (PPB), and most of those reported have focused on human serum albumin (HSA), which is the most abundant protein in plasma although certainly not the only one responsible for PPB [4]. *In silico* approaches to PPB (HSA) prediction have been critically reviewed in the recent past by Lombardo *et al.* [4] and Colmenarejo [67].

One paper to emerge subsequent to these reviews is that of Hajduk *et al.* [68]. This group measured the binding affinity of 889 diverse compounds to the domain-3A of HSA (which contains the diazepam binding site) using heteronuclear NMR correlation and fluorescence spectroscopy. A moderate correlation between affinity and $C\log P$ was observed ($r^2 = 0.56$), which reflects medicinal chemistry experience [69]. A superior model was obtained by adopting a group contribution approach using 74 structural fragments. The resulting model exhibited good statistics ($r^2 = 0.94$, q^2(leave-several-out) $= 0.90$) and a mean error in predicted binding affinity of only 0.11 log units, although no validation on an external test set was reported. The interpretability of the model is attractive inasmuch as the coefficients for the fragments comprising the model should provide a straightforward guide when seeking to modify compounds to modulate their HSA binding. In other recent work, Hall *et al.* [71] used the chromatographic retention data generated by Colmenarejo *et al.* [70] to generate a predictive QSPR comprising various E-state and molecular connectivity indices.

10. METABOLIC STABILITY

A variety of *in vitro* assays for the high-throughput assessment of the metabolic stability of compounds have been developed in recent years, including those based on hepatocytes and various subcellular fractions such as microsomes and S9 [72]. As data from these have become available, publications describing attempts to generate predictive models have begun to appear [73].

Shen *et al.* have reported QSPR models for metabolic stability based on percent turnover (at 30 min) data generated in human liver S9 homogenate for 631 GlaxoSmithKline compounds [74]. The compounds were grouped into four classes according to their percent turnover values, ranging from stable ($<25\%$) to unstable ($>75\%$). The 631 compounds were split into a training set of 572 and a test set of 59 compounds. An additional validation set of 107 compounds became available during the course of the work. Using a *k*-nearest neighbors method, two different QSPR models were generated: one based on atom-pair descriptors, the other on topological indices. Both showed an impressive classification accuracy for the stable and unstable classes of about 80% across training and test sets. The model based on topological descriptors predicted 50 of the additional validation compounds to be stable and 42 of these were verified as being stable by experiment (84% success in prediction).

Other models have been reported by Ekins [75], who used a recursive partitioning procedure to generate a model based on human liver microsomal stability data for 800 compounds, and Bursi *et al.* [76], who developed a CoMFA model to help rationalize the SAR of the microsome-catalyzed hydrolysis rate of some ester prodrugs. Finally, taking a more *a priori* approach, Lewin and Cramer [77] have evaluated various quantum mechanical models for the estimation of C–H bond dissociation energy (BDE), concluding that a method based on the AM1 Hamiltonian should be able to predict BDEs with an error of <4 kcal/mol.

11. INTERACTION WITH CYTOCHROME P450s

As well as studying metabolic stability in a general sense, there has been much interest in recent years in the prediction of the interactions of compounds with individual cytochrome P450 (CYP450) enzymes, which constitute the major drug metabolizing enzyme system in the human body. Two broad approaches have been adopted: those using available X-ray structures to create homology models of important CYP450s and those that are ligand based, studying known inhibitors/substrates in an attempt to generate pharmacophore or QSAR models. However, the approaches should not be considered as mutually exclusive, and are often best used in a complementary fashion [78].

An excellent illustration of the complementary use of homology modeling and QSAR applied to CYP2C9 has been published recently by Afzelius *et al.* [79]. A training set of 22 CYP2C9 inhibitors was used to develop a 3-D QSAR model using the alignment-independent ALMOND descriptors ($r^2 = 0.81$, q^2(leave-one-out) $= 0.62$). When applied to a test set of 12 compounds, the K_i values of nine were predicted with an error of <0.3 log units and the worst error was 0.97 log units. An examination of the key interaction points derived from the 3-D QSAR model showed that they corresponded well with important residues in the active site of a CYP2C9 homology model.

Reviews of comparative modeling of CYP450s by Kirton *et al.* [80] and Lewis [81] give a picture of the field prior to the publication of the first X-ray structure of a human CYP450 [82]. This landmark achievement is likely to give a new impetus to homology modeling efforts. Another major human CYP450 structure (3A4) has been solved but not yet published [83]. Lewis *et al.* [84–88] have published a number of models recently based on the mammalian 2C5 structure. The pharmacophore approach has been reviewed by de Groot and Ekins [89], and Ekins *et al.* [90] have more recently discussed applications to just the CYP3A family. Lewis [91] has compiled data for substrates and inhibitors of the CYP1, CYP2 and CYP3 families and also described a number of QSAR models, particularly in the context of his COMPACT methodology [92]. High-throughput computational filters seeking to identify compounds likely to interact strongly with important CYPs have also been developed. Two recent examples have been published by Ekins *et al.* [93] and Susnow and Dixon [94]. Both pieces of work employed recursive partitioning: the former to develop models for the prediction of CYP2D6 and CYP3A4 inhibition, the latter for just inhibition of CYP2D6.

The next level of sophistication in CYP450 modeling is the prediction of the likely site(s) of metabolism in a compound. Singh *et al.* [95] have recently described a method for predicting sites in compounds that are likely to be metabolized by CYP3A4. Two properties of an atom were found to be important in determining its likelihood of being a site of metabolism: the energy required to abstract a hydrogen atom from it (as computed by AM1) and its accessible surface area.

Finally, while most attention to date has focused on the prediction of CYP450 inhibition, the important issue of CYP450 induction is also now receiving attention from modelers [96].

12. TOXICITY

Lack of space precludes a thorough coverage of the prediction of toxicity. Suffice it to say that it remains a significant challenge, not least because of the plethora of toxicological endpoints, some of which can be ill-defined and the fact that multiple mechanisms can lead to the same endpoint. The progress in computational toxicity prediction has been reviewed by Greene [97] and Dearden [98], and Feng *et al.* [99] have recently published an interesting comparison of statistical methods and molecular descriptors applied to four toxicological data sets. There has also been much interest recently in homology and QSAR models to help predict interaction with the human Ether-a-go-go Related Gene (hERG) channel, due to its implication in cardiac toxicity [57,100].

13. CONCLUSIONS

It is clear from the above that much effort has been, and continues to be, expended in the development of predictive ADMET models. Currently, the major obstacles on the path to improved models are related to data. Primarily, there are simply too few data (at least in the public domain) to allow the creation of robust models that are applicable to a variety of chemotypes. Models for predicting log *P* have over 11,000 experimental data (the Starlist) to draw on [3]. In most cases, ADME models have only a few tens or hundreds of data at most. Thus, it is unreasonable to expect ADME models to perform any better than log *P* models, which can still be erroneous, despite the size of the training set. Secondly, there is a growing realization that the data need to be of the 'right kind' – they should be generated for drug-like compounds that span a diverse structure and physicochemical property space. It is not sufficient just to use the data that happen to be available from past and present projects. Future ADME models may well require bespoke data generation. The complexity – and our current lack of understanding – of many of the underlying physiological processes involved in ADMET also places a limit on the accuracy with which they can be modeled [30]. In some respects, it is surprising that the current generation of models performs as well as it does.

In the future, it is likely that the field of ADME prediction will see an increasing adoption of consensus approaches – already popular in virtual screening – in the hope of generating more robust predictions by combining the outputs of multiple models. Future models should also give better estimates of the errors in their predictions and indicate when a given compound falls outside the training set space [73]. This kind of output will help to engender greater confidence in models in non-expert users. There will also be a growing realization that different models are needed for different tasks. For instance, high-throughput 'filters' may be appropriate for very early stage drug discovery to sift through large compound collections or help design exploratory combinatorial libraries. However, in the lead optimization phase, lower-throughput, but highly interpretable, models will be required to guide molecular design. Models may also be differentiated as being 'global' or 'local' in nature, the former being applicable to a wide range of chemical classes and the latter to perhaps just a single compound series. While there seems to be no shortage of molecular descriptors and statistical methods to apply to ADMET data sets, it may be that as those data sets increase in size, there will be a need for further developments in both these areas. Support vector machines are an example of a recently adopted statistical method that seems to be finding favor with the modeling community [28].

Looking at the full spectrum of ADMET properties, it is evident that there are gaps that are not currently being addressed in detail by predictive methods. In particular, excretion-related properties have been sparsely treated so far [4]. The prediction of active transport and efflux processes should also develop in the coming years as more data become available for modeling. Additionally, our understanding of how and why compounds interact with CYP450s should increase as more co-crystal structures with bound ligands become available. The ultimate goal would be an integrated system capable of predicting the fate of a compound in man, taking into account all the phenomena outlined in this review. It may be that the pharmacologically based pharmacokinetic models that are already available will provide a means to accomplish this [101]. This remains as yet a distant goal, but given a more profound understanding of the processes involved, sufficient data and appropriate modeling techniques, it may not be an unattainable one.

REFERENCES

[1] H. Yu and A. Adedoyin, ADME-Tox in drug discovery: integration of experimental and computational technologies, *Drug Discov. Today*, 2003, **8**, 852–861.
[2] A. P. Beresford, H. E. Selick and M. H. Tarbit, The emerging importance of predictive ADME simulation in drug discovery, *Drug Discov. Today*, 2002, **7**, 109–116.
[3] H. Van de Waterbeemd and E. Gifford, ADMET *in silico* modelling: towards prediction paradise, *Nat. Rev. Drug Discov.*, 2003, **2**, 192–204.
[4] F. Lombardo, E. Gifford and M. Y. Shalaeva, *In silico* ADME prediction: data, models, facts and myths, *Mini-rev. Med. Chem.*, 2003, **3**, 861–875.
[5] S. Winiwarter, N. M. Bonham, F. Ax, A. Hallberg, H. Lennernaes and A. Karlen, Correlation of human jejunal permeability (*in vivo*) of drugs with experimentally and theoretically derived parameters. A multivariate data analysis approach, *J. Med. Chem.*, 1998, **41**, 4939–4949.
[6] S. Winiwarter, F. Ax, H. Lennernaes, A. Hallberg, C. Pettersson and A. Karlen, Hydrogen bonding descriptors in the prediction of human in vivo intestinal permeability, *J. Mol. Graph. Model.*, 2003, **21**, 273–287.
[7] H. Lennernaes, Human intestinal permeability, *J. Pharm. Sci.*, 1998, **87**, 403–410.
[8] A. K. Ghose and G. M. Crippen, Atomic physicochemical parameters for three-dimensional structure-directed quantitative structure–activity relationships. I. Partition coefficients as a measure of hydrophobicity, *J. Comput. Chem.*, 1986, **7**, 565–577.
[9] P. Artursson and S. Tavelin, Caco-2 and emerging alternatives for prediction of intestinal drug transport: a general overview. In *Drug Bioavailability: Estimation of Solubility, Permeability and Bioavailability* (eds P. Artursson, H. Lennernäs and H. van de Waterbeemd), Methods and Principles in Medicinal Chemistry, Wiley-VCH, Weinheim, 2003, Vol. 18, pp. 72–89.
[10] D. E. Clark, Computational prediction of intestinal absorption. In *Encyclopedia of Computational Chemistry* (eds P. v. R. Schleyer, W. L. Jorgensen, H. F. Schaefer, III, P. R. Schreiner, W. Thiel and R. Glen), Wiley, Athens, 2004. URL: http://www.mrw.interscience.wiley.com/ecc/articles/cu0049/frame.html
[11] P. Stenberg, U. Norinder, K. Luthman and P. Artursson, Experimental and computational screening models for the prediction of intestinal drug absorption, *J. Med. Chem.*, 2001, **44**, 1927–1937.
[12] C. A. S. Bergstroem, M. Strafford, L. Lazorova, A. Avdeef, K. Luthman and P. Artursson, Absorption classification of oral drugs based on molecular surface properties, *J. Med. Chem.*, 2003, **46**, 558–570.
[13] C. A. Lipinski, Drug-like properties and the causes of poor solubility and poor permeability, *J. Pharmacol. Toxicol. Methods*, 2000, **44**, 235–249.
[14] C. A. Lipinski, Aqueous solubility in discovery, chemistry, and assay changes. In *Drug Bioavailability: Estimation of Solubility, Permeability and Bioavailability* (eds P. Artursson, H. Lennernäs and H. van de Waterbeemd), Methods and Principles in Medicinal Chemistry, Wiley-VCH, Weinheim, 2003, Vol. 18, pp. 215–231.

[15] C. A. Lipinski, F. Lombardo, B. W. Dominy and P. J. Feeney, Experimental and computational approaches to estimate solubility and permeability in drug discovery and development settings, *Adv. Drug Deliv. Rev.*, 2001, **46**, 3–26.
[16] M. Lobell and V. Sivarajah, *In silico* prediction of aqueous solubility, human plasma protein binding and volume of distribution of compounds from calculated pK_a and A log P98 values, *Mol. Divers.*, 2003, **7**, 69–87.
[17] W. L. Jorgensen and E. M. Duffy, Prediction of drug solubility from structure, *Adv. Drug Deliv. Rev.*, 2002, **54**, 355–366.
[18] J. Huuskonen, Estimation of aqueous solubility in drug design, *Comb. Chem. High Throughput Screen.*, 2001, **4**, 311–316.
[19] D. E. Clark and P. D. J. Grootenhuis, Progress in computational methods for the prediction of ADMET properties, *Curr. Opin. Drug Discov. Dev.*, 2002, **5**, 382–390.
[20] D. Eroes, G. Keri, I. Koevesdi, C. Szantai-Kis, G. Meszaros and L. Oerfi, Comparison of predictive ability of water solubility QSPR models generated by MLR, PLS and ANN methods, *Mini-Rev. Med. Chem.*, 2004, **4**, 167–177.
[21] A. Cheng and K. M. Merz, Jr., Prediction of aqueous solubility of a diverse set of compounds using quantitative structure–property relationships, *J. Med. Chem.*, 2003, **46**, 3572–3580.
[22] Cerius^2, Accelrys Inc., 9685 Scranton Road, San Diego, CA 92121-3752, USA.
[23] A. Yan and J. Gasteiger, Prediction of aqueous solubility of organic compounds based on a 3D structure representation, *J. Chem. Inf. Comput. Sci.*, 2003, **43**, 429–434.
[24] A. Yan and J. Gasteiger, Prediction of aqueous solubility of organic compounds by topological descriptors, *QSAR Comb. Sci.*, 2003, **22**, 821–829.
[25] M. C. Hemmer, V. Steinhauer and J. Gasteiger, Deriving the 3D structure of organic molecules from their infrared spectra, *Vib. Spectrosc.*, 1999, **19**, 151–164.
[26] G. Klopman and H. Zhu, Estimation of the aqueous solubility of organic molecules by the group contribution approach, *J. Chem. Inf. Comput. Sci.*, 2001, **41**, 439–445.
[27] T. J. Hou, K. Xia, W. Zhang and X. J. Xu, ADME evaluation in drug discovery. 4. Prediction of aqueous solubility based on atom contribution approach, *J. Chem. Inf. Comput. Sci.*, 2004, **44**, 266–275.
[28] P. Lind and T. Maltseva, Support vector machines for the estimation of aqueous solubility, *J. Chem. Inf. Comput. Sci.*, 2003, **43**, 1855–1859.
[29] Y. H. Zhao, J. Le, M. H. Abraham, A. Hersey, P. J. Eddershaw, C. N. Luscombe, D. Butina, G. Beck, B. Sherborne, I. Cooper and J. A. Platts, Evaluation of human intestinal absorption data and subsequent derivation of a quantitative structure–activity relationship (QSAR) with the Abraham descriptors, *J. Pharm. Sci.*, 2001, **90**, 749–784.
[30] P. S. Burton, J. T. Goodwin, T. J. Vidmar and B. M. Amore, Predicting drug absorption: how nature made it a difficult problem, *J. Pharmacol. Exp. Ther.*, 2002, **303**, 889–895.
[31] E. Deretey, M. Feher and J. M. Schmidt, Rapid prediction of human intestinal absorption, *Quant. Struct.–Act. Relat.*, 2002, **21**, 493–506.
[32] Molecular Operating Environment (MOE), Chemical Computing Group, Inc., 1010 Sherbrooke St. West, Suite 910, Montreal, Quebec H3A 2R7, Canada.
[33] S. A. Wildman and G. M. Crippen, Prediction of physicochemical parameters by atomic contributions, *J. Chem. Inf. Comput. Sci.*, 1999, **39**, 868–873.
[34] H. Sun, A universal molecular descriptor system for prediction of log P, log S, log BB and absorption, *J. Chem. Inf. Comput. Sci.*, 2004, **44**, 748–757.
[35] Algorithm Builder, Pharma Algorithms, Inc., 591 Indian Road, Toronto, Ontario M6P 2C4, Canada.
[36] D. Zmuidinavicius, R. Didziapetris, P. Japertas, A. Avdeef and A. Petrauskas, Classification structure–activity relations (C-SAR) in prediction of human intestinal absorption, *J. Pharm. Sci.*, 2003, **92**, 621–633.

[37] T. I. Oprea, Property distribution of drug-related chemical databases, *J. Comput.-Aided Mol. Des.*, 2000, **14**, 251–264.
[38] G. Klopman, L. R. Stefan and R. D. Saiakhov, ADME evaluation 2. A computer model for the prediction of intestinal absorption in humans, *Eur. J. Pharm. Sci.*, 2002, **17**, 253–263.
[39] G. Klopman, Artificial intelligence approach to structure–activity studies. Computer automated structure evaluation of biological activity of organic molecules, *J. Am. Chem. Soc.*, 1984, **106**, 7315–7321.
[40] H. van de Waterbeemd and B. C. Jones, Predicting oral absorption and bioavailability, *Prog. Med. Chem.*, 2003, **41**, 1–59.
[41] C. W. Andrews, L. Bennett and L. X. Yu, Predicting human oral bioavailability of a compound: development of a novel quantitative structure–bioavailability relationship, *Pharm. Res.*, 2000, **17**, 639–644.
[42] F. Yoshida and J. G. Topliss, QSAR model for drug human oral bioavailability, *J. Med. Chem.*, 2000, **43**, 2575–2585, see p. 4723 for erratum.
[43] M. Pintore, H. van de Waterbeemd, N. Piclin and J. R. Chretien, Prediction of oral bioavailability by adaptive fuzzy partitioning, *Eur. J. Med. Chem.*, 2003, **38**, 427–431.
[44] J. V. Turner, B. D. Glass and S. Agatonovic-Kustrin, Prediction of drug bioavailability based on molecular structure, *Anal. Chim. Acta*, 2003, **485**, 89–102.
[45] J. V. Turner, D. J. Maddalena and S. Agatonovic-Kustrin, Bioavailability prediction based on molecular structure for a diverse series of drugs, *Pharm. Res.*, 2004, **21**, 68–82.
[46] W. Bains, R. Gilbert, L. Sviridenko, J.-M. Gascon, R. Scoffin, K. Birchall, I. Harvey and J. Caldwell, Evolutionary computational methods to predict oral bioavailability QSPRs, *Curr. Opin. Drug Discov. Dev.*, 2002, **5**, 44–51.
[47] A. K. Mandagere and B. Jones, Prediction of bioavailability. In *Drug Bioavailability: Estimation of Solubility, Permeability and Bioavailability* (eds P. Artursson, H. Lennernäs and H. van de Waterbeemd), Methods and Principles in Medicinal Chemistry, Wiley-VCH, Weinheim, 2003, Vol. 18, pp. 444–460.
[48] D. Herrera-Ruiz and G. T. Knipp, Current perspectives on established and putative mammalian oligopeptide transporters, *J. Pharm. Sci.*, 2003, **92**, 691–714.
[49] P. W. Swaan and J. J. Tukker, Molecular determinants of recognition for the intestinal peptide carrier, *J. Pharm. Sci.*, 1997, **86**, 596–602.
[50] P. W. Swaan, B. C. Koops, E. E. Moret and J. J. Tukker, Mapping the binding site of the small intestinal peptide carrier (PepT1) using comparative molecular field analysis, *Recept. Channels*, 1998, **6**, 189–200.
[51] P. D. Bailey, C. A. R. Boyd, J. R. Bronk, I. D. Collier, D. Meredith, K. M. Morgan and C. S. Temple, How to make drugs orally active: a substrate template for peptide transporter PepT1, *Angew. Chem. Int. Ed.*, 2000, **39**, 506–508.
[52] S. Gebauer, I. Knuetter, B. Hartrodt, M. Brandsch, K. Neubert and I. Thondorf, Three-dimensional quantitative structure–activity relationship analyses of peptide substrates of the mammalian H+/peptide cotransporter PEPT1, *J. Med. Chem.*, 2003, **46**, 5725–5734.
[53] E. Y. Zhang, M. A. Phelps, C. Cheng, S. Ekins and P. W. Swaan, Modeling of active transport systems, *Adv. Drug Deliv. Rev.*, 2002, **54**, 329–354.
[54] J. H. Lin and M. Yamazaki, Role of P-glycoprotein in pharmacokinetics: clinical implications, *Clin. Pharmacokinet.*, 2003, **42**, 59–98.
[55] T. R. Stouch and O. Gudmundsson, Progress in understanding the structure–activity relationships of P-glycoprotein, *Adv. Drug Deliv. Rev.*, 2002, **54**, 315–328.
[56] A. Seelig, E. Landwojtowicz, H. Fischer and X. Li Blatter, Towards P-glycoprotein structure–activity relationships. In *Drug Bioavailability: Estimation of Solubility, Permeability and Bioavailability* (eds P. Artursson, H. Lennernäs and H. van de Waterbeemd), Methods and Principles in Medicinal Chemistry, Wiley-VCH, Weinheim, 2003, Vol. 18, pp. 461–492.

[57] S. Ekins, Predicting undesirable drug interactions with promiscuous proteins *in silico*, *Drug Discov. Today*, 2004, **9**, 276–285.
[58] V. K. Gombar, J. W. Polli, J. E. Humphreys, S. A. Wring and C. S. Serabjit-Singh, Predicting P-glycoprotein substrates by a quantitative structure–activity relationship model, *J. Pharm. Sci.*, 2004, **93**, 957–968.
[59] U. Norinder and M. Haeberlein, Computational approaches to the prediction of the blood–brain distribution, *Adv. Drug Deliv. Rev.*, 2002, **54**, 291–313.
[60] D. E. Clark, *In silico* prediction of blood–brain barrier permeation, *Drug Discov. Today*, 2003, **8**, 927–933.
[61] T. J. Hou and X. J. Xu, ADME evaluation in drug discovery. 3. Modeling blood–brain barrier partitioning using simple molecular descriptors, *J. Chem. Inf. Comput. Sci.*, 2003, **43**, 2137–2152.
[62] M. C. Hutter, Prediction of blood–brain barrier permeation using quantum chemically derived information, *J. Comput.-Aided Mol. Des.*, 2003, **17**, 415–433.
[63] G. Subramanian and D. B. Kitchen, Computational models to predict blood–brain barrier permeation and CNS activity, *J. Comput.-Aided Mol. Des.*, 2003, **17**, 643–664.
[64] M. Adenot and R. Lahana, Blood–brain barrier permeation models: discriminating between potential CNS and non-CNS drugs including P-glycoprotein substrates, *J. Chem. Inf. Comput. Sci.*, 2004, **44**, 239–248.
[65] X. Liu, M. Tu, R. S. Kelly, C. Chen and B. J. Smith, Development of a computational approach to predict blood–brain barrier permeability, *Drug Metab. Dispos.*, 2004, **32**, 132–139.
[66] Advanced Chemistry Development, Inc., 90 Adelaide Street West, Suite 600, Toronto, Ontario M5H 3V9, Canada.
[67] G. Colmenarejo, *In silico* prediction of drug-binding strengths to human serum albumin, *Med. Res. Rev.*, 2003, **3**, 275–301.
[68] P. J. Hajduk, R. Mendoza, A. M. Petros, J. R. Huth, M. Bures, S. W. Fesik and Y. C. Martin, Ligand binding to domain-3 of human serum albumin: a chemometric analysis, *J. Comput.-Aided Mol. Des.*, 2003, **17**, 93–102.
[69] H. van de Waterbeemd, D. A. Smith and B. C. Jones, Lipophilicity in PK design: methyl, ethyl, futile, *J. Comput.-Aided Mol. Des.*, 2001, **15**, 273–286.
[70] G. Colmenarejo, A. Alvarez-Pedraglio and J.-L. Lavandera, Cheminformatic models to predict binding affinities to human serum albumin, *J. Med. Chem.*, 2001, **44**, 4370–4378.
[71] L. M. Hall, L. H. Hall and L. B. Kier, Modeling drug albumin binding affinity with E-state topological structure representation, *J. Chem. Inf. Comput. Sci.*, 2003, **43**, 2120–2128.
[72] E. H. Kerns and D. Li, Pharmaceutical profiling in drug discovery, *Drug Discov. Today*, 2003, **8**, 316–323.
[73] V. K. Gombar, I. S. Silver and Z. Zhao, Role of ADME characteristics in drug discovery and their *in silico* evaluation: *in silico* screening of chemicals for their metabolic stability, *Curr. Top. Med. Chem.*, 2003, **3**, 1205–1225.
[74] M. Shen, Y. Xiao, A. Golbraikh, V. K. Gombar and A. Tropsha, Development and validation of k-nearest-neighbor QSPR models of metabolic stability of drug candidates, *J. Med. Chem.*, 2003, **46**, 3013–3020.
[75] S. Ekins, *In silico* approaches to predicting drug metabolism, toxicology and beyond, *Biochem. Soc. Trans.*, 2003, **31**, 611–614.
[76] R. Bursi, A. Grootenhuis, J. van der Louw, J. Verhagen, M. de Gooyer, P. Jacobs and D. Leysen, Structure–activity relationship study of human liver microsomes-catalyzed hydrolysis rate of ester prodrugs of MENT by comparative molecular field analysis (CoMFA), *Steroids*, 2003, **68**, 213–220.
[77] J. L. Lewin and C. J. Cramer, Rapid quantum mechanical models for the computational estimation of C–H bond dissociation energies as a measure of metabolic stability, *Mol. Pharm.*, 2004, **1**, 128–135.

[78] N. P. E. Vermeulen, Prediction of drug metabolism: the case of cytochrome P450 2D6, *Curr. Top. Med. Chem.*, 2003, **3**, 1227–1239.
[79] L. Afzelius, I. Zamora, C. M. Masimirembwa, A. Karlen, T. B. Andersson, S. Mecucci, M. Baroni and G. Cruciani, Conformer- and alignment-independent model for predicting structurally diverse competitive CYP2C9 inhibitors, *J. Med. Chem.*, 2004, **47**, 907–914.
[80] S. B. Kirton, C. A. Baxter and M. J. Sutcliffe, Comparative modeling of cytochromes P450, *Adv. Drug Deliv. Rev.*, 2002, **54**, 385–406.
[81] D. F. V. Lewis, P450 structures and oxidative metabolism of xenobiotics, *Pharmacogenomics*, 2003, **4**, 387–395.
[82] P. A. Williams, J. Cosme, A. Ward, H. C. Angove, D. M. Vinkovic and H. Jhoti, Crystal structure of human cytochrome P450 2C9 with bound warfarin, *Nature*, 2003, **424**, 464–468.
[83] Astex determines structure of the key drug metabolising enzyme – human cytochrome P450 3A4, press release, October 28, 2002, http://www.astex-technology.com/press_release.jsp?press_release_id=58.
[84] D. F. V. Lewis, M. Dickins, B. G. Lake and P. S. Goldfarb, A molecular model of CYP2D6 constructed by homology with the CYP2C5 crystallographic template: investigation of enzyme–substrate interactions, *Drug Metab. Drug Interact.*, 2003, **19**, 189–210.
[85] D. F. V. Lewis, M. Dickins, B. G. Lake and P. S. Goldfarb, Investigation of enzyme selectivity in the human CYP2C subfamily: homology modelling of CYP2C8, CYP2C9 and CYP2C19 from the CYP2C5 crystallographic template, *Drug Metab. Drug Interact.*, 2003, **19**, 257–285.
[86] D. F. V. Lewis, B. G. Lake, M. Dickins and P. S. Goldfarb, Homology modelling of CYP2A6 based on the CYP2C5 crystallographic template: enzyme–substrate interactions and QSARs for binding affinity and inhibition, *Toxicol. In Vitro*, 2003, **17**, 179–190.
[87] D. F. V. Lewis, B. G. Lake, M. Dickins, Y.-F. Ueng and P. S. Goldfarb, Homology modelling of human CYP1A2 based on the CYP2C5 crystallographic template structure, *Xenobiotica*, 2003, **33**, 239–254.
[88] D. F. V. Lewis, B. G. Lake, M. G. Bird, G. D. Loizou, M. Dickins and P. S. Goldfarb, Homology modelling of human CYP2E1 based on the CYP2C5 crystal structure: investigation of enzyme–substrate and enzyme–inhibitor interactions, *Toxicol. In Vitro*, 2003, **17**, 93–105.
[89] M. J. de Groot and S. Ekins, Pharmacophore modeling of cytochromes P450, *Adv. Drug Deliv. Rev.*, 2002, **54**, 367–383.
[90] S. Ekins, D. M. Stresser and J. A. Williams, In vitro and pharmacophore insights into CYP3A enzymes, *Trends Pharmacol. Sci.*, 2003, **24**, 161–166.
[91] D. F. V. Lewis, Human cytochromes P450 associated with the phase 1 metabolism of drugs and other xenobiotics: a compilation of substrates and inhibitors of the CYP1, CYP2 and CYP3 families, *Curr. Med. Chem.*, 2003, **10**, 1955–1972.
[92] D. F. V. Lewis, Quantitative structure–activity relationships (QSARs) within the cytochrome P450 system: QSARs describing substrate binding, inhibition and induction of P450s, *Inflammopharmacology*, 2003, **11**, 43–73.
[93] S. Ekins, J. Berbaum and R. K. Harrison, Generation and validation of rapid computational filters for CYP2D6 and CYP3A4, *Drug Metab. Dispos.*, 2003, **31**, 1077–1080.
[94] R. G. Susnow and S. L. Dixon, Use of robust classification techniques for the prediction of human cytochrome P450 2D6 inhibition, *J. Chem. Inf. Comput. Sci.*, 2003, **43**, 1308–1315.
[95] S. B. Singh, L. Q. Shen, M. J. Walker and R. P. Sheridan, A model for predicting likely sites of CYP3A4-mediated metabolism on drug-like molecules, *J. Med. Chem.*, 2003, **46**, 1330–1336.

[96] D. C. Mankowski and S. Ekins, Prediction of human drug metabolizing enzyme induction, *Curr. Drug Metab.*, 2003, **4**, 381–391.
[97] N. Greene, Computer systems for the prediction of toxicity: an update, *Adv. Drug Deliv. Rev.*, 2002, **54**, 417–431.
[98] J. C. Dearden, *In silico* prediction of drug toxicity, *J. Comput.-Aided Mol. Des.*, 2003, **17**, 119–127.
[99] J. Feng, L. Lurati, H. Ouyang, T. Robinson, Y. Wang, S. Yuan and S. S. Young, Predictive toxicology: benchmarking molecular descriptors and statistical methods, *J. Chem. Inf. Comput. Sci.*, 2003, **43**, 1463–1470.
[100] D. Fernandez, A. Ghanta, G. W. Kauffman and M. C. Sanguinetti, Physicochemical features of the hERG channel drug binding site, *J. Biol. Chem.*, 2004, **279**, 10120–10127.
[101] I. Nestorov, Whole body pharmacokinetic models, *Clin. Pharmacokinet.*, 2003, **42**, 883–908.

Section 4
Applications of Computational Methods

Section Editor: Heather Carlson
University of Michigan
College of Pharmacy
428 Church Street
Ann Arbor, MI 48109-1065
USA

CHAPTER 11

Filtering in Drug Discovery

Christopher A. Lipinski

Pfizer Global Research and Development, Groton, CT 06340, USA

Contents

1. Drug-likeness 155
2. Drug-likeness and chemistry quality 157
3. Positive desirable chemistry filters 158
4. Lead-likeness 159
5. Oral drug activity 159
6. CNS drugs 160
7. Intestinal permeability 161
8. Aqueous solubility 162
9. Drug metabolism 162
10. Promiscuous compounds 162
11. Agrochemicals 163
References 163

Recent advances in filtering in drug discovery are reviewed. Filtering is used in a global broad sense to include exclusionary as well as inclusionary criteria and encompasses the following topics related to the discovery of drugs: (1) computational definitions of drug-likeness; (2) positive filters for drug-like activity; (3) lead-likeness as a concept; (4) oral activity filters; (5) CNS drug filters; (6) intestinal permeability filters; (7) drug metabolism parameter filters; (8) promiscuous compound filters; and (9) agrochemical filters. This review does not cover what could be termed as local parameters such as pharmacophore models, docking and scoring, etc.

1. DRUG-LIKENESS

To this author there seems to be much more agreement as to what is drug-like than there is as to what is diverse. Defining what is drug-like and non-drug-like requires some type of reference point. The property distributions of commercially available databases have been examined. The available chemicals directory (ACD) seems to be the most common non-drug-like database. The pesticide manual has been used as an alternate standard for non-drug-likeness because it

ANNUAL REPORTS IN COMPUTATIONAL CHEMISTRY, VOLUME 1
ISSN: 1574-1400 DOI 10.1016/S1574-1400(05)01011-X

is composed primarily of compounds designed to cause fatality to the primary organism. The Comprehensive Medicinal Chemistry (CMC), Derwent Word Drug Index (WDI) and Modern Drug Data Report (MDDR) are among the more commonly used drug-like databases [1,2]. In addition, the 10,000 or so Phase II compounds are used to define drug-like compounds [3]. A compound's drug-like index has been calculated based upon the knowledge derived from known drugs selected from the CMC database [4]. The property distributions in combinatorial compounds compared to drugs or natural products largely reflect combinatorial chemistry synthesis constraints such that there are fewer chiral centers and complex ring systems [5]. The distribution of ring systems across multiple databases has been described [6] and a program was written and tested on the MDDR database [7] to identify candidate chemical ring replacements (bioisosteres). From the study of a database of commercially available drugs it is clear that the diversity of molecular framework (ring) shapes is extremely low. The shapes of half of the drugs in the database are described by the 32 most frequently occurring frameworks [8]. The diversity that side chains provide to drug molecules is quite low since only 20 side chains account for over 70% of the side chains [9]. Defining drug-like by what exists in databases leads to the criticism that most of chemistry space will be undefined and that discovery opportunities in unexplored chemistry space will be limited. A solution is to populate chemistry space with non-drug-like markers akin to the way point in a GPS navigation system [10].

Multiple filters (properties) may be incorporated into a definition of drug-likeness and this leads to trade-offs among compound properties in compounds intended for screening [11]. Optimization of compound properties may require some type of multi-parameter optimization scheme in library design [12]. Fingerprint algorithms can be used to guide diversity [13]. Filters also need to be employed in the chemistry synthesis planning process so that good quality compounds are made [14]. Differences in property ranges between oral and injectable drugs have been summarized [15]. Oral drugs are lower in MWT and have fewer H-bond donors, acceptors and rotatable bonds. Property profiles of oral drugs are independent of the year in which the drug was approved to market and to some degree independent of target. Polar surface area (PSA) in one definition is the solvent accessible surface covered by oxygen, nitrogen and the hydrogens attached to oxygen and nitrogen. As a compound progresses through clinical trials there is a steady change in properties, e.g., MWT, Log *P* and PSA all decline with a MWT of about 340 found for marketed drugs [16,17]. The reason for this pattern is unclear since properties related to oral absorption would be expected to have reached a plateau by Phase II and hence selection pressure for properties related to oral absorption should have disappeared by then [18]. Pulmonary drugs tend to have higher PSA because pulmonary permeability is less sensitive to polar hydrogen-bonding functionality [19]. Anatomically this makes sense since lungs are a closed compartment and any accumulating fluid and compounds in terminal alveoli must be cleared. Discrimination between antibacterial and non-antibacterial activity has been achieved based on 3D molecular descriptors. The overall classification rate was around 90% on a data set of 661 compounds using 2–3 variables selected from log *P*, charged-weighted negative surface area, positive surface area of heavy atoms and maximum donor delocalizability. Three-dimensional geometry variations had little impact on the discriminatory performance [20].

2. DRUG-LIKENESS AND CHEMISTRY QUALITY

Descriptors for drug-like are most effective if they have physical meaning so as to facilitate chemists designing in drug-likeness [21]. Drug-likeness in the design of combinatorial libraries [22,23] involves the use of rule-based filters like the rule of 5 [24], the use of exclusionary filters to remove undesired chemistry functionality [25] and the capture of privileged structure information, e.g., from natural product collections [26] or from retro synthetic analysis of collections of bioactive molecules [27]. Natural product structural features are particularly well represented in the cancer chemotherapy and infectious disease areas [28]. Exclusionary filters have been described that remove reactive chemical functionality based on the premise that compounds having covalent chemistry possibilities have no place in drug discovery [29]. Filters are also necessary to remove cross reactivity in pooled compounds [30]. Pooling is a procedure in which single well-characterized compounds are deliberately mixed to speed screening. Components of the mixture must neither contain structural features causing assay false positives nor must they contain common substructural elements that would confuse the deconvolution of activities of the individual components. The magnitude of the number of poor quality screening compounds is emphasized by the report that only 37% of 1.6 million unique commercially available compounds are drug-like [31]. A very similar result was found in a virtual screen for SARS-CoV protease against commercially available and academic compounds. Of the 0.07% virtual hits against 3.6 million compounds, 47% failed three or more of 13 druggability criteria [32]. The criteria were based on physical, chemical and structural properties. Providing high-quality chemistry subject matter is now supported under the NIH molecular library small molecule repository initiative which aims to collect one million drug-like molecules from commercial, industry and government sources [33]. Emphasizing the point that drugs must contain adequate functionality to achieve acceptable receptor interactions, a single filter separates drug-like from non-drug-like compounds based on the observation that non-drugs are often under-functionalized [34].

3. POSITIVE DESIRABLE CHEMISTRY FILTERS

Privileged structures, e.g., benzodiazepines are recurring structures active against targets unrelated by target family. They can be viewed as molecular filters selecting for desirable chemistry subject matter. As such they are rich sources for screening libraries and have recently been reviewed [35]. Privileged structure features have been employed in the combinatorial design of GPCR libraries [36], in the combinatorial synthesis of privileged bicyclic structures [37] and in the combinatorial synthesis of cyclic peptides [38]. Homology modeling suggests a parallelism between common privileged GPCR ligand features and complementary deeply buried protein features in class A GPCRs [39]. Grouping by target family is also another method helping focus on particular target-directed privileged structures [40]. The idea is that structurally similar target family members will bind structurally similar small molecule ligands [41]. NMR screening helps identify privileged protein binding elements albeit of smaller size [42]. Although not strictly speaking a privileged structure, privileged structural elements such as the hydroxamate moiety found in many metalloprotease inhibitors can be identified [43]. Discernment of privileged structures has historically largely been a data mining exercise. However, very similar recurring structural motifs, so-called 'molecular anchors' have been described based on structure-based ligand binding considerations [44,45]. Rigid small molecule ligands (the molecular anchors) are incapable of hydrophobic collapse and a single non-collapsible ligand conformer binds at a protein cavity site which is also often incapable of hydrophobic collapse. This concept explains the frequent occurrence of non-collapsible spiro structures in privileged structures/molecular anchors. Chemistry design principles directed to the very difficult goal of small molecule interference with protein–protein interactions *via* an allosteric interaction have been described [46]. An intriguing aspect is the hypothesis that chemistry emphasis should be placed on compound cores capable of interacting with relatively fixed protein hinge regions rather than on elaboration of lipophilic side chains attached to the core. The thermodynamic penalty attendant to ligand binding to a non-lowest energy protein conformer suggests that screening should allow for slow binding with adequate assay equilibration time. An implication is that for this type of target it is better to make larger numbers of smaller libraries than fewer numbers of large libraries. This trend to smaller libraries is now well documented [47]. Taken to its extreme this approach takes the typical dense chemistry space coverage of the traditional combinatorial library (target-oriented synthesis) towards the direction of the diversity-oriented synthesis approach to chemistry lead generation which populates diverse single molecules broadly through chemistry space [48]. This direction is of course in the direction of less efficient, more difficult chemistry. The focus on biological information content richness suggests natural products as combinatorial library starting points [49]. Chemical content richness is found in compounds produced by multicomponent reactions (MCR) which are chemical transformations in which as many as four components form a new compound in a single chemistry reaction step. The Ugi reaction of a carboxylic acid, amine, aldehyde and isonitrile is a classic example. In theory while offering an efficient approach to synthesis of diverse compounds, MCR in high-throughput mode currently suffers from significant chemistry limitations [50].

4. LEAD-LIKENESS

The difference between drug-like and lead-like has been described [51]. Leads are less complex in most parameters than drugs, which is understandable in that medicinal chemistry optimization almost invariably increases MWT and Log *P* [52]. However, the structural resemblance between a starting lead and a drug is marked [53]. The implication is that a quality lead as opposed to a flawed lead is far likely to lead to a real drug [54]. Lead-like discovery also refers to the screening of small molecule libraries with detection of weak affinities in the high micromolar to millimolar range. The process usually by itself does not lead to an acceptable chemistry starting point. Something else has to be added after the primary screen. Generally, multiple small molecules do not bind to non-adjacent target sites [55], so the screening is that of small MWT singletons. However, binding of two components to the same receptor site is possible as attested by the discovery of sub-nanomolar ligands in what is termed click chemistry [56]. In this process an acetylene and azide terminus from two receptor site independently bound molecules cyclize to a single compound with the two components linked *via* a 1,2,3 triazole ring. Filtering in the context of lead-like small molecule screening implies control of the properties of drug starting points that eventually result from this process. A rule of three [57] has been coined for small molecule fragment screening libraries; MWT < 300; Log $P < 3$; H-bond donors and acceptors < 3 and rotatable bonds < 3. Small fragment screening can be by NMR [58–60], by X-ray [61,62], or in theory by any method capable of detecting weak interactions.

5. ORAL DRUG ACTIVITY

The topic of filtering in human therapeutic drug discovery has received numerous frequent reviews [23,63–65] as well as criticism if fundamental medicinal chemistry principles are neglected [66]. The 'rule of 5' describes four simple parameters associated with improved prospects for oral activity. Poor solubility or poor permeability are more likely if there are >5 H-bond donors (expressed as sum of OH and NH); >10 H-bond acceptors (expressed as sum of $O + N$); MWT > 500 and Log $P > 5$. There are only four rules. The 5 in rule of 5 arises from the frequent appearance of a 5 in the cutoff parameters. Compounds classes such as natural products, infectious disease drugs, etc. where transporter affinity is prevalent are exceptions [24]. Rotatable bond count is now a widely used filter following the finding that greater than 10 rotatable bonds correlates with decreased rat oral bioavailability [67]. The mechanistic basis for the rotatable bond filter is unclear since the rotatable bond count does not correlate with *in vivo* clearance rate in the rat but the filter is reasonable from an *in vitro* screening viewpoint since ligand affinity on average decreases 0.5 kcal for each two rotatable bonds [68]. Compounds indexed in medicinal chemistry journals show the recent trend towards poor properties. Over 50% of medicinal chemistry compounds with activities above 1 nM have $MW > 425$, $\text{Log}\, P > 4.25$ and $\text{Log}\, Sw < -4.75$, indicating that these compounds are larger, more hydrophobic and less soluble when compared to time-tested quality leads [52]. The concept of the importance of compound properties (e.g., rule of 5 compliance) beyond potency is widely accepted [69]. although there are notable occasional exceptions where an orally bioavailable compound is found that lies well outside the rule of 5 limits [70]. Can the rule of 5 be bypassed by delivering drug by a non-oral route, e.g., pulmonary, intra-nasal or dermal? The answer depends very much on the dose. If the total dose is 20 mg or less then alternative delivery routes begin to be feasible. However, a limitation is that only about 10% of current clinical candidates have sufficient potency in the 0.1 mg/kg range to result in such a low dose and finding such very potent compounds seems to be mostly a matter of luck [71]. Beyond chemistry-based features, oral drugs can also be defined by their biological target. It is striking that the 100 best selling (mostly oral) drugs are ligands for proteins encoded by only a very small subset of genes and that a very considerable portion of the targets for orally active drugs may have already been discovered [72]. The term 'druggable genome' has been coined to describe the severe restriction that chemistry considerations related to oral activity superimpose on possible biology target space [73].

6. CNS DRUGS

A scheme for separating CNS from non-CNS active drugs in the WDI allowed discovery of simple parameters relating to passive blood brain barrier (BBB) permeability and prediction of p-glycoprotein (PGP) affinity [74]. The PGP transporter is a major barrier to the entry of compounds to the CNS [75]. Appropriately determined PGP efflux ratios can be used as a measure of compound affinity to PGP. However, the value of filters based on PGP efflux ratios from the commonly used high-throughput mode Caco-2 colonic cell permeability cell culture assay have been questioned as efflux ratios do not correlate with *in vivo* rat brain penetration [76]. A PSA value of less than 60–70 Å^2 tends to identify

CNS active compounds [77]. A very simple set of two rules predicts CNS activity: If N + O (the number of nitrogen and oxygen atoms) in a molecule is less than or equal to 5, it has a high chance of entering the brain. The second rule predicts that if log $P - (N + O)$ is positive then the compound is CNS active [78]. More complex commercially available software programs have been compared as to their ability to predict CNS log BBB ratio [79]. Experimental and theoretical reasons support the belief that surface tension measurements can be predictors for blood brain permeability [80]. Predictors for absorption, distribution, metabolism and excretion (ADME) currently appear most useful in global models. Limitations in local models likely reflect a lack of quality experimental data sets [81] and user dissatisfaction may result from unrealistic expectations given the magnitude of experimental ADME errors [82]. An additional limitation to schemes for separating CNS from non-CNS compounds is the complexity of the BBB. Compounds with affinity to transporters are exceptions to physicochemically based filters like the rule of 5. This is a problem for the CNS since it is estimated that about 15% of all genes selectively expressed at the BBB encode for transporter proteins and that only about 50% of BBB transporters are currently known [83].

7. INTESTINAL PERMEABILITY

PSA in rather simple models is a commonly used parameter to predict intestinal permeability [84]. Its rule-based calculation (TPSA) is very fast and does not require 3D structure [85]. A better prediction of intestinal permeability has been reported when PSA is partitioned into smaller molecularly based components [86]. Using molecular surface properties compounds selected from the World Health Organization's (WHO) list of essential drugs could be classified with 87% accuracy as to permeability and solubility using a six bin scheme similar to that in the FDA biopharmaceutical classification system [87]. Pharmacokinetic parameters including permeability can also be generated for filtering or ligand affinity prediction through the Volsurf software [88]. An analysis of small drug-like molecules suggests a filter of log $D > 0$ and < 3 enhances the probability of good permeability [89]. A collection of 222 commercially available drugs was used to determine the exclusion criteria that differentiate poorly absorbed drugs from well-absorbed drugs. Similar to the rule of 5, MWT < 500 and log $P < 5$ were associated with better absorbed compounds. Exceptions to the MWT criteria were compounds with a sugar moiety, high atomic weight and large cyclic structure [90] suggesting the involvement of absorptive biological transporter systems. Based on the intestinal absorption of 158 drug and drug-like compounds in rats there is a significant relationship between rat intestinal absorption, and by extrapolation human absorption, to drug hydrogen-bond acidity and basicity [91].

8. AQUEOUS SOLUBILITY

Poor aqueous solubility is a wide spread problem in combinatorial libraries as opposed to poor intestinal permeability which is much less of a problem. About one-half of poor solubility is due to large size/lipophilicity. Log $P > 5$ identifies 75% of these compounds. The other 50% of poor aqueous solubility is due to crystal packing considerations for which there is no computational filter [92]. Melting point is an experimental indicator of crystal packing. Aqueous solubility decreases about 10 × for each 100 °C rise in melting point and so melting point, if available, is a valuable parameter in solubility prediction [93]. Progress toward a computational melting point is suggested by the 63% success in qualitative ranking of compounds into low-medium and high-solubility bins. Descriptors for hydrophilicity, polarity, partial atom charge and molecular rigidity were found to be positively correlated with melting point whereas non-polar atoms and high flexibility within the molecule were negatively correlated [94].

9. DRUG METABOLISM

Volume of distribution (VD) is a key pharmacokinetic parameter. A low VD of less than 1 l/kg identifies drugs residing in the plasma compartment. A VD greater than 1 l/kg identifies compounds accessing tissue compartments outside the plasma compartment, e.g., many CNS drugs have VD values in the tens or higher. A recently developed computational approach to predict VD for neutral and basic drugs works as well as the *in vivo* experimental measurement provided that accurate experimental compound log D and pK_a are available. Predictivity is retained if computed log D and pK_a are used but accuracy declines somewhat [95,96]. Approximately 80% of drugs are oxidized by the cytochrome P450 (CYP) family of enzymes; hence a decision tree for CYP substrate affinity is important. This has been described in that characteristics of CYP substrates, such as lipophilicity, MWT and hydrogen-bonding potential, govern selectivity towards individual CYPs [97].

10. PROMISCUOUS COMPOUNDS

Compounds with a marked propensity to bind to multiple targets, so-called nuisance compounds, are of little value in drug discovery. Such compounds can be experimentally identified by their binding to fetal calf serum [98]. It has long been known that compounds could be identified as reproducible actives in HTS screens that could not be optimized in chemistry. Such compounds often appear active in multiple screens that have no biological relationship. An analysis of such promiscuous compounds from HTS hits led to the conclusion that colloidal aggregates in the 50–1000 nm size range were responsible. The apparent HTS screen activity was due to a biophysical effect rather than due to a normal ligand receptor affinity and hence the hits were unoptimizable in chemistry [99]. This promiscuous aggregation effect was found among 8 of 15 kinase inhibitors widely used in biology screening [100] emphasizing the importance of exclusionary filters to prevent wasting of biology research time by testing compounds with flawed properties. The aggregation phenomenon has been found among known drugs, albeit only when tested at high non-physiological concentrations and a predictive model was developed [101].

11. AGROCHEMICALS

Filtering has also been applied to agrochemicals. Compared to drugs intended for human use, agrochemicals tend to have fewer hydrogen-bond donors [102]. For agrochemical screening computationally intensive surface area parameters offered no advantage over the rule of 5 [103]. Analogous to drug-likeness, agrochem-likeness for large compound collections has been explored using support vector machines (SVM). In this study SVM performed better than neural networks [104].

REFERENCES

[1] T. I. Oprea, Property distribution of drug-related chemical databases, *J. Comput.-Aided Mol. Des.*, 2000, **14**, 251–264.
[2] M. P. Bradley, An overview of the diversity represented in commercially-available databases, *Mol. Divers.*, 2002, **5**, 175–183.
[3] C. A. Lipinski, Drug-like properties and the causes of poor solubility and poor permeability, *J. Pharmacol. Toxicol. Methods*, 2001, **44**, 235–249.
[4] J. Xu and J. Stevenson, Drug-like index: a new approach to measure drug-like compounds and their diversity, *J. Chem. Inf. Comput. Sci.*, 2000, **40**, 1177–1187.
[5] M. Feher and J. M. Schmidt, Property distributions: differences between drugs natural products, and molecules from combinatorial chemistry, *J. Chem. Inf. Comput. Sci.*, 2003, **43**, 218–227.
[6] X. Q. Lewell, A. C. Jones, C. L. Bruce, G. Harper, M. M. Jones, I. M. McLay and J. Bradshaw, Drug rings database with web interface. A tool for identifying alternative chemical rings in lead discovery programs, *J. Med. Chem.*, 2003, **46**, 3257–3274.
[7] R. P. Sheridan, The most common chemical replacements in drug-like compounds, *J. Chem. Inf. Comput. Sci.*, 2002, **42**, 103–108.
[8] G. W. Bemis and M. A. Murcko, The properties of known drugs. 1. Molecular frameworks, *J. Med. Chem.*, 1996, **39**, 2887–2893.
[9] G. W. Bemis and M. A. Murcko, Properties of known drugs. 2. Side chains, *J. Med. Chem.*, 1999, **42**, 5095–5099.
[10] T. I. Oprea, I. Zamora and A.-L. Ungell, Pharmacokinetically based mapping device for chemical space navigation, *J. Comb. Chem.*, 2002, **4**, 258–266.

[11] T. Wright, V. J. Gillet, D. V. S. Green and S. D. Pickett, Optimizing the size and configuration of combinatorial libraries, *J. Chem. Inf. Comput. Sci.*, 2003, **43**, 381–390.
[12] D. K. Agrafiotis, Multiobjective optimization of combinatorial libraries, *J. Comput.-Aided Mol. Des.*, 2002, **16**, 335–356.
[13] F. L. Stahura, L. Xue, J. W. Godden and J. Bajorath, Methods for compound selection focused on hits and application in drug discovery, *J. Mol. Graph. Model.*, 2002, **20**, 439–446.
[14] T. I. Oprea, J. Gottfries, V. Sherbukhin, P. Svensson and T. C. Kuhler, Chemical information management in drug discovery: optimizing the computational and combinatorial chemistry interfaces, *J. Mol. Graph. Model.*, 2000, **18**, 512–524.
[15] M. Vieth, M. G. Siegel, R. E. Higgs, I. A. Watson, D. H. Robertson, K. A. Savin, G. L. Durst and P. A. Hipskind, Characteristic physical properties and structural fragments of marketed oral drugs, *J. Med. Chem.*, 2004, **47**, 224–232.
[16] M. C. Wenlock, R. P. Austin, P. Barton, A. M. Davis and P. D. Leeson, A comparison of physiochemical property profiles of development and marketed oral drugs, *J. Med. Chem.*, 2003, **46**, 1250–1256.
[17] J. F. Blake, Examination of the computed molecular properties of compounds selected for clinical development, *BioTechniques*, 2003, **Suppl.**, 16–20.
[18] J. Askenash, The rule of five revisited, *Preclinica*, 2004, **2**, 92.
[19] A. Tronde, B. Norden, H. Marchner, A.-K. Wendel, H. Lennernaes and U. H. Bengtsson, Pulmonary absorption rate and bioavailability of drugs in vivo in rats: structure-absorption relationships and physicochemical profiling of inhaled drugs, *J. Pharm. Sci.*, 2003, **92**, 1216–1233.
[20] A. O. Aptula, R. Kuehne, R. Ebert, M. T. D. Cronin, T. I. Netzeva and G. Schueuermann, Modeling discrimination between antibacterial and non-antibacterial activity based on 3D molecular descriptors, *QSAR Comb. Sci.*, 2003, **22**, 113–128.
[21] T. Mitchell and G. A. Showell, Design strategies for building drug-like chemical libraries, *Curr. Opin. Drug Discov. Dev.*, 2001, **4**, 314–318.
[22] A. Polinsky and G. B. Shaw, High-speed chemistry libraries: assessment of drug-likeness. In *Practice of Medicinal Chemistry* (ed. C. G. Wermuth), 2nd edition, Elsevier, London, UK, 2003, pp. 147–157.
[23] H. Matter, K.-H. Baringhaus, T. Naumann, T. Klabunde and B. Pirard, Computational approaches towards the rational design of drug-like compound libraries, *Comb. Chem. High Throughput Screening*, 2001, **4**, 453–475.
[24] C. A. Lipinski, F. Lombardo, B. W. Dominy and P. J. Feeney, Experimental and computational approaches to estimate solubility and permeability in drug discovery and development settings, *Adv. Drug Deliv. Rev.*, 1997, **23**, 3–25.
[25] W. P. Walters, Ajay and M. A. Murcko, Recognizing molecules with drug-like properties, *Curr. Opin. Chem. Biol.*, 1999, **3**, 384–387.
[26] G. Muller and H. Giera, Protein secondary structure templates derived from bioactive natural products, *J. Comput.-Aided Mol. Des.*, 1998, **12**, 1–6.
[27] X. Q. Lewell, D. Judd, S. Watson and M. Hann, RECAP – Retrosynthetic combinatorial analysis procedure: a powerful new technique for identifying privileged molecular fragments with useful applications in combinatorial chemistry, *J. Chem. Inf. Comput. Sci.*, 1998, **38**, 511–522.
[28] D. J. Newman, G. M. Cragg and K. M. Snader, Natural products as sources of new drugs over the period 1981–2002, *J. Nat. Prod.*, 2003, **66**, 1022–1037.
[29] G. M. Rishton, Reactive compounds and in vitro false positives in HTS, *Drug Discov. Today*, 1997, **2**, 382–384.
[30] M. Hann, B. Hudson, X. Lewell, R. Lifely, L. Miller and N. Ramsden, Strategic pooling of compounds for high-throughput screening, *J. Chem. Inf. Comput. Sci.*, 1999, **39**, 897–902.

[31] N. Baurin, R. Baker, C. Richardson, I. Chen, N. Foloppe, A. Potter, A. Jordan, S. Roughley, M. Parratt, P. Greaney, D. Morley and R. E. Hubbard, Drug-like annotation and duplicate analysis of a 23-supplier chemical database totaling 2.7 million compounds, *J. Chem. Inf. Comput. Sci.*, 2004, **44**, 643–651.
[32] S. Sirois, D.-Q. Wei, Q. Du and K.-C. Chou, Virtual screening for SARS-CoV protease based on KZ7088 pharmacophore points, *J. Chem. Inf. Comput. Sci.*, 2004, **44**, 1111–1122.
[33] http://nihroadmap.nih.gov/molecularlibraries/index.asp.
[34] I. Muegge, S. L. Heald and D. Brittelli, Simple selection criteria for drug-like chemical matter, *J. Med. Chem.*, 2001, **44**, 1841–1846.
[35] A. A. Patchett and R. P. Nargund, Chapter 26. Privileged structures – an update, *Annu. Rep. Med. Chem.*, 2000, **35**, 289–298.
[36] T. Guo and D. W. Hobbs, Privileged structure-based combinatorial libraries targeting G protein-coupled receptors, *Assay Drug Dev. Technol.*, 2003, **1**, 579–592.
[37] D. A. Horton, G. T. Bourne and M. L. Smythe, The combinatorial synthesis of bicyclic privileged structures or privileged substructures, *Chem. Rev.*, 2003, **103**, 893–930.
[38] D. A. Horton, G. T. Bourne and M. L. Smythe, Exploring privileged structures: the combinatorial synthesis of cyclic peptides, *Mol. Divers.*, 2002, **5**, 289–304.
[39] K. Bondensgaard, M. Ankersen, H. Thogersen, B. S. Hansen, B. S. Wulff and R. P. Bywater, Recognition of privileged structures by G-protein coupled receptors, *J. Med. Chem.*, 2004, **47**, 888–899.
[40] B. R. Roberts, Target information in lead discovery, *Targets*, 2003, **2**, 14–18.
[41] A. Schuffenhauer, P. Floersheim, P. Acklin and E. Jacoby, Similarity metrics for ligands reflecting the similarity of the target proteins, *J. Chem. Inf. Comput. Sci.*, 2003, **43**, 391–405.
[42] P. J. Hajduk, M. Bures, J. Praestgaard and S. W. Fesik, Privileged molecules for protein binding identified from NMR-based screening, *J. Med. Chem.*, 2000, **43**, 3443–3447.
[43] G. Muller, Medicinal chemistry of target family-directed masterkeys, *Drug Discov. Today*, 2003, **8**, 681–691.
[44] P. A. Rejto and G. M. Verkhivker, Unraveling principles of lead discovery: from unfrustrated energy landscapes to novel molecular anchors, *Proc. Natl Acad. Sci. USA*, 1996, **93**, 8945–8950.
[45] P. A. Rejto and G. M. Verkhivker, Molecular anchors with large stability gaps ensure linear binding free energy relationships for hydrophobic substituents. In *Pacific Symposium on Biocomputing'98* (ed. R. B. Altman), Maui, Hawaii, January 4–9, 1998, pp. 362–373.
[46] S. J. Teague, Implications of protein flexibility for drug discovery, *Nat. Rev. Drug Discov.*, 2003, **2**, 527–541.
[47] R. Goodnow, Small molecule lead generation processes for drug discovery, *Drug Future*, 2002, **27**, 1165–1180.
[48] M. D. Burke and S. L. Schreiber, A planning strategy for diversity-oriented synthesis, *Angew. Chem. Int. Ed.*, 2004, **43**, 46–58.
[49] R. Breinbauer, I. R. Vetter and H. Waldmann, From protein domains to drug candidates: natural products as guiding principles in the design and synthesis of compound libraries, *Angew. Chem. Int. Ed.*, 2002, **41**, 2878–2890.
[50] C. Hulme and V. Gore, Multi-component reactions: emerging chemistry in drug discovery from xylocain to crixivan, *Curr. Med. Chem.*, 2003, **10**, 51–80.
[51] T. I. Oprea, A. M. Davis, S. J. Teague and P. D. Leeson, Is there a difference between leads and drugs? A historical perspective, *J. Chem. Inf. Comput. Sci.*, 2001, **41**, 1308–1315.
[52] T. I. Oprea, Current trends in lead discovery: are we looking for the appropriate properties?, *Mol. Divers.*, 2002, **5**, 199–208.

[53] J. R. Proudfoot, Drugs, leads, and drug-likeness: an analysis of some recently launched drugs, *Bioorg. Med. Chem. Lett.*, 2002, **12**, 1647–1650.
[54] C. A. Lipinski, Compound properties and drug quality. In *Practice of Medicinal Chemistry* (ed. C. G. Wermuth), 2nd edition, 2003, pp. 341–349.
[55] C. W. Murray and M. L. Verdonk, The consequences of translational and rotational entropy lost by small molecules on binding to proteins, *J. Comput.-Aided Mol. Des.*, 2002, **16**, 741–753.
[56] H. C. Kolb and K. B. Sharpless, The growing impact of click chemistry on drug discovery, *Drug Discov. Today*, 2003, **8**, 1128–1137.
[57] M. Congreve, R. Carr, C. Murray and H. Jhoti, A 'rule of three' for fragment-based lead discovery?, *Drug Discov. Today*, 2003, **8**, 876–877.
[58] E. Jacoby, J. Davies and M. J. J. Blommers, Design of small molecule libraries for NMR screening and other applications in drug discovery, *Curr. Top. Med. Chem.*, 2003, **3**, 11–23.
[59] C. A. Lepre, J. Peng, J. Fejzo, N. Abdul-Manan, J. Pocas, M. Jacobs, X. Xie and J. M. Moore, Applications of SHAPES screening in drug discovery, *Comb. Chem. High Throughput Screen*, 2002, **5**, 583–590.
[60] C. A. Lepre, *Strategies for NMR Screening and Library Design. Methods and Principles in Medicinal Chemistry*, BioNMR in Drug Research, 2003, Vol. 16, pp. 391–415.
[61] R. Carr and H. Jhoti, Structure-based screening of low-affinity compounds, *Drug Discov. Today*, 2002, **7**, 522–527.
[62] A. M. Davis, S. J. Teague and G. J. Kleywegt, Application and limitations of X-ray crystallographic data in structure-based ligand and drug design, *Angew. Chem. Int. Ed.*, 2003, **42**, 2718–2736.
[63] D. Fattori, Molecular recognition: the fragment approach in lead generation, *Drug Discov. Today*, 2004, **9**, 229–238.
[64] P. S. Charifson and W. P. Walters, Filtering databases and chemical libraries, *Mol. Divers.*, 2002, **5**, 185–197.
[65] V. J. Gillet and P. Willett, *Computational Methods for the Analysis of Molecular Diversity. Pharmacochemistry Library*, Trends in Drug Research III, 2002, Vol. 32, pp. 125–133.
[66] H. Kubinyi, Opinion: drug research: myths, hype and reality, *Nat. Rev. Drug Discov.*, 2003, **2**, 665–668.
[67] D. F. Veber, S. R. Johnson, H.-Y. Cheng, B. R. Smith, K. W. Ward and K. D. Kopple, Molecular properties that influence the oral bioavailability of drug candidates, *J. Med. Chem.*, 2002, **45**, 2615–2623.
[68] P. R. Andrews, D. J. Craik and J. L. Martin, Functional group contributions to drug-receptor interactions, *J. Med. Chem.*, 1984, **27**, 1648–1657.
[69] T. I. Oprea, Virtual screening in lead discovery: a viewpoint, *Molecules*, 2002, **7**, 51–62.
[70] P. Sauerberg, P. S. Bury, J. P. Mogensen, H.-J. Deussen, I. Pettersson, J. Fleckner, J. Nehlin, K. S. Frederiksen, T. Albrektsen, N. Din, L. A. Svensson, L. Ynddal, E. M. Wulff and L. Jeppesen, Large dimeric ligands with favorable pharmacokinetic properties and peroxisome proliferator-activated receptor agonist activity in vitro and in vivo, *J. Med. Chem.*, 2003, **46**, 4883–4894.
[71] K. R. Horspool and C. A. Lipinski, Advancing new drug delivery concepts to gain the lead, *Drug Deliv. Technol.*, 2003, **3**, 34–46.
[72] B. P. Zambrowicz and A. T. Sands, Knockouts model the 100 best-selling drugs – will they model the next 100?, *Nat. Rev. Drug Discov.*, 2003, **2**, 38–51.
[73] A. L. Hopkins and C. R. Groom, Opinion: the druggable genome, *Nat. Rev. Drug Discov.*, 2002, **1**, 727–730.
[74] M. Adenot, R. S. Lahana and F. Nimes, Blood–brain barrier permeation models: discriminating between potential CNS and non-CNS drugs including P-glycoprotein substrates, *J. Chem. Inf. Comput. Sci.*, 2004, **44**, 239–248.

[75] K. M. M. Doan, J. E. Humphreys, L. O. Webster, S. A. Wring, L. J. Shampine, C. J. Serabjit-Singh, K. K. Adkison and J. W. Polli, Passive permeability and P-glycoprotein-mediated efflux differentiate central nervous system (CNS) and non-CNS marketed drugs, *J. Pharmacol. Exp. Ther.*, 2002, **303**, 1029–1037.
[76] F. Faassen, G. Vogel, H. Spanings and H. Vromans, Caco-2 permeability, P-glycoprotein transport ratios and brain penetration of heterocyclic drugs, *Int. J. Pharm.*, 2003, **263**, 113–122.
[77] J. Kelder, P. D. J. Grootenhuis, D. M. Bayada, L. P. C. Delbressine and J.-P. Ploemen, Polar molecular surface as a dominating determinant for oral absorption and brain penetration of drugs, *Pharm. Res.*, 1999, **16**, 1514–1519.
[78] U. Norinder and M. Haeberlein, Computational approaches to the prediction of the blood–brain distribution, *Adv. Drug Deliv. Rev.*, 2002, **54**, 291–313.
[79] M. Lobell, L. Molnar and G. M. Keseru, Recent advances in the prediction of blood–brain partitioning from molecular structure, *J. Pharm. Sci.*, 2003, **92**, 360–370.
[80] P. Suomalainen, C. Johans, T. Soederlund and P. K. J. Kinnunen, Surface activity profiling of drugs applied to the prediction of blood–brain barrier permeability, *J. Med. Chem.*, 2004, **47**, 1783–1788.
[81] F. Lombardo, E. Gifford and M. Y. Shalaeva, In silico ADME prediction: data, models, facts and myths, *Mini-Rev. Med. Chem.*, 2003, **3**, 861–875.
[82] T. R. Stouch, J. R. Kenyon, S. R. Johnson, X.-Q. Chen, A. Doweyko and Y. Li, In silico ADME/Tox: why models fail, *J. Comput.-Aided Mol. Des.*, 2003, **17**, 83–92.
[83] W. M. Pardridge, Blood–brain barrier genomics and the use of endogenous transporters to cause drug penetration into the brain, *Curr. Opin. Drug Discov. Dev.*, 2003, **6**, 683–691.
[84] P. Stenberg, U. Norinder, K. Luthman and P. Artursson, Experimental and computational screening models for the prediction of intestinal drug absorption, *J. Med. Chem.*, 2001, **44**, 1927–1937.
[85] P. Ertl, B. Rohde and P. Selzer, Fast calculation of molecular polar surface area as a sum of fragment-based contributions and its application to the prediction of drug transport properties, *J. Med. Chem.*, 2000, **43**, 3714–3717.
[86] P. Artursson and C. A. S. Bergstroem, *Intestinal Absorption: the Role of Polar Surface Area. Methods and Principles in Medicinal Chemistry*, Drug Bioavailabilty, 2003, Vol. 18, pp. 341–357.
[87] C. A. S. Bergstroem, M. Strafford, L. Lazorova, Al. Avdeef, K. Luthman and P. Artursson, Absorption classification of oral drugs based on molecular surface properties, *J. Med. Chem.*, 2003, **46**, 558–570.
[88] I. Zamora, T. Oprea, G. Cruciani, M. Pastor and A.-L. Ungell, Surface descriptors for protein–ligand affinity prediction, *J. Med. Chem.*, 2003, **46**, 25–33.
[89] T. Fichert, M. Yazdanian and J. R. Proudfoot, A structure-permeability study of small drug-like molecules, *Bioorg. Med. Chem. Lett.*, 2003, **13**, 719–722.
[90] T. Sakaeda, N. Okamura, S. Nagata, T. Yagami, M. Horinouchi, K. Okumura, F. Yamashita and M. Hashida, Molecular and pharmacokinetic properties of 222 commercially available oral drugs in humans, *Biol. Pharm. Bull.*, 2001, **24**, 935–940.
[91] Y. H. Zhao, M. H. Abraham, A. Hersey and C. N. Luscombe, Quantitative relationship between rat intestinal absorption and Abraham descriptors, *Eur. J. Med. Chem.*, 2003, **38**, 939–947.
[92] C. A. Lipinski, Integration of physicochemical property considerations into the design of combinatorial libraries, *Pharm. News*, 2002, **9**, 195–202.
[93] Y. Ran and S. H. Yalkowsky, Prediction of drug solubility by the general solubility equation (GSE), *J. Chem. Inf. Comput. Sci.*, 2001, **41**, 354–357.
[94] C. A. S. Bergstroem, U. Norinder, K. Luthman and P. Artursson, Molecular descriptors influencing melting point and their role in classification of solid drugs, *J. Chem. Inf. Comput. Sci.*, 2003, **43**, 1177–1185.

[95] F. Lombardo, R. S. Obach, M. Y. Shalaeva and F. Gao, Prediction of human volume of distribution values for neutral and basic drugs. 2. Extended data set and leave-class-out statistics, *J. Med. Chem.*, 2004, **47**, 1242–1250.

[96] F. Lombardo, R. S. Obach, M. Y. Shalaeva and F. Gao, Prediction of volume of distribution values in humans for neutral and basic drugs using physicochemical measurements and plasma protein binding data, *J. Med. Chem.*, 2002, **45**, 2867–2876.

[97] D. F. V. Lewis and M. Dickins, Substrate SARs in human P450s, *Drug Discov. Today*, 2002, **7**, 918–925.

[98] K. M. Comess, M. J. Voorbach, M. L. Coen, H. Tang, L. Gao, X. Cheng, M. E. Schurdak, B. A. Beutel and D. J. Burns, Affinity-based high-throughput screening of orphan targets: practical solutions for removing promiscuous binders, Abstracts of Papers, 224th ACS National Meeting, Boston, MA, United States, August 18–22, 2002, COMP-220.

[99] S. L. McGovern, B. T. Helfand, B. Feng and B. K. Shoichet, A specific mechanism of nonspecific inhibition, *J. Med. Chem.*, 2003, **46**, 4265–4272.

[100] S. L. McGovern and B. K. Shoichet, Kinase inhibitors: not just for kinases anymore, *J. Med. Chem.*, 2003, **46**, 1478–1483.

[101] J. Seidler, S. L. McGovern, T. N. Doman and B. K. Shoichet, Identification and prediction of promiscuous aggregating inhibitors among known drugs, *J. Med. Chem.*, 2003, **46**, 4477–4486.

[102] C. M. Tice, Selecting the right compounds for screening: does Lipinski's rule of 5 for pharmaceuticals apply to agrochemicals?, *Pest Manage. Sci.*, 2001, **57**, 3–16.

[103] C. M. Tice, Selecting the right compounds for screening: use of surface-area parameters, *Pest Manage. Sci.*, 2002, **58**, 219–233.

[104] V. V. Zernov, K. V. Balakin, A. A. Ivaschenko, N. P. Savchuk and I. V. Pletnev, Drug discovery using support vector machines. The case studies of drug-likeness, agrochemical-likeness, and enzyme inhibition predictions, *J. Chem. Inf. Comput. Sci.*, 2003, **43**, 2048–2056.

CHAPTER 12

Structure-Based Lead Optimization

Diane Joseph-McCarthy

Wyeth Research, 200 CambridgePark Drive, Cambridge, MA 02140, USA

Contents

1. Introduction 169
2. Lead discovery 171
   2.1. Compound equity 171
   2.2. High-throughput screening 171
   2.3. Virtual screening 172
3. Lead modification 175
   3.1. Structure visualization 175
   3.2. Fragment positioning 175
   3.3. Molecular simulation 177
   3.4. Library enumeration and docking 178
   3.5. Ligand–target complex evaluation 178
4. Application to a specific target 179
   4.1. Acyl carrier protein synthase 179
5. Conclusions 180
References 180

1. INTRODUCTION

Structure-based drug design is an increasingly integral part of the drug discovery process. As the number of therapeutic targets with structural information dramatically increases [1,2], computational drug design methods continue to advance [3,4] to make it possible for the pharmaceutical industry to achieve greater efficiency. Structure-based drug design (Fig. 1) encompasses both lead generation (through virtual screening, including molecular docking) and lead optimization/*de novo* design. While structure-based virtual screening is emerging as an increasingly powerful tool for lead discovery, structure-based lead optimization has played a crucial role in drug discovery over the last 10 years. A typical computational chemist/molecular modeler in an industrial setting spends the majority of their time working on lead modification. Although true *de novo* design – starting with an apo protein structure and no ligand information – is rare, when structural information is available it is always used to guide the synthesis of new compounds.

ANNUAL REPORTS IN COMPUTATIONAL CHEMISTRY, VOLUME 1
ISSN: 1574-1400 DOI 10.1016/S1574-1400(05)01012-1

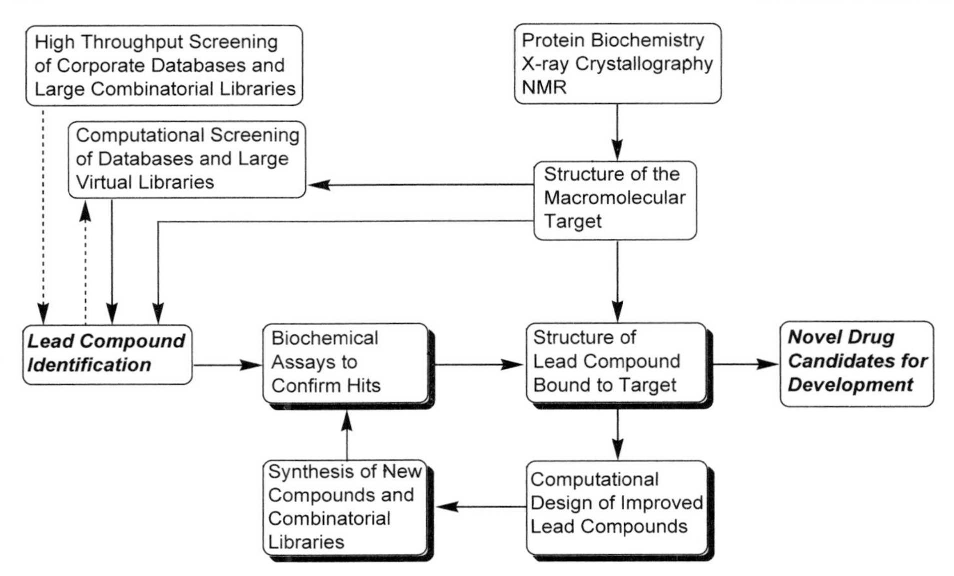

Fig. 1. Structure-based lead generation and optimization.

Given a high-resolution (ideally <2.5 Å) X-ray structure of a protein–ligand complex, computational methods are used to design potentially improved compounds that are predicted to make better or similar interactions with the target structure. A high-quality NMR structure or homology model may also be used, provided the binding mode of the initial lead can be identified with a high level of confidence. For use of a homology model, a high-resolution structure of a closely related protein with the same lead or a very similar compound bound may be required. Additional biological data (competition assay data, NMR binding data, etc.) to help confirm the binding mode in the actual target would also be desirable. With that level of structural information, an iterative process of designing, synthesizing, and experimentally testing new compounds can ensue. Much of the computational work is not published or eventually appears as a paragraph in an experimental paper describing a few molecules. Leads are optimized not only in terms of potency but often to improve the physical properties of the compounds to address issues such as bioavailability, metabolism, and protein binding, and sometimes to engineer novelty into the compounds to avoid patent infringement.

This report includes a brief section on lead generation and how the origin of a lead may affect the choice of lead optimization methods used, a discussion of the process an industrial chemist typically goes through for lead modification and some of the computational methods used and, finally, a specific example of the application of the approach to a given target.

2. LEAD DISCOVERY

New leads for a therapeutic target can be generated *via* (i) known inhibitors for the target or a very related target, (ii) high-throughput screening of corporate compound collections or screening of natural product libraries, or (iii) virtual screening of corporate or other large vendor databases. The way that leads are generated and the characteristics of the leads will determine to an extent the optimization strategy that is subsequently employed.

2.1. Compound equity

Occasionally, if nothing novel has been found, a published inhibitor or even an existing drug is taken as a 'lead' for a project. If a published compound is taken as a starting point, the goal is to modify the chemical entity without dramatically reducing its potency, although some decrease in potency will be tolerated. Then, the resulting new lead will be further optimized. More often compound equity is taken advantage of by testing all in-house leads or hits for a related target against the target of interest. In addition, family-focused compound plates may be screened. Kinases are an example of a protein family for which the latter two approaches can be productive. If the most closely related kinase (or kinases) in the company's portfolio can be identified by sequence alignment or possibly active site alignment, all hits for that kinase can be tested for the new target kinase. If a lead is generated in this manner, the immediate objective of lead modification is to design more selective analogs. Also, since these new leads may be fully optimized compounds for the related target, another challenge is to improve the potency of the compounds for the target of interest without significantly increasing their molecular weight and lipophilicity.

2.2. High-throughput screening

Today a significant percentage of novel leads for a target is found through high-throughput screening (HTS) [5]. Companies routinely and rapidly screen their corporate libraries each year for hits against a variety of targets. The advantage of HTS is that several hundred thousand compounds can be experimentally screened for binding to the target in a matter of weeks. The disadvantages are that it is expensive and the resulting data are typically very noisy. A commitment of resources to transfer the biological assay to HTS format, run the screen, confirm statistically significant hits, analyze the results, and follow up with secondary assays must be in place. Often throughput limits dictate the number of compounds passed from one stage to the next, so true hits may be overlooked. Also, the more potent confirmed hits are often known hits for related targets or in the worst case nonspecific binders that hit for many targets. Unless all statistically significant hits in the primary HTS are confirmed, less potent and more lead-like hits may be missed. Ideally, a lead has molecular weight less than 400, clogP less than 5, fewer than seven rotatable bonds, and a polar surface area less than 140 $Å^2$ [6,7]. These filters allow for the optimization process to increase the complexity of the molecules, as is common, and still yield drug-like molecules [8,9].

2.3. Virtual screening

Increasingly new leads for a target are identified through virtual screening [10]. Both ligand-based [11,12] and structure-based [13,14] approaches have been successfully used. If for a given therapeutic project, a set of active ligand molecules is known for the macromolecular target, but little or no structural information exists on the target, ligand-based screening methods can be employed. These include pharmacophore searches (e.g., [15–18]) and shape searches (e.g., the program ROCS [19]; OpenEye, Santa Fe, NM, 2003). Traditionally, a pharmacophore is the set of features common to a series of active molecules, where features can include acceptors, donors, ring centroids, hydrophobes, etc. A three-dimensional (3D) pharmacophore specifies the spatial relationship between the groups or features, often defining distances or distance ranges between groups, angles between groups or planes, and exclusion spheres. Deriving a pharmacophore model involves aligning or overlaying the structures of the known ligands and identifying the common features. Once generated, programs like Catalyst (Accelrys, San Diego, CA, 2003) and UNITY (Tripos, St. Louis, MO, 2003) can search large 2D or 3D molecular databases for additional molecules that possess the pharmacophore. Given just one active ligand known to bind to the target, a shape search can be performed, whereby 3D molecular databases are searched for other compounds that have the same shape. Knowledge of the bound conformation of the ligand is highly desirable. With certain shape search methods [19], some chemical matching can also be specified in addition to shape fit. The assumptions behind ligand-based methods are that the set of known ligands all bind at the same site in the target structure, that the bioactive conformer of each known ligand can be accurately predicted (in the case of 3D methods), and that any utilized alignment of the set is biologically relevant.

Structure-based virtual screening or molecular docking uses a heuristic to orient each database ligand in the binding site of the target structure and then scores each orientation. In this manner, poses for a given ligand are ranked relative to one another and to those of all other ligands in the database. In general, molecular docking involves searching a database for compounds that fit into the binding site of the target structure in terms of shape and electronic complementarity. To be a viable technique for screening commercial and corporate databases as well as virtual libraries of several hundred thousand molecules, search times on the order of seconds per molecule are desired. With molecular docking the 3D structure of the target binding site is known, potential key interactions with the target are identifiable, and the search is for ligands that can bind to the specified site. One typical assumption is that the protein structure used is competent to bind all active compounds; i.e., generally, the protein structure is held rigid during the docking. In reality, a given protein structure may only be competent to bind, say, 60% of active compounds [20,21]. As a result, some true actives will be missed. With structure-based virtual screening compared to ligand-based, however, additional experimental information is taken into account and expected to make the results more predictive. Furthermore, greater novelty may be expected in the ligands that are identified.

Conformational flexibility of the database ligands, however, must be considered in docking. Ligand flexibility can be incorporated either by flexing the ligand as it is incrementally built into the binding site or by docking rigid, precomputed conformers from conformationally expanded databases. The first category includes methods that employ Monte Carlo sampling [22], genetic and evolutionary algorithms [23–25], simulated annealing [26], and incremental construction [27–29]. The second group includes rigid docking of flexibases, in which individual conformers are separately docked and scored [30], and rigid docking of conformational ensembles [31–34] generated by overlaying related conformers. Some of the many docking programs in use include FlexX [28], GOLD [23], ICM [35], FRED (OpenEye, Santa Fe, NM, 2003), GLIDE [36,37], DOCK [38], and our in-house PhDock [31].

FlexX [28] is an incremental construction method that first decomposes the ligand into fragments by breaking all single acyclic and nonterminal bonds. The ligand is incrementally built up starting from the position of an anchor fragment. The set of allowed interaction types and the empirical scoring function are defined as in the program LUDI [39] with slight modifications. This model of discrete conformational flexibility for the ligand, with finite sets of allowed torsional angles for single acyclic bonds and precomputed conformations for ring systems, allows the docking to be fast.

The program DOCK [40] uses a heuristic that matches ligand atoms to pre-defined site points in the binding site of the target structure. Our pharmacophore-based docking method, PhDock [31], is implemented in DOCK4.0 [38] and allows for pharmacophore-based docking of ensembles of precomputed conformers. In a PhDock database, precomputed conformers of the same or different molecules are overlaid based on their largest 3D pharmacophore. The pharmacophore points (and not all of the ligand atoms) are matched to the predefined site points to determine the allowed orientations for the ensemble in the binding site. Each individual conformer within the ensemble is scored in the binding site and the best scoring conformer (or a set number of best scoring conformers) for each molecule is saved. Docking pharmacophoric ensembles provides an efficient strategy for rapidly docking large databases of molecules. Theoretical pharmacophore points which represent 'hot spots' in the target binding site can be used as the matching site points to orient the candidate ligands more rapidly. More specifically, chemically labeled site points can be generated in an automated fashion using the script MCSS2SPTS [20]. MCSS2SPTS employs the program MCSS (Multiple Copy Simultaneous Search) [41–43] to determine the target-based theoretical pharmacophores. MCSS functional group maps (or preferred binding sites) are calculated for nine different groups, spanning ionizable, polar, and hydrophobic atom types. Chemically labeled site points are automatically extracted from selected low energy functional-group minima and clustered together. Chemically labeling the site points can significantly reduce the search time by restricting the search space to areas relevant to the target, thereby reducing the combinatorial problem.

A number of research groups are developing docking methods that allow for some degree of target flexibility. This has largely been limited to a few side chain motions or water displacements (e.g., SLIDE [44]). Others try to account for protein flexibility by using multiple scoring grids, representing multiple protein structures, or a scoring function with values averaged over or derived from a number of protein structures (e.g., AutoDock [45], FlexE [46]).

Once a virtual screen is completed and some visual inspection or further prioritization ([47]; http://www.metaphorics.com/products/magnet) of the top ranked hits has occurred, compounds (10s–100s) are selected for experimental testing. Ideally, docking hits are molecules with novel scaffolds for binding the target, which may initially be relatively weak binders; these hits serve as new starting points for synthetic optimization. Thus, virtual screening should be initiated as early in the project as possible. Often virtual screening is done prior to or concurrently with experimental HTS. It can also be carried out at the later stages of a project to search for backup or follow-on series. Very large commercial databases, such as the Available Chemicals Directory (ACD) database, as well as corporate databases can be rapidly screened. When a virtual screen is carried out for a corporate database in parallel with experimental HTS, virtual screening can identify false negatives from the HTS or compounds not present in the HTS plates. During the setup or generation of these large databases in a suitable docking format, filters are applied to eliminate nonlead-like and undesirable molecules [47,48]. If a target-focused or family-focused library is being screened, less stringent filters may be applied because the resulting hits are expected to require less optimization.

In a recent example using PhDOCK and MCSS2SPTS, our corporate database was screened for binding to a particular kinase target (D. Joseph-McCarthy, unpublished). Out of 20 compounds selected and experimentally tested, the screen identified two novel scaffolds for the target not found through the HTS. Both were reversible, μM inhibitors; one was also relatively selective. These compounds are currently being further characterized and serve as backup series for the program.

3. LEAD MODIFICATION

3.1. Structure visualization

A lead optimization project involves interaction and collaboration between the modeler and the medicinal chemists on the project, on a weekly or regular basis. In today's global economy with multisite, international, or collaborative alliances this communication is often required to be *via* the Internet and teleconference. An Internet meeting software such as WebEx (WebEx, San Jose, CA, 2003) can be used with a pc-based graphics or visualization program such as Weblab Viewer (Accelrys, San Diego, CA, 2002), VIDA (OpenEye, Santa Fe, NM, 2003), or MOE (Chemical Computing Group, Montreal, Quebec, 2003). At the present time, often the pc-based graphical user interface is not the one associated with the programs that the modeler used to generate the models; a number of commonly used modeling packages are still largely Unix based although this is expected to change in the near future. These Internet meetings are still limited by the speed of the connection, and one would ideally like to have visualization software which transmits the commands (e.g., the translation/rotation matrices for a move) then applies them locally in another instance of the program. Whether the meetings are face to face or *via* the Internet, an iterative process ensues and can last for many months. The simplest part of the collaboration involves viewing existing structures and newly calculated models to generate additional design ideas. Structures are overlaid, surfaces are displayed on the protein or sometimes on the ligand to get a sense of how well the ligand is filling and complementing (in terms of electrostatic interactions, lipophilicity, etc.) the protein binding site, and distances between ligand and protein atoms are measured to help to determine where different or additional substituents could be modeled.

3.2. Fragment positioning

Fragment positioning methods are often employed for lead optimization (Fig. 2). These methods determine energetically favorable binding site positions for

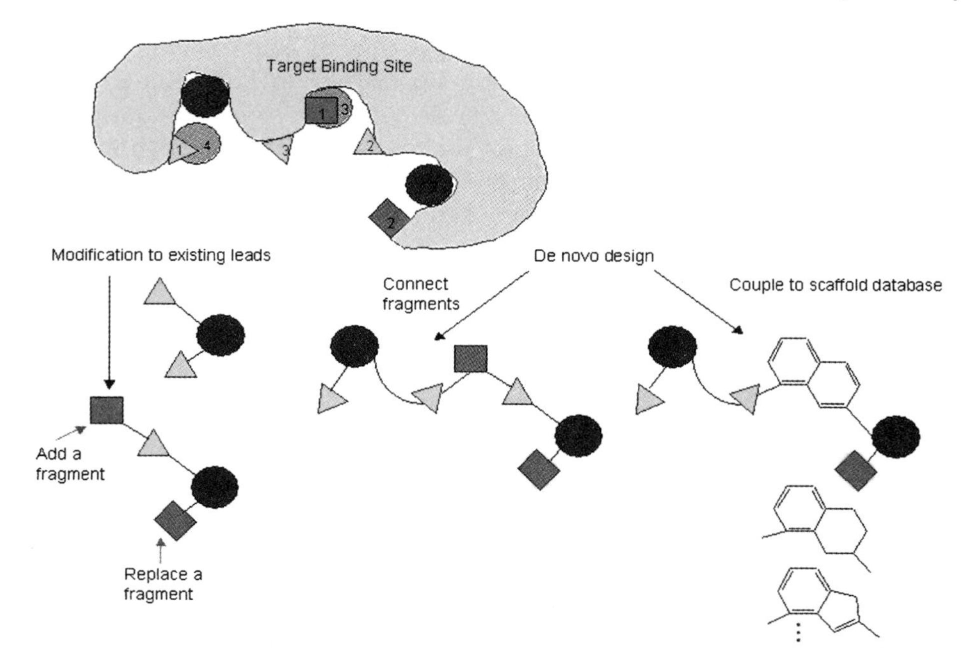

Fig. 2. Optimal fragment positions in the binding site (e.g., the light grey triangles represent a given chemical group and the labels 1, 2, 3, etc. specify their position rank) identified for the modification of an existing lead or the design of entirely new molecules.

various functional group types or chemical fragments. Several well-known programs are GRID [49], MCSS [41–43], MUSIC [47], LUDI [51,52], and Superstar [53]. With the MCSS program, probe groups are fully flexible and individual atoms are represented using the CHARMM [54] potential energy function. Very briefly, several thousand copies of a given functional group are randomly distributed in the binding site. The functional group copies are then simultaneously minimized in the time-dependent Hartree approximation such that each copy of the group feels the full force field of the protein but the group copies do not interact with each other. In general, using any fragment positioning method, chemical group sites or 'hot spots' can be calculated for a wide range of groups (typically 10s of groups) either for a protein structure with any bound ligands removed or for a protein structure with a ligand or the scaffold part of the ligand bound. Even when the functional group maps are calculated for the protein structure alone, structures or models of existing leads bound to the target can be superimposed with the maps; this overlay is used to predict substitutions that are likely to improve the potency of the lead. More detailed calculations, like the ones described in the next section, can then be carried out on the proposed improved ligands, or if a small combinatorial library of ligands is being synthesized, the functional group site information on its own can be utilized to focus the library [55]. Since we typically use MCSS2SPTS with PhDOCK to setup a virtual screen, functional group maps for nine diverse groups are automatically calculated for the target protein structure in the process and can be simultaneously or subsequently used for lead optimization.

3.3. Molecular simulation

There are a number of ways to model a series of specific analogs into the binding site of a target structure. The approach taken can depend on the number of analogs (specific molecules with substitutions to the existing lead structure) to be modeled (10s *vs.* 100s or 1000s), the time allotted for the calculations, and the character of the binding site. If, for example, the existing lead makes a specific interaction with the protein that is to be maintained in any candidate compounds, then the modeling method may need to allow for constraints to be setup (e.g., distance constraints specifying a hydrogen bond or a salt bridge). A very flat, narrow binding pocket as in the ATP-binding site of kinase structures may allow only for minimization unless constraints are applied to keep the scaffold portion of the molecule in the binding site. The first step in any lead optimization effort involves determining which methods and scoring schemes can best reproduce the known binding modes of existing leads. A method has to be able to reproduce accurately any X-ray complex structures of the leads bound to the target to be useful for lead optimization. The results are generally system dependent in that an approach that works well for kinases may not work as well for metalloproteases.

If 10s of substitutions are to be modeled for a given lead, the binding mode is well established, and the substituents are small, simply building the molecule into the binding site with the substituents in all plausible orientations and energy minimizing with a molecular mechanics force field such as CHARMM [56], AMBER [57], or MMFF94s [58] may be sufficient. Adding an implicit solvation term while minimizing may help. Minimizing using the MMFF94s force field with a solvation correction can readily be done within the MOE software package (Chemical Computing Group, Montreal, Quebec, 2003), for example. A better approach may be to do either a Monte Carlo or a simulated annealing simulation for each ligand in the binding site. These methods provide more sampling and therefore it is not generally necessary to build the ligand into the binding site in multiple orientations. The FLO99 [22] algorithm involves Monte Carlo (MC) perturbation (wide-angle torsional Metropolis perturbation as well as translation and rotation of ligand atoms) followed by energy minimization in Cartesian space for flexible ligand binding to a target structure based upon a modified AMBER [57] potential. Using the graphical interface of FLO99, individual ligand molecules can be easily built and docked into the protein site. On an SGI R12000 (400 MHz) processor run times for 1000 cycles of MC are approximately 30 min per ligand and therefore this is a medium throughput approach.

3.4. Library enumeration and docking

If 100s or 1000s of candidate ligands are to be modeled, a relatively high throughput approach is desirable. With FLO99, for example, if the scaffold position is known and a particular R group position is to be varied using a specified chemistry up to a thousand or more substituents can be modeled relatively quickly (typically in a day or two). In that case, the attachment position is labeled on the scaffold, an SDF format file containing the set of substituents (a list of amines for example) is read and the chemistry for attachment is indicated. FLO99 automatically generates each library ligand and docks it into the site; because the scaffold position is known, MC sampling can be limited to the R group speeding up the calculations and reducing the number of cycles required. A library of analogs can be enumerated in a number of different ways. Packages such as Sybyl (Tripos, St. Louis, MO, 2003), MOE (Chemical Computing Group, Montreal, Quebec, 2003), and Cerius2 (Accelrys, San Diego, CA, 2003), for example, are capable of constructing the library ligands. The ligands can then be docked into the target binding site using any of the medium to high throughput methods discussed above in Section 2.3.

If the scaffold position is known, an anchor and growth method such as FlexX which places the anchor and then grows the rest of the molecule as the docking proceeds may work well. If a complex structure of the lead bound to the target exists, some part of the ligand molecule can be defined as the scaffold or anchor. It is generally assumed that the binding mode of the scaffold portion of a lead is to be maintained during the optimization. An alternative approach that we are developing involves using DOCK to search databases for suitable R groups that can attach to a given scaffold bound to the target (J. Zou, I.J. McFadyen, J.C. Alvarez, D. Joseph-McCarthy, in preparation). A ranked list of acceptable (and presumably synthetically accessible) substituents is generated for each R group position. Select R group substituents are automatically recombined to generate novel analogs that are predicted to bind to the target and are not found in the original database.

3.5. Ligand–target complex evaluation

Once a model for a ligand–protein complex is obtained, it can be evaluated using any number of scoring functions. If a series of closely related analogs are modeled, molecular mechanics energies may be sufficient to rank them or at least to separate out the binders *vs.* the nonbinders. To account for desolvation effects, a solvation correction can be added using a Poisson–Boltzmann continuum model [59,60] or a generalized Born implicit solvent model [61]. Physicochemical scoring functions [62–65] and knowledge-based potentials [66] can also be used to evaluate complexes. Consensus scoring is an effective method that requires a complex to score well by several different metrics to be predicted as a good ligand [67,68].

4. APPLICATION TO A SPECIFIC TARGET

4.1. Acyl carrier protein synthase

Acyl carrier protein synthase (AcpS) catalyzes the conversion of apo-acyl carrier protein (apo-Acp) to holo-Acp through the transfer of the 4′-phosphopantetheinyl group of coenzyme A to a serine residue of apo-Acp. This transfer is an important step in the fatty acid biosynthesis pathway and therefore AcpS is of interest as an antibacterial target [69]. An HTS lead was the starting point for a lead optimization project that involved designing and modeling improved analogs for individual synthesis as well as focusing, or biasing, a relatively large combinatorial library for binding to the target. The initial lead was a novel anthranilic acid inhibitor of AcpS with a μM IC_{50} value. Modifications to the lead were proposed based on structure-based modeling using FLO99 and a small number of (less than 10) individual compounds were synthesized resulting in a 20-fold lower IC_{50} value. In addition, the binding mode of the final inhibitor bound to AcpS has been confirmed by X-ray crystallography (K.D. Parris, D. Joseph-McCarthy *et al.*, in preparation). Concurrently, modeling was used to identify positions off a related scaffold that could be varied for optimization (Fig. 3).

Modeling constraints aided in the design of an initial focused library that led through the use of D-optimal design to the synthesis of a second library resulting in a 0.27 μM inhibitor [70].

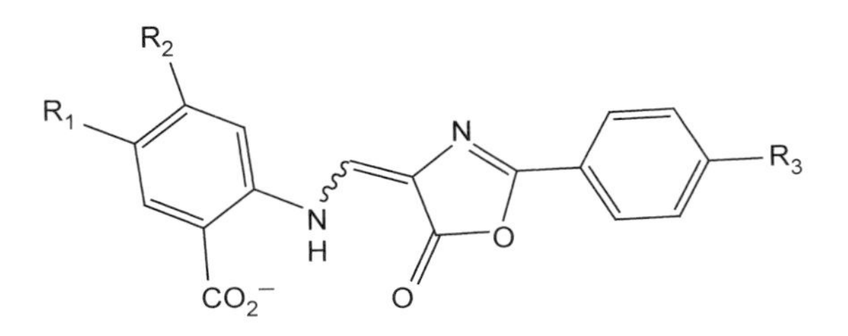

Fig. 3. Schematic of the scaffold.

5. CONCLUSIONS

Structure-based lead optimization continues to make the drug discovery process more efficient. As computer power increases, scoring functions improve, and computational methods for lead optimization are more fully automated, an even greater impact of the technology will be realized.

REFERENCES

[1] H. M. Berman, T. N. Bhat, P. E. Bourne, Z. Feng, G. Gilliland, H. Weissig and J. Westbrook, The Protein Data Bank and the challenge of structural genomics, *Nat. Struct. Biol.*, 2000, **7** (Suppl.), 957–959.
[2] B. Rost, Marrying structure and genomics, *Structure*, 1998, **6**, 259–263.
[3] D. Joseph-McCarthy, Computational approaches to structure-based ligand design, *Pharmacol. Ther.*, 1999, **84**, 179–191.
[4] G. Klebe, Recent developments in structure-based drug design, *J. Mol. Med.*, 2000, **78**, 269–281.
[5] R. P. Hertzberg and A. J. Pope, High-throughput screening: new technology for the 21st century, *Curr. Opin. Chem. Biol.*, 2000, **4**, 445–451.
[6] C. A. Lipinski, F. Lombardo, B. W. Dominy and P. J. Feeney, Experimental and computational approaches to estimate solubility and permeability in drug discovery and development settings, *Adv. Drug Deliv. Rev.*, 1997, **23**, 3–25.
[7] T. I. Oprea, Property distribution of drug-related chemical databases, *J. Comput.-Aided Mol. Des.*, 2000, **14**, 251–264.
[8] T. I. Oprea, A. M. Davis, S. J. Teague and P. D. Leeson, Is there a difference between leads and drugs? A historical perspective, *J. Chem. Inf. Comput. Sci.*, 2001, **41**, 1308–1315.
[9] D. F. Veber, S. R. Johnson, H.-Y. Cheng, B. R. Smith, K. W. Ward and K. D. Kopple, Molecular properties that influence the oral bioavailability of drug candidates, *J. Med. Chem.*, 2002, **45**, 2615–2623.
[10] G. Schneider and H. J. Bohm, Virtual screening and fast automated docking methods, *Drug Discov. Today*, 2002, **7**, 64–70.
[11] A. C. Good, S. R. Krystek and J. S. Mason, High-throughput and virtual screening: core lead discovery technologies move towards integration, *Drug Discov. Today*, 2000, **5**, 61–69.
[12] T. Lengauer, C. Lemmen, M. Rarey and M. Zimmermann, Novel technologies for virtual screening, *Drug Discov. Today*, 2004, **9**, 27–34.
[13] N. Brooijmans and I. D. Kuntz, Molecular recognition and docking algorithms, *Annu. Rev. Biophys. Biomol. Struct.*, 2003, **32**, 335–373.
[14] P. D. Lyne, Structure-based virtual screening: an overview, *Drug Discov. Today*, 2002, **7**, 1047–1055.
[15] O. F. Guner, History and evolution of the pharmacophore concept in computer-aided drug design, *Curr. Top. Med. Chem.*, 2002, **2**, 1321–1332.
[16] J. S. Mason, A. C. Good and E. J. Martin, 3-D pharmacophores in drug discovery, *Curr. Pharm. Des.*, 2001, **7**, 567–597.
[17] J. Van Drie, Strategies for the determination of pharmacophoric 3D database queries, *J. Comput.-Aided Mol. Des.*, 1997, **11**, 39–52.
[18] Y. C. Martin, Computer design of bioactive compounds based on 3-D properties of ligands, *NIDA Res. Monogr.*, 1993, **134**, 84–102.
[19] J. A. Grant, M. A. Gallardo and B. T. Pickup, *J. Comput. Chem.*, 1996, **17**, 1653–1666.

[20] D. Joseph-McCarthy and J. C. Alvarez, Automated generation of MCSS-derived pharmacophoric DOCK site points for searching multiconformation databases, *Proteins*, 2003, **51**, 189–202.

[21] J. A. Erickson, M. Jalaie, D. H. Robertson, R. A. Lewis and M. Vieth, Lessons in molecular recognition: the effects of ligand and protein flexibility on molecular docking accuracy, *J. Med. Chem.*, 2004, **47**, 45–55.

[22] C. McMartin and R. S. Bohacek, QXP: powerful, rapid computer algorithms for structure-based drug design, *J. Comput.-Aided Mol. Des.*, 1997, **11**, 333–344.

[23] G. Jones, P. Willett, R. C. Glen, A. R. Leach and R. Taylor, Development and validation of a genetic algorithm for flexible docking, *J. Mol. Biol.*, 1997, **267**, 727–748.

[24] D. K. Gehlhaar, G. M. Verkhivker, P. A. Rejto, C. J. Sherman, D. B. Fogel, L. J. Fogel and S. T. Freer, Molecular recognition of the inhibitor AG-1343 by HIV-1 protease: conformationally flexible docking by evolutionary programming, *Chem. Biol.*, 1995, **2**, 317–324.

[25] G. M. Morris, D. S. Goodsell, R. S. Halliday, R. Huey, W. E. Hart, R. K. Belew and A. J. Olson, Automated docking using a Lamarckian genetic algorithm and an empirical binding free energy function, *J. Comput. Chem.*, 1998, **19**, 1639–1662.

[26] G. M. Morris, D. S. Goodsell, R. Huey and A. J. Olson, Distributed automated docking of flexible ligands to proteins: parallel applications of AutoDock 2.4, *J. Comput.-Aided Mol. Des.*, 1996, **10**, 293–304.

[27] B. Kramer, M. Rarey and T. Lengauer, CASP2 experiences with docking flexible ligands using FLEXX, *Proteins*, 1997, **Suppl. 1**, 221–225.

[28] M. Rarey, B. Kramer, T. Lengauer and G. Klebe, A fast flexible docking method using an incremental construction algorithm, *J. Mol. Biol.*, 1996, **261**, 470–489.

[29] S. Makino and I. D. Kuntz, Automated flexible ligand docking method and its application for database search, *J. Comput. Chem.*, 1997, **18**, 1812–1825.

[30] S. K. Kearsley, D. J. Underwood and M. D. Miller, Flexibases: a way to enhance the use of molecular docking methods, *J. Comput.-Aided Mol. Des.*, 1994, **8**, 565.

[31] D. Joseph-McCarthy, B. E. Thomas, IV, M. Belmarsh, D. Moustakas and J. C. Alvarez, Pharmacophore-based molecular docking to account for ligand flexibility, *Proteins*, 2003, **51**, 172–188.

[32] B. E. Thomas, IV, D. Joseph-McCarthy and J. C. Alvarez, Pharmacophore-based molecular docking. In *Pharmacophore Perception, Development, and Use in Drug Design* (ed. O. F. Guner), International University Press, La Jolla, CA, 2000, pp. 351–367.

[33] A. I. Su, D. M. Lorber, G. S. Weston, W. A. Baase, B. W. Matthews and B. K. Shoichet, Docking molecules by families to increase the diversity of hits in database screens: computational strategy and experimental evaluation, *Proteins*, 2001, **42**, 279–293.

[34] D. M. Lorber and B. K. Shoichet, Flexible ligand docking using conformational ensembles, *Protein Sci.*, 1998, **7**, 938–950.

[35] R. A. Abagyan, M. M. Totrov and D. N. Kuznetsov, ICM – a new method for protein modeling and design: applications to docking and structure prediction from the distorted native conformation, *J. Comput. Chem.*, 1994, **15**, 488–506.

[36] R. A. Friesner, J. L. Banks, R. B. Murphy, T. A. Halgren, J. J. Klicic, D. T. Mainz, M. P. Repasky, E. H. Knoll, M. Shelley, J. K. Perry, D. E. Shaw, P. Francis and P. S. Shenkin, Glide: a new approach for rapid, accurate docking and scoring. 1. Method and assessment of docking accuracy, *J. Med. Chem.*, 2004, **47**, 1739–1749.

[37] T. A. Halgren, R. B. Murphy, R. A. Friesner, H. S. Beard, L. L. Frye, W. T. Pollard and J. L. Barks, Glide: a new approach for rapid, accurate docking and scoring. 2. Enrichment factors in database screening, *J. Med. Chem.*, 2004, **47**, 1750–1759.

[38] T. Ewing, S. Makino, A. Skillman and I. Kuntz, DOCK 4.0: search strategies for automated molecular docking of flexible molecule databases, *J. Comput.-Aided Mol. Des.*, 2001, **15**, 411–428.

[39] H. J. Bohm, The development of a simple empirical scoring function to estimate the binding constant for a protein ligand complex of known 3-dimensional structure, *J. Comput.-Aided Mol. Des.*, 1994, **8**, 243–256.
[40] I. D. Kuntz, J. M. Blaney, S. J. Oarley, R. Langridge and T. E. Ferrin, A geometric approach to macromolecule–ligand interactions, *J. Mol. Biol.*, 1982, **161**, 269–288.
[41] A. Miranker and M. Karplus, Functionality maps of binding sites: a multiple copy simultaneous search method, *Proteins*, 1991, **11**, 29–34.
[42] E. Evensen, D. Joseph-McCarthy and M. Karplus, *MCSSv2*, 2.1 edition, Harvard University, Cambridge, MA, 1998.
[43] D. Joseph-McCarthy, Structure-based combinatorial library design and screening: application of the Multiple Copy Simultaneous Search method. In *Combinatorial Library Design and Evaluation for Drug Discovery: Principles, Methods, Software Tools and Applications* (eds A. K. Ghose and V. N. Viswanadhan), Marcel Dekker, New York, 2000.
[44] V. Schnecke and L. A. Kuhn, Virtual screening with solvation and ligand-induced complementarity, *Perspect. Drug Discov. Des.*, 2000, **20**, 171–190.
[45] F. Osterberg, G. Morris, M. Sanner, A. Olson and D. Goodsell, Automated docking to multiple target structures: incorporation of protein mobility and structural water heterogeneity in AutoDock, *Proteins*, 2002, **46**, 34–40.
[46] H. Claussen, C. Buning, M. Rarey and T. Lengauer, FlexE: efficient molecular docking considering protein structure variations, *J. Mol. Biol.*, 2001, **308**, 377–395.
[47] D. Joseph-McCarthy, I. J. McFadyen, J. Zou, G. Walker and J. C. Alvarez, Pharmacophore-based molecular docking: a practical guide, *Virtual Screening in Drug Discovery*, Marcel Dekker, New York, NY, 2004.
[48] W. P. Walters, M. T. Stahl and M. A. Murcko, Virtual screening – an overview, *Drug Discov. Today*, 1998, **3**, 160–178.
[49] P. Goodford, A computational procedure for determining energetically favorable binding sites on biologically important macromolecules, *J. Med. Chem.*, 1985, **28**, 849–857.
[50] H. A. Carlson, K. M. Masukawa, K. Rubins, F. D. Bushman, W. L. Jorgensen, R. D. Lins, J. M. Briggs and J. A. McCammon, Developing a dynamic pharmacophore model for HIV-1 integrase, *J. Med. Chem.*, 2000, **43**, 2100–2114.
[51] H. J. Bohm, Ludi – rule-based automatic design of new substituents for enzyme-inhibitor leads, *J. Comput.-Aided Mol. Des.*, 1992, **6**, 593–606.
[52] H. J. Bohm, On the use of Ludi to search the Fine Chemicals Directory for ligands of proteins of known 3-dimensional structure, *J. Comput.-Aided Mol. Des.*, 1994, **8**, 623–632.
[53] M. L. Verdonk, J. C. Cole and R. Taylor, SuperStar: a knowledge-based approach for identifying interaction sites in proteins, *J. Mol. Biol.*, 1999, **289**, 1093–1108.
[54] B. R. Brooks, R. E. Bruccoleri, B. D. Olafson, D. J. States, S. Swaminathan and M. Karplus, CHARMM: a program for macromolecular energy, minimization, and dynamics calculations, *J. Comput. Chem.*, 1983, **4**, 187–217.
[55] D. Joseph-McCarthy, S. K. Tsang, D. J. Filman, J. M. Hogle and M. Karplus, Use of MCSS to design small targeted libraries: application to picornavirus ligands, *J. Am. Chem. Soc.*, 2001, **123**, 12758–12769.
[56] A. D. MacKerell, D. Bashford, M. Bellott, R. L. Dunbrack, J. D. Evanseck, M. J. Field, S. Fischer, J. Gao, H. Guo, S. Ha, D. Joseph-McCarthy, L. Kuchnir, K. Kuczera, F. T. K. Lau, C. Mattos, S. Michnick, T. Ngo, D. T. Nguyen, B. Prodhom, W. E. Reiher, B. Roux, M. Schlenkrich, J. C. Smith, R. Stote, J. Straub, M. Watanabe, J. Wiorkiewicz-Kuczera, D. Yin and M. Karplus, All-atom empirical potential for molecular modeling and dynamics studies of proteins, *J. Phys. Chem. B*, 1998, **102**, 3586–3616.
[57] W. D. Cornell, P. Cieplak, C. I. Bayly, I. R. Gould, K. M. Merz, D. M. Ferguson, D. C. Spellmeyer, T. Fox, J. W. Caldwell and P. A. Kollman, A 2nd generation force-field for

the simulation of proteins, nucleic-acids, and organic-molecules, *J. Am. Chem. Soc.*, 1995, **117**, 5179–5197.

[58] T. A. Halgren, MMFF VI. MMFF94s option for energy minimization studies, *J. Comput. Chem.*, 1999, **20**, 720–729.

[59] B. Honig and A. Nicholls, Classical electrostatics in biology and chemistry, *Science*, 1995, **268**, 1144–1149.

[60] J. Srinivasan, T. E. Cheatham, III, P. Cieplak, P. A. Kollman and D. A. Case, Continuum solvent studies of the stability of DNA, RNA, and phosphoramidate-DNA helices, *J. Am. Chem. Soc.*, 1998, **120**, 9401–9409.

[61] D. Bashford and D. A. Case, Generalized born models of macromolecular solvation effects, *Annu. Rev. Phys. Chem.*, 2000, **51**, 129–152.

[62] H. J. Bohm and G. Klebe, What can we learn from molecular recognition in protein–ligand complexes for the design of new drugs, *Angew. Chem. Int. Ed. Engl.*, 1996, **35**, 2588–2614.

[63] M. D. Eldridge, C. W. Murray, T. R. Auton, G. V. Paolini and R. P. Mee, Empirical scoring functions. 1. The development of a fast empirical scoring function to estimate the binding affinity of ligands in receptor complexes, *J. Comput.-Aided Mol. Des.*, 1997, **11**, 425–445.

[64] A. N. Jain, Scoring noncovalent protein–ligand interactions: a continuous differentiable function tuned to compute binding affinities, *J. Comput.-Aided Mol. Des.*, 1996, **10**, 427–440.

[65] D. K. Gehlhaar, G. M. Verkhivker, P. A. Rejto, C. J. Sherman, D. B. Fogel, L. J. Fogel and S. T. Freer, Molecular recognition of the inhibitor AG-1343 by HIV-1 protease: conformationally flexible docking by evolutionary programming, *Chem. Biol.*, 1995, **2**, 317–324.

[66] W. A. Koppensteiner and M. J. Sippl, Knowledge-based potentials – back to the roots, *Biochemistry (Mosc.)*, 1998, **63**, 247–252.

[67] P. S. Charifson, J. J. Corkery, M. A. Murcko and W. P. Walters, Consensus scoring: a method for obtaining improved hit rates from docking databases of three-dimensional structures into proteins, *J. Med. Chem.*, 1999, **42**, 5100–5109.

[68] R. D. Clark, A. Strizhev, J. M. Leonard, J. F. Blake and J. B. Matthew, Consensus scoring for ligand/protein interactions, *J. Mol. Graph. Model.*, 2002, **20**, 281–295.

[69] K. D. Parris, L. Lin, A. Tam, R. Mathew, J. Hixon, M. Stahl, C. C. Fritz, J. Seehra and W. S. Somers, Crystal structures of substrate binding to *Bacillus subtilis* holo-(acyl carrier protein) synthase reveal a novel trimeric arrangement of molecules resulting in three active sites, *Struct. Fold. Des.*, 2000, **8**, 883–895.

[70] A. M. Gilbert, M. Kirisits, P. Toy, D. S. Nunn, A. Failli, E. G. Dushin, E. Novikova, P. J. Petersen, D. Joseph-McCarthy, I. McFadyen and C. C. Fritz, Anthranilate 4H-oxazol-5-ones: novel small molecule antibacterial acyl carrier protein synthase (AcpS) inhibitors, *Bioorg. Med. Chem. Lett.*, 2004, **14**, 37–41.

CHAPTER 13

Targeting the Kinome with Computational Chemistry

Michelle L. Lamb

AstraZeneca R&D Boston, Cancer Research, 35 Gatehouse Drive, Waltham, MA 02451, USA

Contents

1. Introduction 185
2. The kinome 186
   2.1. Background 186
   2.2. ATP site recognition elements 187
3. Methodology for kinase targets 188
   3.1. Homology models 188
   3.2. Docking and scoring 189
   3.3. Selectivity 190
   3.4. Structure-based hybridization 191
4. Applications across the kinome 192
   4.1. CMGC group 192
   4.2. TK group 194
   4.3. CAMK group 196
   4.4. AGC group 196
   4.5. Other kinase groups 196
5. Conclusions 197
References 198

1. INTRODUCTION

Despite their importance in cellular signal trafficking, protein kinases were initially considered poor drug discovery targets due to the highly conserved protein fold and ATP binding site. This conservation was expected to result in a lack of selectivity against other members of this large target class. Recently, the kinase family has garnered significantly more interest with the clinical success in cancer therapeutics of small-molecule inhibitors of tyrosine kinases. These compounds include Gleevec (imatinib, STI-571), acting through the Bcr–Abl kinase, and Iressa (gefitinib), which inhibits the epidermal growth factor receptor kinase. Medicinal chemistry efforts in this field have benefited much from structural biology insights into inhibitor binding interactions and selectivity.

ANNUAL REPORTS IN COMPUTATIONAL CHEMISTRY, VOLUME 1
ISSN: 1574-1400 DOI 10.1016/S1574-1400(05)01013-3

Computational chemistry has flourished at the interface of these two disciplines. This review will outline the kinase structural motif, computational methods that have been adapted for kinase applications, and many of the most recent applications of computational chemistry techniques to the design of potent and selective kinase inhibitors.

2. THE KINOME

2.1. Background

The kinases are a family of enzymes that catalyze the transfer of the γ-phosphate group of ATP in the presence of metal ion cofactors (i.e., Mg^{2+}) to substrates, primarily protein sequences with hydroxyl-containing residues serine, threonine, or tyrosine, using the machinery of a catalytic domain with a conserved fold. More than 500 protein kinases encoded in the human genome, the kinome, have been classified into groups and families by sequence alignment [1], expanding upon the fundamental work in this area by Hanks and Hunter [2]. The cognate kinases for the CMGC group are the computationally well-studied cyclin-dependent kinases (CDKs), the mitogen-activated protein kinases (MAP kinases), glycogen synthase kinase 3 (GSK3), and CDK-like kinases, all of which phosphorylate serine and threonine residues in their appropriate substrates. The tyrosine kinase group (TK) can be further organized into families of receptor TKs, such as epidermal growth factor receptors (EGFR, ErbB) and vascular endothelial growth factor receptors (VEGFR), and nonreceptor TKs, of which Abl, Src, and Lck are prime examples. Other groups include CAMK, with kinases related to the family of calmodulin-dependent kinases, and AGC, named for the cyclic-nucleotide-dependent kinases such as protein kinase A, protein kinase G, and protein kinase C which form its basis. As will be seen, fewer modeling studies have been reported for other groups of kinases, for example, the tyrosine-kinase-like (TKL) or homologs of yeast sterile kinase (STE) groups, and a number of kinases fall outside the major classifications described here.

Protein crystallography has made a tremendous impact on elucidating the mechanisms by which kinases can be inhibited, as is described in a number of excellent reviews [3–5]. Although crystal structures are static 'snapshots' of protein conformation, the variety of conformations reported for kinases is an indication of the importance of conformational dynamics in the binding process. Kinases in their active states have similar structures, but kinases crystallized in their unphosphorylated, inactive forms reveal the complexity of the situation [6]. Small-molecule kinase inhibitors have been reviewed [7,8], and medicinal chemistry efforts are introducing compounds at a rapid rate, resulting in entire journal issues devoted to the topic [9]. Most reported inhibitors are ATP competitive, and historically these have been shown to inhibit the active form of a given kinase. However, compounds have also been identified that bind to the inactive form, most notably Gleevec (with Abl) and inhibitors of isoforms of p38 kinase of the MAPK family [10]. Most recently, Pfizer has disclosed inhibitors of MAP Kinase/Erk Kinase (MEK) that bind concurrently with ATP [11].

The regulation of kinase signaling can be controlled through a number of mechanisms [12]. Among those identified for the tyrosine kinase family are extracellular ligand binding and dimerization of receptor TKs, autophosphorylation, recognition of additional domains such as Src-homology SH2 and SH3 domains, and N-terminal myristoylation, in the case of c-Abl [13,14]. While some computational work has focused on these binding events, for example, the binding of ephrin ligands to ephrin receptor TKs [15], and diacylglycerol and diacylglycerol lactones which initiate translocation to the membrane of the protein kinase C [16], as the majority of recent effort has been directed toward ATP-competitive compounds, these will be the focus of this report. For earlier work, the reader is referred to the review written by Woolfrey and Weston [17].

2.2. ATP site recognition elements

The kinase fold consists of two domains connected by a linker (or hinge) region. The N-terminal domain is primarily β-sheet in structure with a single α-helix, while the C-terminal domain is largely composed of α-helices. The ATP binding site is a narrow channel located at the hinge between the domains. A flexible, glycine-rich loop is responsible for phosphate binding, while the activation loop regulates kinase activity through its phosphorylation state and conformation. This loop in most kinases contains a well-conserved Asp-Phe-Gly motif; in the active conformation, the aspartic acid can interact with an inhibitor or ATP and the phenylalanine side chain is directed out of the ATP site. In an inactive conformation, the Phe residue is directed into the site and can either block ATP binding or stack against aromatic rings of inhibitors such as Gleevec.

The ATP site (scheme below) can be divided into a number of subsites including the adenine-binding region, the sugar-binding pocket, the phosphate-binding region, a hydrophobic specificity pocket not exploited by ATP, and a hydrophobic surface patch en route to the entrance of the site (the 'solvent channel'). Molecular recognition of adenine in 68 ATP-binding proteins (11 kinases) through hydrogen-bonding, $\pi-\pi$ stacking, and cation$-\pi$ interactions has been characterized in these crystal structures [18,19]. *Ab initio* calculations on model systems were used to interpret the contributions of each interaction to binding. The adenine-recognition motif includes hydrogen bonds to backbone atoms in the hinge region. One nitrogen of the pyrimidine is a hydrogen-bond acceptor interacting with the amide NH of residue i, while the acyclic amine donates to the backbone carbonyl oxygen of residue $i-2$.

In the scheme, these interactions are indicated with solid arrows. Beyond the adenine site, the hydrophobic selectivity pocket has been the target of many kinase inhibitors, to pick up interactions not accessible to ATP. The size of this pocket has been related to the residue at position $i-3$ in the hinge; this residue may be termed the 'gatekeeper' of the selectivity pocket [5]. The contribution of heterocyclic CH interactions with protein carbonyl oxygen atoms (a CH· · ·O hydrogen bond) has also been studied [20]. Relative to a general set of aromatic ligand–protein interactions in crystal structures, these authors found contacts between kinase inhibitors and the hinge backbone carbonyl oxygens (with distances similar to those in ATP to the carbonyl of residue i, indicated with a dotted arrow) that provided additional favorable interactions. The computed quantum mechanical interaction energies and geometries for 19 heterocycle–water complexes elucidate the magnitude of the effect for different inhibitors and suggest ideas for future ligand design. In addition to these studies, within the phosphate site, quantum mechanical calculations have been used in two laboratories recently to investigate the catalytic mechanism, in particular, supporting the participation of a conserved aspartic acid residue in the phosphoryl transfer reaction in cAMP-dependent protein kinase (Protein Kinase A) [21,22].

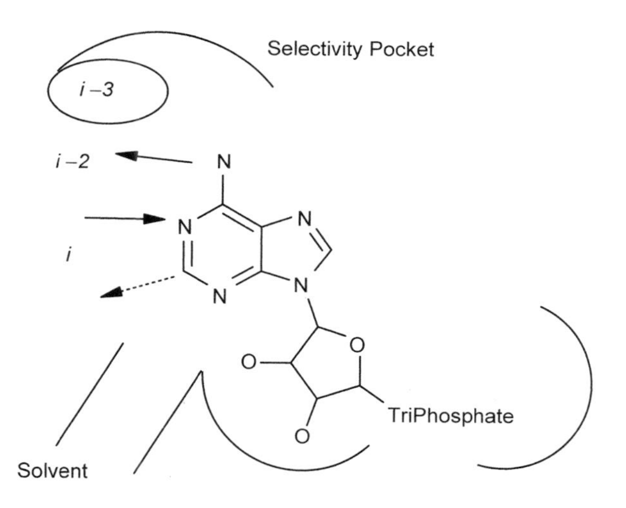

3. METHODOLOGY FOR KINASE TARGETS

3.1. Homology models

Models for kinase targets based on sequence homology to those with known structure have provided insights useful for the development of potent and selective inhibitors for this protein family, as will be seen in the applications below. Commonly used homology modeling approaches [23] have been employed successfully in many cases. In addition, a new algorithm has been described for kinases that focuses on the conformations of loops near the active site [24]. This approach, using dipeptide conformations from a database of proteins, was validated on an 'open' structure of Lck and a 'closed' structure of c-Src, and applied to modeling of the open conformations of c-Src and Jak-2 based on the Lck structure.

Two studies have considered if homology models are accurate enough for molecular docking and scoring algorithms to correctly identify kinase inhibitors from within a larger set of compounds [25,26]. Diller *et al.* created homology models for four growth factor receptor kinase domains and also p38 and Src [25]. More than 1000 compounds gathered from the literature were seeded within 32,000 random compounds from a corporate collection with similar calculated physical properties, and these compounds were docked into the models (and the corresponding crystal structure, if known). The authors discuss model building with apo, ATP-analog-, and inhibitor-bound templates where induced fit can reduce performance of the model in docking. For kinases excluding the fibroblast growth factor receptor (FGFR1), the docking protocol demonstrated an enrichment of ~5 compared to random compounds and an enrichment of ~2 with respect to other kinase inhibitors. With FGFR1, the choice of another template improved the docking outcome, illustrating the importance of the template-selection step in model building. In a more limited study, with CDK2 as the representative kinase, Oshiro *et al.* addressed the question of when might a homology model improve enrichment of active molecules relative to its template kinase structure (i.e., the structure upon which the model was based), and to the actual crystal structure [26]. It was suggested that when sequence homology is greater than 50%, it was worth building a model for a specific target – otherwise, docking to the template structure was just as effective.

3.2. Docking and scoring

A number of the challenges in applying molecular docking and scoring techniques (virtual screening) to kinase targets have recently been highlighted through enrichment analyses for docking methods applied retrospectively to a literature CDK2 indenopyrazole and in-house Src kinase data sets [27]. To improve docking results for kinases, constraints that incorporate the canonical hydrogen bonds within the ATP site or other key pharmacophore features can be employed. Programs such as FlexX-Pharm [28], Glide [29,30], DOCK [31], and GOLD [32] have this facility, to either focus the search phase of the docking calculation, or to filter the molecular poses to report only those which satisfy the constraint. Other commonly used docking methods have been further developed to obtain better performance when approaching the kinase family. For example, results from AutoDock 2.4 have been improved by the augmentation of hydrogen bond potentials (an alternative to explicit hydrogen-bond constraints) and the addition of terms reflecting solvation within the scoring functions [33]. Cavasotto and Abagyan have expanded the internal coordinate mechanics (ICM) docking approach to include receptor flexibility in an effort to account for conformational changes in kinase structures [34]. Occasionally, accounting for crystallographic water molecules or their displacement has been necessary to reproduce binding mode of a known kinase–inhibitor complex [35]; a method to carry out more detailed analysis of active site hydration has been reported for CDK2 and may be useful in such cases [36].

Tominaga and Jorgensen have developed a scoring function for kinases using Monte Carlo simulations with explicit solvent and an extended linear response analysis of the sampled configurations [37]. Fifteen descriptors were considered in a multiple linear regression approach following simulations of ~150 CDK2, p38, and Lck ligands and their protein–ligand complexes. Key factors in the derived models were the protein–ligand interaction energy, the change in hydrogen bonds to the ligand, the total change in ligand solvent accessible surface area, and L_{corr}, an indicator variable that is 1 if the compound is an Lck inhibitor. This final variable was initially included to accommodate variations in assay conditions (IC_{50} *vs.* K_d, etc.) but may also compensate for aryl CH···O interactions and protein–ligand interactions bridged by water molecules, which are not accounted for in the hydrogen-bonding term.

3.3. Selectivity

The promiscuity of many kinase inhibitor scaffolds has been exploited for the design of kinase 'gene family' libraries, expected to provide enhanced hit rates against targets in this family [38,39]. However, selectivity, while demonstrably achievable (at least for a subset of kinases profiled with a given inhibitor) and theoretically preferable to avoid off-target toxicity, has been difficult to predict to date. In the Monte Carlo simulation study described above, the average electrostatic and van der Waals components of the protein–ligand interaction energy were shown to emphasize residues in the kinase domain that should contribute to specificity amongst the kinases [37]. Earlier work by Rockey and Elcock demonstrated that starting from a given inhibitor and its crystal structure, average binding energies for correctly docked poses in homology models of related targets could generally predict which other kinases would be successfully inhibited by the compound [40].

Within the Janus kinase family, alpha shape molecular similarity analysis and multiple-copy simultaneous search techniques were used to map protein binding sites and explore functional groups that may contribute to selectivity of tyrphostin inhibitors such as AG-490 **1** between Jak2 and Tyk2 kinases [41]. A similar principal components (PC) analysis was employed to classify and explore selectivity more broadly across kinase subfamilies [42]. Probe interactions across binding-site grids from 26 aligned kinases were analyzed. The first principal component, associated with the adenine-binding pocket when visualized using hydrophobic probe contours, largely discriminates between the PKA structures and those of other kinases, while the second PC separates the CDK group from the other structures and appears to be associated with differences in the phosphate-recognition site and hydrophobic pockets not occupied by ATP. Interactions highlighted for CDKs through the protein-structural analysis were interpreted using a 3D-QSAR analysis of 2,6,9-substituted purines aligned within a CDK2 crystal structure.

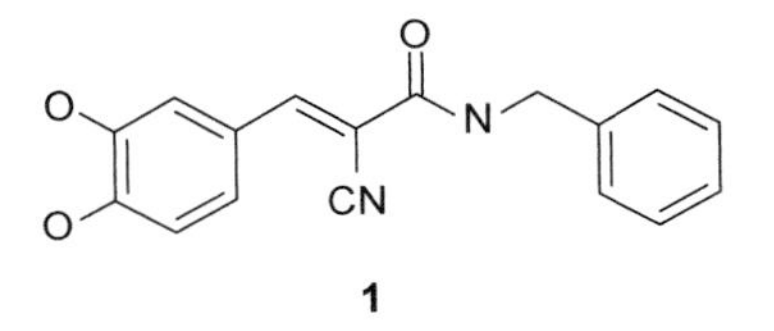

1

In another study, the selectivity of STI-571 (Gleevec) for Abl and platelet-derived growth factor receptor (PDGFR) α and β isoforms over the closely related Flt3 receptor tyrosine kinase was traced to the gatekeeper residue (threonine in Abl and PDGFRs, phenylalanine in Flt3) through homology modeling and docking. This finding was confirmed when the Flt3 F691T mutant was successfully inhibited by the compound [43]. An analogous mutation to threonine was not sufficient to render FGFR1 or the insulin receptor kinase sensitive to the compound, however. Analysis of the ϕ and ψ backbone angles within the Asp-Phe-Gly motif of the activation loops in crystal structures of seven tyrosine kinases (active and inactive forms) suggested the conformation of this loop is essential for binding STI-571. Interactions that stabilize this loop conformation were indicated, as were residues that might be targeted for selectivity by corresponding inhibitor design.

3.4. Structure-based hybridization

A staple of modern medicinal chemistry is the use of multiple crystallographic structures and compound SAR to suggest combinations of known inhibitors from different chemical classes to improve potency, selectivity, or novelty. One recent application demonstrated optimization of a series of imidazo[1,2-*a*]pyridine inhibitors of CDK2 (or CDK4), incorporating information from overlaid structures of bisanilinopyrimidine compounds bound in a similar mode [44]. Pierce *et al.* at Vertex have reported a simple but powerful automated approach to structure-based hybridization, called BREED, to create novel inhibitors from crystal structures of known ligands [45]. Target structures are aligned, resulting in inhibitor superposition. Acyclic bonds between pairs of ligands that match in terms of bond order, atom proximity at each end of the bond within 1 Å, and an angle between the two bonds of $\leq 15°$ indicate fragments that might beneficially be exchanged between the molecules. New molecules are created for each matching bond as the fragments are 'swapped' and are output to a file, retaining the coordinates of the proposed binding mode based on the original superposition. These molecules/poses can then be evaluated visually within the aligned active sites, or scored by any external approach. Interesting molecules generated from known CDK2 and p38 ligands were reported, and the statistics for the full experiment based on 10 kinase–inhibitor complexes (including proprietary structures for GSK3, Src, and Aurora2) were described as well. These 10 ligands resulted in 119 compound combinations, including known inhibitors (not contained in the input set) and novel kinase structures. The authors suggest that the pool of solved crystal structures could be increased further with the addition of docked structures, to provide additional new ideas through this approach.

4. APPLICATIONS ACROSS THE KINOME

4.1. CMGC group

To identify new inhibitors for the CDK family of kinases, docking of ~50,000 commercially available compounds, requiring a match of four site points from within the ATP site of the CDK2–staurosporine complex, has been performed [46]. In the first round of screening, 120 top-ranked compounds resulted in six hits with IC_{50} less than 20 μM. In a second round, following modification of van der Waals parameters, the best-scoring 28 compounds, along with 28 randomly selected compounds, were screened. In this case, the hit rate at greater than 30% inhibition (compound concentration, 30 μM) was 29% for the docked compounds and 7% for those selected at random. The crystal structures of leads based on scaffolds such as **2** and **3** revealed their binding modes, and the role of inhibitor binding on the stability of the Lys33-Asp145 salt-bridge interaction as a determinant of activation loop conformation was investigated.

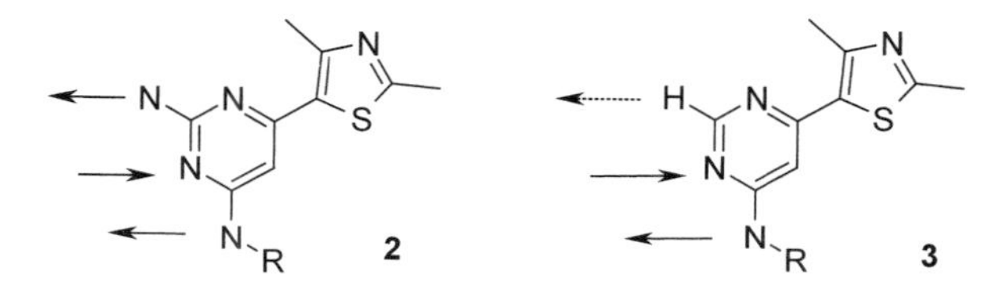

An alternative approach to select compounds for screening, iterative informative library design, was illustrated with a retrospective analysis of data generated during a CDK2 project [47]. The dataset of more than 17,500 compounds, composed of a general screening library as well as CDK2-focused synthetic libraries, was released with this report. An initial 3- and 4-point pharmacophore design space was derived from the general screening library. For subsequent rounds of 'synthesis', compounds were selected such that the sampling of the pharmacophore space was optimized for refinement with subsequent screening data. The model in each round was then refined to retain pharmacophores that best discriminate between active and inactive molecules. For the final round of selection, the 100 remaining compounds most similar to the actives identified in earlier rounds were selected. The cumulative enrichment for this protocol compared to a baseline diversity/similarity process was significant, and the informative approach identified 11 of 14 active scaffolds, compared to seven with the diversity/similarity technique. The same group previously reported a structure-based approach to informative design, using pharmacophores derived from the ATP sites of one or more CDK2 structures [48]; others [49] have utilized Catalyst [50] to create higher-order pharmacophore models of CDK activity and screened large combinatorial libraries.

The McCammon group has reported free energy calculations using a continuum solvent model to understand the binding of flavopiridol **4**-based inhibitors of CDK2 [51]. Probing the sensitivity of the binding energy to small changes in atomic partial charges highlighted regions of the ligand that should be optimized and the electrostatic characteristics that would be preferred, as well as regions of the kinase site that contributed most to ligand binding. Based on this analysis, the rationale for synthesis of a number of additional compounds was provided. Hybrid homology models for other CDK family members, where only active site residues were 'mutated' from the template structure, did not yield quantitative results with this approach. However, an analysis of the CDK2 active site in terms of the residue conservation of ~400 kinase sequences provided other opportunities for selectivity; Glu12, Phe82, and Lys89 were proposed to have a role in selectivity *vs.* CDK4.

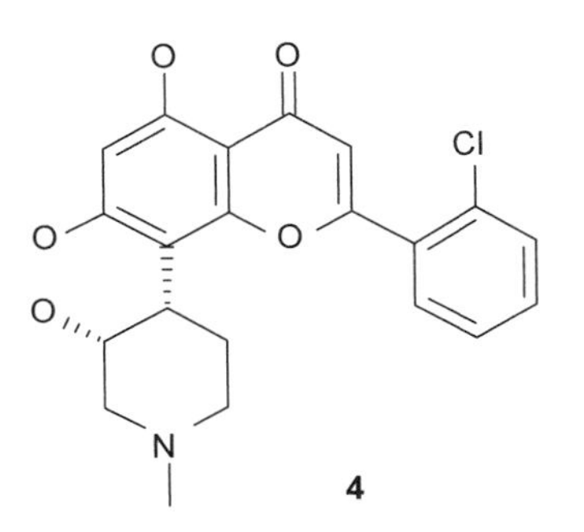

Fifty-two paullone analogs (alsterpaullone, **5**) formed the basis of a comparative molecular similarity analysis of GSK-3β inhibition and the related CDK1 and CDK5 serine/threonine kinases [52]. GSK-3β is of interest as an Alzheimer's disease target. Compounds were docked into a model of CDK1/cyclin B and minimized within the kinase ATP site to generate alignments for the 3D-QSAR model, which yielded cross-validated q^2 values of 0.699 (CDK1), 0.652 (CDK5), and 0.554 (GSK-3β). The models were further tested with 21 newly synthesized paullones. Optimization of the electrostatics within this series was suggested for future selective GSK-3β inhibitor design. With similar goals in mind, indirubin **6** analogs that inhibit GSK-3β selectively over CDKs have been pursued, although in this case 23 training set compounds were minimized within the active site and their binding affinities estimated using PrGen [53]. The binding affinities of 15 additional compounds were then predicted, with an RMS error of 1.74 kcal/mol. Another recent report of selective inhibitors based on scaffold **7** developed out of a structure-based design approach [54].

The MAP kinases, in particular the p38 kinases, have received much attention, resulting in compounds in the clinic for inflammatory diseases [10]. Most recently, however, two groups have explored the conformational preferences of inhibitors to rationalize structure–activity relationships in indole-5-carboxamides **8** [55] and aminobenzophenone **9** [56] series.

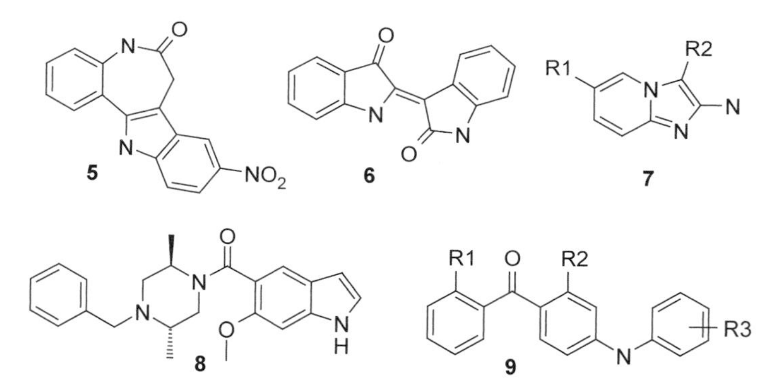

4.2. TK group

In the EGFR tyrosine kinase family, Muegge and Eneydy reported identification of a novel inhibitor of the ErbB2 (HER2) kinase [27]. A homology model for the catalytic domain was built from an insulin receptor kinase domain in the active conformation and was relaxed with molecular dynamics simulations that included explicit water molecules. DOCK was used for a preliminary shape-based screen of the ATP site, followed by more a rigorous docking and scoring protocol that was applied to only the top 20,000 compounds. After visual inspection of the top-ranked compounds at this stage, 141 were chosen for the biochemical assay. The most potent hit **10** had an IC_{50} of 1.5 μM and was selective *vs.* EGFR. Interestingly, it was shown to be an irreversible inhibitor, forming a covalent bond with Cys805.

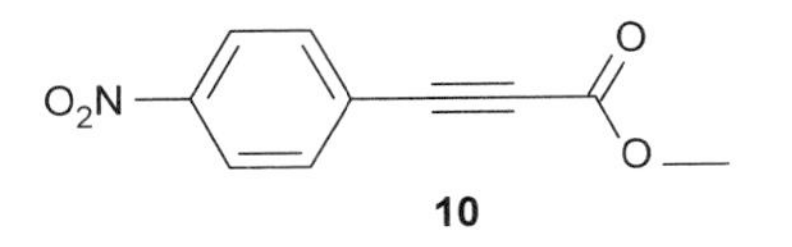

Three recent reports have applied 3D-QSAR methods to develop models for EGFR inhibition. In the Buolamwini laboratory, comparative molecular field and similarity analyses of 172 anilinoquinazoline and anilinoquinoline inhibitors were performed, using a variety of ligand alignment protocols [35]. Receptor-guided alignments with docking or dynamics based on the crystal structure of erlotinib (Tarceva) with EGFR were included. A similar study was carried out to explore the selectivity and potential binding modes of a series of tyrphostins for EGFR or HER2 and to guide future synthesis efforts [57]. In addition, a pseudoreceptor approach to 3D-QSAR has been applied to 27 pyrrolo[2,3-d]pyrimidine and 1H-pyrazolo[3,4-d]pyrimidine inhibitors of EGFR [58].

Within the VEGFR family, inhibitors in a number of chemical series have been discovered [59]. Homology models and docking have contributed to understanding their binding modes, leading to improved aqueous solubility for a set of KDR (VEGFR-2) inhibitors by directing solubilizing groups appropriately toward the solvent channel [60]. When applied to the Src family of nonreceptor tyrosine kinases, this approach led to improved physical properties for Lck inhibitors [61] and selectivity in targeting bone tissue with the addition of phosphonomethyl-phosphonic acid warheads to Src inhibitors [62]. In another study, conformational and molecular electrostatic potential analysis of a potent phthalazine-based inhibitor **11** of KDR ($IC_{50} = 37$ nM) and Flt-1 (VEGFR-1, $IC_{50} = 77$ nM) lead to a substructure search for anthranilic acids that would mimic the phthalazine ring with an intramolecular hydrogen bond. Compound **12** from that search was tested (3.7 μM *vs.* KDR), providing impetus for synthesis of **13**, which provided a similar activity profile to the original lead (20 nM *vs.* KDR, 180 nM *vs.* Flt-1) [63].

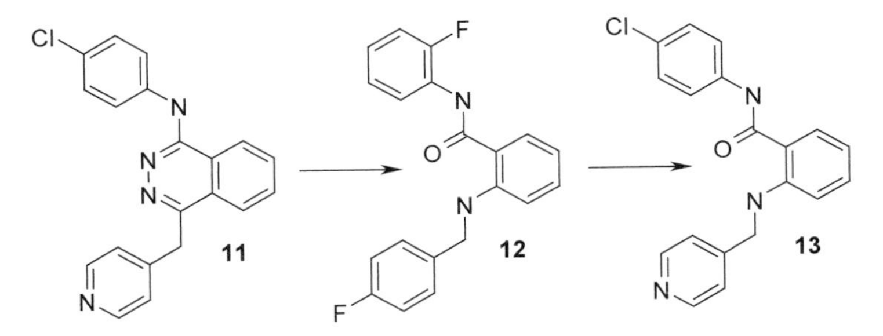

4.3. CAMK group

Associated by sequence similarity to the CAMK group, checkpoint kinase-1 (Chk-1) blocks the activity of CDKs and causes cells with damaged DNA to arrest at the G2/M cell cycle checkpoint. To focus a corporate collection into a subset to be screened against Chk-1 and thereby identify compounds that would allow damaged cells to proceed into mitosis and death, the collection was filtered for compounds that contained a simple kinase recognition pharmacophore (a hydrogen bond donor and acceptor pair separated by 1.35–2.40 Å) [64]. In a subsequent virtual screen with FlexX-Pharm, docked compounds were required to match their donor/acceptor pair to the backbone carbonyl of Glu85 and the backbone NH of Cys87. Compounds were then re-scored with a consensus score validated in a pilot study using CDK2. Of the 250 compounds inspected visually, 103 were screened against Chk-1, and 36 compounds were identified with IC_{50}s between 110 nM and 68 μM. Example hits came from bisanilinopyrimidine and quinazoline scaffolds.

4.4. AGC group

Inhibitors of the serine/threonine Rho kinase were developed from weak screening hits through docking to a homology model [65]. Five scaffold ideas derived from the docked poses of the pyridine-containing compounds lead to the docking of a small virtual library. Following synthesis and testing, compounds with IC_{50} values less than 1 μM were reported from four cores: pyridines (best $IC_{50} = 200$ nM), 1*H*-indazoles (20 nM), isoquinolines (100 nM), and phthalimides (900 nM). For the aurora2 kinase, a cancer target with homology to the cyclic-AMP-dependent kinases in this group, docking of similar scaffolds led to prioritization of quinazoline and isoquinoline scaffolds for further molecular design [66].

4.5. Other kinase groups

A pharmacophore and shape-based virtual screening approach was taken to identify inhibitors of a TKL family member, Type I TGFβ Receptor kinase, based on a literature triarylimidazole inhibitor [67]. The Catalyst query incorporated a requirement for two hydrogen bond acceptors and three aromatic centers, and filtered a commercial 200,000 compound database to 87 hits for testing against this serine/threonine kinase. Crystallography confirmed that compound **14** ($K_d = 5$ nM) bound as predicted by the query.

The compound was found to stabilize the inactive conformation of the target, as does the protein inhibitor of this pathway, FKBP12.

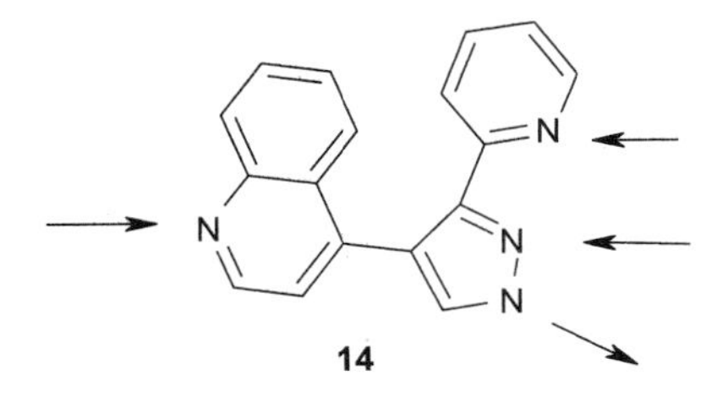

14

Finally, an inhibitor of CK2 (casein kinase 2) was discovered through a virtual screening protocol employing DOCK [68]. Compounds with an energy score better than −35 kcal/mol against the homology model were further required to make two hydrogen bonding interactions to either Val116 alone or in combination with Glu114. After consensus scoring, ~1600 compounds remained for visual inspection. Of 12 compounds tested, four hits were achieved in four chemical classes. The most active, **15** (IC_{50} = 80 nM), was found to be selective against a panel of 20 kinases. The authors suggested that this results from a salt-bridge interaction to Arg43, which is not conserved in other kinases.

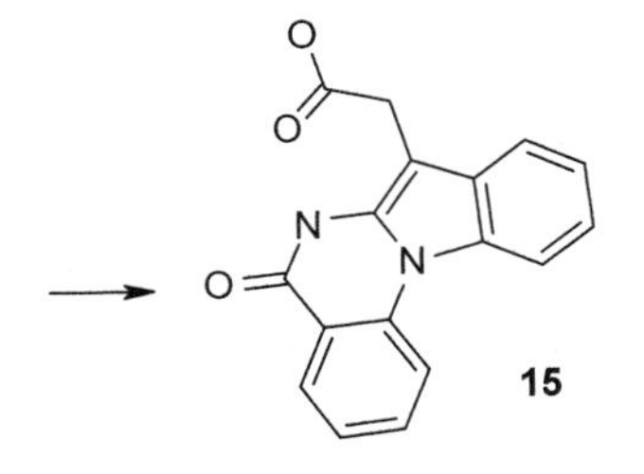

15

5. CONCLUSIONS

Clearly, computational chemistry has had a large impact on kinase inhibitor design and analysis of structure–activity relationships. Crystal structures have assisted many studies, but homology models have played an important role as well. Thus far, the promiscuity of the active site has been exploited to find hits, and certain subsites of the ATP site have provided the selectivity necessary to optimize leads. However, accurate predictions of kinase selectivity for individual compounds or chemical classes remain elusive. Additional efforts to improve modeling of the flexibility of the activation loop should enable a fuller understanding of which compounds will bind to active or inactive kinase conformations. Even more challenging, perhaps, will be the development of methods to identify and adapt inhibitors to mutations within a given kinase, for example, the Gleevec-resistant mutations reported in Bcr–Abl [69] and the recently reported activating mutations in EGFR [70,71] that may point the way toward patient targeted therapies.

REFERENCES

[1] G. Manning, D. B. Whyte, R. Martinez, T. Hunter and S. Sudarsanam, The protein kinase complement of the human genome, *Science*, 2002, **298**, 1912–1934.
[2] S. K. Hanks and T. Hunter, The eukaryotic protein kinase superfamily: kinase (catalytic) domain structure and classification, *FASEB J.*, 1995, **9**, 576–596.
[3] G. Scapin, Structural biology in drug design: selective protein kinase inhibitors, *Drug Discov. Today*, 2002, **7**, 601–611.
[4] M. Cherry and D. H. Williams, Recent kinase and kinase inhibitor X-ray structures: mechanisms of inhibition and selectivity insights, *Curr. Med. Chem.*, 2004, **11**, 663–673.
[5] M. E. M. Noble, J. A. Endicott and L. N. Johnson, Protein kinase inhibitors: insights into drug design from structure, *Science*, 2004, **303**, 1800–1805.
[6] M. Huse and J. Kuriyan, The conformational plasticity of protein kinases, *Cell*, 2002, **109**, 275–282.
[7] A. J. Bridges, Chemical inhibitors of protein kinases, *Chem. Rev.*, 2001, **101**, 2541–2571.
[8] P. Cohen, Protein kinases – the major drug targets of the twenty-first century?, *Nat. Rev. Drug Discov.*, 2002, **1**, 309–315.
[9] W. A. Metz, A perspective on protein kinase inhibitors, *Bioorg. Med. Chem. Lett.*, 2003, **13**, 2953.
[10] J. Regan, S. Breitfelder, P. Cirillo, T. Gilmore, A. G. Graham, E. Hickey, B. Klaus, J. Madwed, M. Moriak, N. Moss, C. Pargellis, S. Pav, A. Proto, A. Swinamer, L. Tong and C. Torcellini, Pyrazole urea-based inhibitors of p38 map kinase: from lead compound to clinical candidate, *J. Med. Chem.*, 2002, **45**, 2994–3008.
[11] H. Chen, A. Delaney, D. T. Dudley, C. A. Hasemann, Jr., P. Kuffa, P. C. McConnell, J. F. Ohren, A. G. Pavlovsky, H. Tecle, C. E. Whitehead, C. Yan and E. Zhang, Modified MEK1 and MEK2, crystal of a peptide:ligand:cofactor complex containing such modified MEK1 or MEK2, and methods or use thereof, European Patent Application 1321518A1.
[12] S. R. Hubbard, Protein tyrosine kinases: autoregulation and small-molecule inhibition, *Curr. Opin. Struct. Biol.*, 2002, **12**, 735–741.
[13] O. Hantschel, B. Nagar, S. Guettler, J. Kretzschmar, K. Dorey, J. Kuriyan and G. Superti-Furga, A myristoyl/phosphotyrosine switch regulates C-Abl, *Cell*, 2003, **112**, 845–857.
[14] B. Nagar, O. Hantschel, M. A. Young, K. Scheffzek, D. Veach, W. Bornmann, B. Clarkson, G. Superti-Furga and J. Kuriyan, Structural basis for the autoinhibition of C-Abl tyrosine kinase, *Cell*, 2003, **112**, 859–871.
[15] E. Myshkin and B. Wang, Chemometrical classification of ephrin ligands and Eph kinases use GRID/CPCA approach, *J. Chem. Inf. Comput. Sci.*, 2003, **43**, 1004–1010.
[16] D. M. Sigano, M. L. Peach, K. Nacro, Y. Choi, N. E. Lewin, M. C. Nicklaus, P. M. Blumberg and V. E. Marquez, Differential binding modes of diacylglycerol (DAG) and DAG lactones to protein kinase C (PK-C), *J. Med. Chem.*, 2003, **46**, 1571–1579.
[17] J. R. Woolfrey and G. S. Weston, The use of computational methods in the discovery and design of kinase inhibitors, *Curr. Pharm. Des.*, 2002, **8**, 1527–1545.
[18] L. Mao, Y. Wang, Y. Liu and X. Hu, Multiple intermolecular interaction modes of positively charged residues with adenine in ATP-binding proteins, *J. Am. Chem. Soc.*, 2003, **125**, 14216–14217.
[19] L. Mao, Y. Wang, Y. Liu and X. Hu, Molecular determinants for ATP-binding in proteins: a data mining and quantum chemical analysis, *J. Mol. Biol.*, 2004, **336**, 787–807.
[20] A. C. Pierce, K. L. Sandretto and G. W. Bemis, Kinase inhibitors and the case for CH···O hydrogen bonds in protein–ligand binding, *Proteins*, 2002, **49**, 567–576.

[21] M. Valiev, R. Kawai, J. A. Adams and J. H. Weare, The role of the putative catalytic base in the phosphoryl transfer reaction in a protein kinase: first-principles calculations, *J. Am. Chem. Soc.*, 2003, **125**, 9926–9927.
[22] N. Diaz and M. J. Field, Insights into the phosphoryl-transfer mechanism of CAMP-dependent protein kinase from quantum chemical calculations and molecular dynamics simulations, *J. Am. Chem. Soc.*, 2004, **126**, 529–542.
[23] S. Goldsmith-Fischman and B. Honig, Structural genomics: computational methods for structure analysis, *Protein Sci.*, 2003, **12**, 1813–1821.
[24] A. Rayan, E. Noy, D. Chema, A. Levitzki and A. Goldblum, Stochastic algorithm for kinase homology model construction, *Curr. Med. Chem.*, 2004, **11**, 674–692.
[25] D. J. Diller and R. Li, Kinases, homology models, and high throughput docking, *J. Med. Chem.*, 2003, **46**, 4638–4647.
[26] C. Oshiro, E. K. Bradley, J. E. Eksterowicz, E. Evensen, M. L. Lamb, J. K. Lanctot, S. Putta, R. Stanton and P. D. J. Grootenhuis, Performance of 3D-database molecular docking studies into homology models, *J. Med. Chem.*, 2004, **47**, 764–767.
[27] I. Muegge and I. J. Enyedy, Virtual screening for kinase targets, *Curr. Med. Chem.*, 2004, **11**, 693–707.
[28] S. A. Hindle, M. Rarey, C. Buning and T. Lengauer, Flexible docking under pharmacophore type constraints, *J. Comput.-Aided Mol. Des.*, 2002, **16**, 129–149.
[29] R. A. Friesner, J. L. Banks, R. B. Murphy, T. A. Halgren, J. J. Klicic, D. T. Mainz, M. P. Repasky, E. H. Knoll, M. Shelley, J. K. Perry, D. E. Shaw, P. Francis and P. S. Shenkin, Glide: a new approach for rapid, accurate docking and scoring. 1. Method and assessment of docking accuracy, *J. Med. Chem.*, 2004, **47**, 1739–1749.
[30] T. A. Halgren, R. B. Murphy, R. A. Friesner, H. S. Beard, L. L. Frye, W. T. Pollard and J. L. Banks, Glide: a new approach for rapid, accurate docking and scoring. 2. Enrichment factors in database screening, *J. Med. Chem.*, 2004, **47**, 1750–1759.
[31] T. J. A. Ewing and I. D. Kuntz, Critical evaluation of search algorithms for automated molecular docking and database screening, *J. Comput. Chem.*, 1997, **18**, 1175–1189.
[32] G. Jones, P. Willett, R. C. Glen, A. R. L. Leach and R. Taylor, Development and validation of a genetic algorithm for flexible docking, *J. Mol. Biol.*, 1997, **267**, 727–748.
[33] O. V. Buzko, A. C. Bishop and K. M. Shokat, Modified autodock for accurate docking of protein kinase inhibitors, *J. Comput.-Aided Mol. Des.*, 2002, **16**, 113–127.
[34] C. N. Cavasotto and R. A. Abagyan, Protein flexibility in ligand docking and virtual screening to protein kinases, *J. Mol. Biol.*, 2004, **337**, 209–225.
[35] H. Assefa, S. Kamath and J. K. Buolamwini, 3-D QSAR and docking studies on 4-anilinoquinazoline and 4-anilinoquinoline epidermal growth factor receptor (EGFR) tyrosine kinase inhibitors, *J. Comput.-Aided Mol. Des.*, 2003, **17**, 475–493.
[36] Z. Kříž, M. Otyepka, I. Bártová and J. Koča, Analysis of CDK2 active-site hydration: a method to design new inhibitors, *Proteins*, 2004, **55**, 258–274.
[37] Y. Tominaga and W. L. Jorgensen, General model for estimation of the inhibition of protein kinases using Monte Carlo simulations, *J. Med. Chem.*, 2004, **47**, 2534–2549.
[38] D. T. Mallanack, W. R. Pitt, E. Gancia, J. G. Montana, D. J. Livingstone, M. G. Ford and D. C. Whitley, Selecting screening candidates for kinase and G protein-coupled receptor targets using neural networks, *J. Chem. Inf. Comput. Sci.*, 2002, **42**, 1256–1262.
[39] L. Kissau, P. Stahl, R. Mazitschek, A. Giannis and H. Waldmann, Development of natural product-derived receptor tyrosine kinase inhibitors based on conservation of protein domain fold, *J. Med. Chem.*, 2003, **46**, 2917–2931.
[40] W. M. Rockey and A. H. Elcock, Progress toward virtual screening for drug side effects, *Proteins*, 2002, **48**, 664–671.

[41] K. Tøndel, E. Anderssen and F. Drabløs, Protein alpha shape similarity analysis (PASSA): a new method for mapping protein binding sites. Application in the design of a selective inhibitor of tyrosine kinase 2, *J. Comput.-Aided Mol. Des.*, 2002, **16**, 831–840.
[42] T. Naumann and H. Matter, Structural classification of protein kinases using 3d molecular interaction field analysis of their ligand binding sites: target family landscapes, *J. Med. Chem.*, 2002, **45**, 2366–2378.
[43] F. D. Böhmer, L. Karagyozov, A. Uecker, H. Serve, A. Botzki, S. Mahboobi and S. Dove, A single amino acid exchange inverts susceptibility of related receptor tyrosine kinases for the ATP site inhibitor STI-571, *J. Biol. Chem.*, 2003, **278**, 5148–5155.
[44] M. Anderson, J. F. Beattie, G. A. Breault, J. Breed, K. F. Byth, J. D. Culshaw, R. P. A. Ellston, S. Green, C. A. Minshull, R. A. Norman, R. A. Pauptit, J. Stanway, A. P. Thomas and P. J. Jewsbury, Imidazo[1,2-*a*]pyridines: a potent and selective class of cyclin-dependent kinase inhibitors identified through structure-based hybridisation, *Bioorg. Med. Chem. Lett.*, 2003, **13**, 3021–3026.
[45] A. C. Pierce, G. Rao and G. W. Bemis, BREED: generating novel inhibitors through hybridization of known ligands, *J. Med. Chem.*, 2004, **47**, 2768–2775.
[46] S. Y. Wu, I. McNae, G. Kontopidis, S. J. McClue, C. McInnes, K. J. Stewart, S. Wang, D. I. Zheleva, H. Marriage, D. P. Lane, P. Taylor, P. M. Fischer and M. D. Walkinshaw, Discovery of a novel family of CDK inhibitors with the program LIDEAUS: structural basis for ligand-induced disordering of the activation loop, *Structure*, 2003, **11**, 399–410.
[47] E. K. Bradley, J. L. Miller, E. Saiah and P. D. J. Grootenhuis, Informative library design as an efficient strategy to identify and optimize leads: application to cyclin-dependent kinase 2 antagonists, *J. Med. Chem.*, 2003, **46**, 4360–4364.
[48] J. E. Eksterowicz, E. Evensen, C. Lemmen, G. P. Brady, J. K. Lanctot, E. K. Bradley, E. Saiah, L. A. Robinson, P. D. J. Grootenhuis and J. M. Blaney, Coupling structure-based design with combinatorial chemistry: application of active site derived pharmacophores with informative library design, *J. Mol. Graph. Model.*, 2002, **20**, 469–477.
[49] E. A. Hecker, C. Duraiswami, T. A. Andrea and D. J. Diller, Use of catalyst pharmacophore models for screening of large combinatorial libraries, *J. Chem. Inf. Comput. Sci.*, 2002, **42**, 1204–1211.
[50] Catalyst, Accelrys, Inc., San Diego, CA.
[51] P. A. Sims, C. F. Wong and J. A. McCammon, A computational model of binding thermodynamics: the design of cyclin-dependent kinase 2 inhibitors, *J. Med. Chem.*, 2003, **46**, 3314–3325.
[52] C. Kunick, K. Lauenroth, K. Wieking, X. Xie, C. Schultz, R. Gussio, D. Zaharevitz, M. Leost, L. Meijer, A. Weber, F. S. Jørgensen and T. Lemcke, Evaluation and comparison of 3D-QSAR CoMSIA models for CDK1, CDK5, and GSK-3 inhibition by paullones, *J. Med. Chem.*, 2004, **47**, 22–36.
[53] P. Polychronopoulos, P. Magiatis, A.-L. Skaltsounis, V. Myrianthopoulos, E. Mikros, A. Tarricone, A. Musacchio, S. M. Roe, L. Pearl, M. Leost, P. Greengard and L. Meijer, Structural basis for the synthesis of indirubins as potent and selective inhibitors of glycogen synthase kinase-3 and cyclin-dependent kinases, *J. Med. Chem.*, 2004, **47**, 935–946.
[54] C. Hamdouchi, H. Keyser, E. Collins, C. Jaramillo, J. E. de Diego, C. D. Spencer, J. A. Dempsey, B. D. Anderson, T. Leggett, N. B. Stamm, R. M. Schultz, S. A. Watkins, K. Cocke, S. Lemke, T. F. Burke, R. P. Beckmann, J. T. Dixon, T. M. Gurganus, N. B. Rankl, K. A. Houck, F. Zhang, M. Vieth, J. Espinosa, D. E. Timm, R. M. Campbell, B. K. R. Patel and H. B. Brooks, The discovery of a new structural class of cyclin-dependent kinase inhibitors, aminoimidazo[1,2-*a*]pyridines, *Mol. Cancer Ther.*, 2004, **3**, 1–9.

[55] B. J. Mavunkel, S. Chakravarty, J. J. Perumattam, G. R. Luedtke, X. Liang, D. Lim, Y.-j. Xu, M. Laney, D. Y. Liu, G. F. Schreiner, J. A. Lewicki and S. Dugar, Indole-based heterocyclic inhibitors of p38α MAP kinase: designing a conformationally restricted analogue, *Bioorg. Med. Chem. Lett.*, 2003, **13**, 3087–3090.
[56] E. R. Ottosen, M. D. Sørensen, F. Björkling, T. Skak-Nielsen, M. S. Fjording, H. Aaes and L. Binderup, Synthesis and structure–activity relationship of aminobenzophenones. A novel class of p38 map kinase inhibitors with high antiinflammatory activity, *J. Med. Chem.*, 2003, **46**, 5651–5662.
[57] S. Kamath and J. K. Buolamwini, Receptor-guided alignment-based comparative 3D-QSAR studies of benzylidene malonitrile tyrphostins as EGFR and HER-2 kinase inhibitors, *J. Med. Chem.*, 2003, **46**, 4657–4668.
[58] T. Peng, J. Pei and J. Zhou, 3D-QSAR and receptor modeling of tyrosine kinase inhibitors with flexible atom receptor model (FLARM), *J. Chem. Inf. Comput. Sci.*, 2003, **43**, 298–303.
[59] T. L. Underiner, B. Ruggeri and D. E. Gingrich, Development of vascular endothelial growth factor receptor (VEGFR) kinase inhibitors as anti-angiogenic agents in cancer therapy, *Curr. Med. Chem.*, 2004, **11**, 731–745.
[60] M. E. Fraley, W. F. Hoffman, K. L. Arrington, R. W. Hungate, G. D. Hartman, R. C. McFall, K. E. Coll, K. Rickert, K. A. Thomas and G. B. McGaughey, Property-based design of KDR kinase inhibitors, *Curr. Med. Chem.*, 2004, **11**, 709–719.
[61] D. F. C. Moffat, R. A. Allen, S. E. Rapecki, P. D. Davis, J. O'Connell, M. C. Hutchings, M. A. King, B. A. Boyce and M. J. Perry, 4-Thiophenoxy-*N*-(3,4,5-trialkoxyphenyl) pyrimidine-2-amines as potent and selective inhibitors of the T-cell tyrosine kinase p56lck, *Curr. Med. Chem.*, 2004, **11**, 747–753.
[62] R. Sundaramoorthi, W. C. Shakespeare, T. P. Keenan, C. A. Metcalf III, Y. Wang, U. Mani, M. Taylor, S. Liu, R. S. Bohacek, S. S. Narula, D. C. Dalgarno, M. R. van Schravendijk, S. M. Violette, S. Liou, S. Adams, M. K. Ram, J. A. Keats, M. Weigele and T. K. Sawyer, Bone-targeted Src kinase inhibitors: novel pyrrolo- and pyrazolopyrimidine analogues, *Bioorg. Med. Chem. Lett.*, 2003, **13**, 3063–3066.
[63] P. Furet, G. Bold, F. Hofmann, P. Manley, T. Meyer and K.-H. Altmann, Identification of a new chemical class of potent angiogenesis inhibitors based on conformational considerations and database searching, *Bioorg. Med. Chem. Lett.*, 2003, **13**, 2967–2971.
[64] P. D. Lyne, P. W. Kenny, D. A. Cosgrove, C. Deng, S. Zabludoff, J. J. Wendoloski and S. Ashwell, Identification of compounds with nanomolar binding affinity for checkpoint kinase-1 using knowledge-based virtual screening, *J. Med. Chem.*, 2004, **47**, 1962–1968.
[65] A. Takami, M. Iwakubo, Y. Okada, T. Kawata, H. Odai, N. Takahashi, K. Shindo, K. Kimura, Y. Tagami and M. Miyake, Design and synthesis of Rho kinase inhibitors (I), *Bioorg. Med. Chem.*, 2004, **12**, 2115–2137.
[66] H. Vankayalapati, D. J. Bearss, J. W. Saldanha, R. M. Muñoz, S. Rojanala, D. D. von Hoff and D. Mahadevan, Targeting aurora2 kinase in oncogenesis: a structural bioinformatics approach to target validation and rational drug design, *Mol. Cancer Ther.*, 2003, **2**, 283–294.
[67] J. Singh, C. E. Chuaqui, P. A. Boriack-Sjodin, W.-C. Lee, T. Pontz, M. J. Corbley, H.-K. Cheung, R. M. Arduini, J. N. Mead, M. N. Newman, J. L. Papadatos, S. Bowes, S. Josiah and L. E. Ling, Successful shape-based virtual screening: the discovery of a potent inhibitor of the type I TGFβ receptor kinase (TβRI), *Bioorg. Med. Chem. Lett.*, 2003, **13**, 4355–4359.
[68] E. Vangrevelinghe, K. Zimmermann, J. Schoepfer, R. Portmann, D. Fabbro and P. Furet, Discovery of a potent and selective protein kinase CK2 inhibitor by high-throughput docking, *J. Med. Chem.*, 2003, **46**, 2656–2662.
[69] M. Azam, R. R. Latek and G. Q. Daley, Mechanisms of autoinhibition and STI-571/imatinib resistance revealed by mutagenesis of *BCR–ABL*, *Cell*, 2003, **112**, 831–843.

[70] T. J. Lynch, D. W. Bell, R. Sordella, S. Gurubhagavatula, R. A. Okimoto, B. W. Brannigan, P. L. Harris, S. M. Haserlat, J. G. Supko, F. G. Haluska, D. N. Louis, D. C. Christiani, J. Settleman and D. A. Haber, Activating mutations in the epidermal growth factor receptor underlying responsiveness of non-small-cell lung cancer to gefitinib, *N. Engl. J. Med.*, 2004, **350**, 2129–2139.
[71] J. G. Paez, P. A. Jänne, J. C. Lee, S. Tracy, H. Greulich, S. Gabriel, P. Herman, F. J. Kaye, N. Lindeman, T. J. Boggon, K. Naoki, H. Sasaki, Y. Fujii, M. J. Eck, W. R. Sellers, B. E. Johnson and M. Meyerson, EGFR mutations in lung cancer: correlation with clinical response to gefitinib therapy, *Science*, 2004, **304**, 1497–1500.

Section 5
Chemical Education

Section Editor: Theresa Zielinski
Department of Chemistry
Medical Technology, and Physics
Edison Science Hall, Room E245
Monmouth University
400 Cedar Avenue
West Long Branch
NJ 07764-1898
USA

CHAPTER 14

Status of Research-Based Experiences for First- and Second-Year Undergraduate Students

Jeffrey D. Evanseck[1] and Steven M. Firestine[2]

[1] *Department of Chemistry and Biochemistry, Center for Computational Sciences, Duquesne University, 600 Forbes Avenue, Pittsburgh, PA 15282, USA*
[2] *Division of Pharmaceutical Sciences, Duquesne University, 600 Forbes Avenue, Pittsburgh, PA 15282, USA*

Contents

1. Introduction 205
2. Current status 207
3. Council on undergraduate research 208
4. National science foundation 209
5. Undergraduate research program 209
6. Conclusions 212
Acknowledgements 213
References 213

1. INTRODUCTION

It is well known and accepted that the United States provides a world-premiere postsecondary educational system. According to the 2000 US census, 66% of Americans enter college after high school. In contrast, the United Kingdom has 43%, France 33%, and Switzerland 15% of their population entering college (based on 1998 numbers). Despite the obvious successes and scientific advances since World War II, it is clear that a number of modern day issues have begun to place considerable stress on the US educational system's ability to provide the superb educational infrastructure that has flourished in the past.

Two statistics indicate a growing problem for the future. First, there has been a decrease in the percentage of domestic students entering college as science majors since 1970. To compensate for the decline in domestic students applying to graduate school, the American system has relied heavily upon quality international graduate students to support the research and educational mission of the university. However, this option is quickly becoming unavailable as foreign universities have become more advanced and competitive, and student visas have become more restricted due to recent political and world events.

ANNUAL REPORTS IN COMPUTATIONAL CHEMISTRY, VOLUME 1
ISSN: 1574-1400 DOI 10.1016/S1574-1400(05)01014-5

Over the last few years, the trend has become clear to institutions outside the window of the top 50 elite universities; fewer quality international students are willing to travel abroad or simply cannot obtain visas to enter the United States. A recent report has indicated that 64% of the 320 institutions surveyed have experienced declines in international student applications and visa requests [1]. The major reason for the decline was the increase in visa denials. The dwindling number of quality graduate students will become a major issue in the near future for all universities. One method to address this problem is to focus efforts towards recruiting domestic undergraduates into science.

Secondly, only 3% of the institutions that grant bachelor's degrees per year are classified as research universities [2]. This statistic underscores a dramatic departure from the 'golden years' in the American higher educational system. As a greater percentage of the population enters college, a great number of the students will experience 2-year colleges, or schools not equipped with the infrastructure to carry out research [3]. This is of concern since an active research experience is considered one of the most effective ways to attract talented undergraduates and retain them in careers in science and engineering, including careers in teaching [2]. Consequently, the inspiration and motivation for studying chemistry or biochemistry will not be experienced by a vast majority of college students. Unless we find a way to provide research opportunities for college-bound students early in their academic experience, there will be fewer students, domestic or international, available for research and teaching in chemistry.

The Boyer Commission on educating undergraduates in the research university was created in 1995 under the Carnegie Foundation for the advancement of teaching [2,4]. The commission released its report on undergraduate education in 1998, which described 10 ways to reinvent the undergraduate experience to strengthen science education in the United States. One focus of the commission is research- and inquiry-based learning for freshmen. The belief is that undergraduate education at research universities should be fully integrated with the research faculty. As the report states "...thousands of students graduate without ever seeing the world-famous professors or tasting genuine research." This sentiment is echoed in the National Academies report entitled 'Improving Undergraduate Instruction in Science, Technology, Engineering, and Mathematics: Report of a Workshop' which indicated that "all undergraduates [need to] have learning experiences that motivate them to persist in their studies and consider careers in these fields" [5]. As a result of these reports, more efforts have been placed into the development of undergraduate research experiences.

The two ideas described within this annual report touch upon several of the Boyer strategic points and are reflected by a number of organizations and agencies, such as the Council on Undergraduate Research (CUR) and the National Science Foundation (NSF). First, how can we provide active and engaging modes of learning involving research opportunities of significant pedagogical value? Specifically, how do we approach a larger number of students earlier in their undergraduate careers in a manner that will attract and retain them as science majors? Secondly, how can we successfully influence a broad spectrum of students at institutions with no history of research and the increasingly important 2-year colleges? It is clear that academic 'business as usual' is no longer sufficient to educate the modern student body and instill a desire for graduate education.

2. CURRENT STATUS

From 1977 to 1999, the total number of college students in America increased 250%, from about 6 million in 1970 to about 15 million in 1999 [6]. However, the US Census Bureau indicates that between 1985 and 1998, the number of graduate students in the physical sciences stayed approximately constant at 29,000 [7]. The number of chemistry majors has not kept pace with the large influx of college students, which indicates that as a percentage of college students, chemistry and biochemistry is losing significant ground. The apparent loss of students from the chemistry profession is less troublesome than the missed opportunity to attract the brightest and most talented students that will ultimately have the greatest impact on chemical and biochemical advances in the future.

The higher number of students entering college has been attributed to the desire of students to earn higher wage jobs. However, students have primarily focused on technology or health-related fields. For example, between 1970 and 1998 computer and information sciences, agriculture and health-related fields saw an average of a 250% increase [8]. During this same time period, Snyder observed a loss of students in the physical sciences and Hudson noted "...the shift in the past three decades... away from the humanities and hard [physical] sciences toward business, technical, and health fields" [9]. Studies conducted by Seymour and Hewitt on the reason why students have moved away from the physical sciences has indicated that over half of the students who enter college intending to pursue majors in the natural sciences change majors within 2 years of taking their first college science or mathematics classes [10]. Seymour and Hewitt reported that students were dissatisfied with what they perceived as poor teaching and other negative experiences in 'weed-out' science courses and the reliance upon memorization rather than problem and research-based testing [10].

Several investigators have recently reported on the importance of undergraduate research [11–18]. The belief is that undergraduate education should be fully integrated with research faculty. However, the proper method and timing of introducing undergraduates to meaningful research experiences have become a matter of debate and controversy. Too early of an introduction could lead to student confusion and alienation, yet a delay to research exposure could fail to pique student interest during critical decision times concerning career choices. Due to a number of issues, fundamental research skills are rarely taught to undergraduates by using a structured classroom setting. Instead, traditional research opportunities, primarily reserved for upper-level juniors and seniors, have been carried out under the guidance of a faculty member; well after a commitment to study chemistry has been made.

Educators have recently reported their attempts to create active learning environments in lower level university classes by invoking research experiences through less regulated projects and laboratory assignments [11–38]. For example, pre-defined industrial-based mini-projects have been used to better prepare second-year students for more sophisticated third-year research [29]. Similarly, second-year students have gained valuable experience and skills by optimizing problematic analytical procedures commonly used in pharmaceutical and industrial production [32]. Sophomore organic and spectroscopy laboratories have been redesigned to provide more project-based and independent research experiences [33,34]. Representative of most upper-level classes, Whipple-VanPatter [27] required second-year students to design, conduct, analyze, and present a final 5-week research project based upon the analytical techniques taught in class. Finally, Weisshaar [37] exposed first-year students to an 8-week independent laboratory project based upon a genuine research project. One current view of the more successful pedagogies is in the development of 'research-supportive curriculum' to create time in the already tight curriculum schedules for the students and alleviate increasing pressures on faculty members [11].

3. COUNCIL ON UNDERGRADUATE RESEARCH

The CUR and its affiliated colleges, universities, and individuals share a focus on providing undergraduate research opportunities for faculty and students at predominantly undergraduate institutions [39]. Many of the recent publications involving undergraduate research are associated with CUR members [20–28]. CUR believes that faculty members enhance their teaching and contribution to society by remaining active in research and by involving undergraduates in research. CUR's publications and outreach activities are designed to share successful models and strategies for establishing and institutionalizing Undergraduate Research Programs (URPs). Serving faculty and administrators at primarily undergraduate institutions, CUR has 3000 members representing over 870 institutions in eight academic divisions. CUR's wide range of services has had a positive impact on URPs at their individual institutions and increases their connections with funding agencies and colleagues.

4. NATIONAL SCIENCE FOUNDATION

The NSF has been proactive in promoting research experiences for undergraduates. Two of the NSF's principal efforts have been through the Research Experiences for Undergraduates (REU) and Undergraduate Research Center (URC) programs [40,41]. There are currently 67 REU sites across the nation. The REU program aims to provide appropriate and valuable educational experiences for undergraduate students through research participation. REU projects involve students in meaningful ways in ongoing research programs or in research projects specially designed for the purpose. REU projects feature high-quality interaction of students with faculty and/or other research mentors and access to appropriate facilities and professional development opportunities. REU opportunities are an excellent way to reach broadly into the student talent pool of our nation. The NSF provides a complete list of REU sites and contact information for undergraduate research opportunities [42]. The URC program seeks new models and partnerships with the potential to expand the reach of undergraduate research to include first- and second-year college students; and to enhance the research capacity, infrastructure, and culture of participating institutions, thereby strengthening the nation's research enterprise.

5. UNDERGRADUATE RESEARCH PROGRAM

Several excellent university programs have been established to promote the involvement of undergraduates into the research laboratory. CIRRUS is a chemistry Internet resource that identifies opportunities for undergraduate research [43]. Duquesne University is one such example where a tradition of undergraduate research has been a priority for the faculty and school [44]. The Bayer School of Natural and Environmental Sciences at Duquesne University officially established its URP in 1998. However, the School has been providing REU since 1951. Initially the research experiences were limited to the biological sciences, chemistry and biochemistry, and physics. Over the last 4 years, the URP has expanded to provide research opportunities in the pharmaceutical, health, and forensic sciences.

To assist in a successful and meaningful 10-week experience in research, the URP coordinates faculty and staff in the training of students in using the library for effective literature searches [45], reading and writing scientific papers [46], presenting their research findings, and treatment of ethical issues in science [47,48]. In addition, the URP administers web-based advertising, online application processing, program assessment, scheduling of weekly scientific seminars, community outreach activities, group outings, and coordination of combined ethics meetings with local REU sites in the greater Pittsburgh area.

Chemistry and biochemistry faculty members have extensive experience in mentoring undergraduate researchers (Table 1). Specifically, over the last 11 years, the faculty members have provided mentoring for 199 undergraduates, resulting in 229 professional presentations and 63 peer-reviewed publications by undergraduates. The faculty members have provided research experiences for over 15 students per summer, limited only by available scholarships and stipends. Collaborative efforts have been on the rise over the last 4 years, where chemistry faculty members have worked with pharmaceutical sciences and biological faculty. In 2004 URP, six students worked on interdisciplinary projects with more than one research mentor. Currently the students are supported through a variety of sources derived from the National Science Foundation, National Institutes of Health, Department of Defense, Department of Education, and Duquesne University.

Acceptance into the URP is competitive. Each year over 160 completed applications are considered for approximately 35 positions. The URP has a flexible recruitment plan that has three principal target areas involving students from regional colleges without research capacity, commonly overlooked first- and second-year undergraduates, and underrepresented students. Sixty percent of the students come from colleges or universities that do not have the resources to carry out research. Not only does the university consider this a form of service learning, but also the faculty members find it to be an effective tool for graduate student recruitment. Over the last 3 years, one of every eight entering graduate students in chemistry has had a URP experience; 90% have benefited from a previous research experience. On the other hand, there is immense pressure on faculty members to increase or maintain the highest level of scholarship. As such, some faculty members struggle with the idea of training a student only to lose them after 10 weeks.

Analysis of the student demographics reveals that 17% of the successful candidates are first-year students. The URP committee intentionally keeps the percentage of first year undergraduates at a high rate, since they know that the first year is a critical time in the student's career choices. The faculty members at Duquesne University have come up with a number of innovative ideas to provide more research exposure for freshmen. Our philosophy is to give freshmen the necessary skills and motivation to carry out research, and not let them 'sink or swim'. The development of our new educational model of introducing freshmen into research revolves about honors freshman chemistry. Since its inception in 1998, the class size is approximately 20 students per year. The students are accepted into the class based upon SAT scores or from TOLEDO examination performance. We have also honored requests from students who display the necessary motivation and desire to be challenged.

The students meet once a week in the first semester to be trained in the use of the library for literature searching, safety procedures and guidelines, reading and writing of scientific papers, and application of the scientific method. A university

Table 1. Bayer School's Undergraduate Research Program (URP) statistics over the last 7 years

1998	1999	2000	2001	2002	2003	2004	Average	URP Statistics
8	8	8	8	7	9	8	8	Funded by Bayer School
2	8	9	6	4	0	0	6	Funded by NSF REU (biology)
0	0	0	0	0	0	10	10	Funded by NSF REU (chemistry)
14	9	12	3	6	14	9	10	Funded by departments
0	1	0	4	4	4	3	3	Funded by other schools
6	10	4	7	10	8	6	7	Funded by external sources
30	**36**	**33**	**28**	**31**	**35**	**36**	**33±3**	**Total participants**
6	7	6	4	5	7	2	5	Freshmen
10	12	14	10	15	14	13	13	Sophomores
12	16	12	13	9	13	19	13	Juniors
2	1	1	1	2	1	2	1	Seniors
13	21	16	11	14	15	9	14	Worked in biological sciences
16	14	16	11	13	15	20	15	Worked in chemistry/biochemistry
0	0	0	1	0	0	0	1	Worked in health sciences
0	0	0	2	1	1	2	2	Worked in physics
0	0	3	3	3	4	5	4	Worked in pharmacy

Entries are the numbers of students.

librarian trains the students how to use the library's electronic and printed resources. Faculty members assist by providing examples on how to search the literature exhaustively. The university's safety officer trains the students to use fire extinguishers and gas cylinders, treat chemical spills, store compounds, and dispose of hazardous waste. The honors chemistry faculty members share the responsibility of teaching the students how to read and write scientific papers. The process starts with the reading of short papers or communications and discussion as a group under the direction of a faculty member. The students are also given specific writing assignments on mock research projects. The students are then exposed to the research faculty by short research seminars. The semester-long experience culminates in the student selection of their research mentor. A research project is decided upon and becomes the focus of the following semester. The faculty members feel that this program results in a well-trained student who has an appreciation of how science is conducted and whose success within the sciences are increased.

Another goal of the URP is to improve upon the number of underrepresented students (currently 14%) and maintain the number of women participants (currently 60%) in the program. We have made significant advances in the last 3 years. A strong connection with Florida Memorial College (FMC), a historically black college with 2242 students and 38 undergraduate degree programs, has been established to create research opportunities for under-represented students. Prof. Ayivi Huisso from FMC has become a crucial member of the URP team in terms of mentoring and recruitment. Over the last 3 years, Duquesne has awarded 12 research fellowships to FMC and Mississippi State College for Women. Five of these students have won research awards by presenting their research at subsequent research symposia.

6. CONCLUSIONS

The United States continues to provide a world-premiere postsecondary educational system. However, a number of modern day issues have begun to deteriorate the superb educational infrastructure that has flourished over the last century. Difficult times lie ahead for a majority of research universities. The problems stem from decades of neglect in the recruitment of domestic students into the sciences and from the lack of training of the entering college population by the top research universities. Agencies and organizations such as the National Science Foundation and the Council on Undergraduate Research have been active in reversing trends that are detriment to the higher education system. Novel ideas and programs are starting to make a difference. Programs that promote the involvement of undergraduate students in research and incorporate these ideas not just in the laboratory, but also in the classroom should provide one mechanism to help recruit talented students into the physical sciences.

ACKNOWLEDGEMENTS

We thank Carmen Rios, Director of Student Services, for providing information regarding undergraduate research participation from Duquesne University. We also thank the National Science Foundation for grants CHE-0321147 and CHE-0354052.

REFERENCES

[1] (a) Survey of Applications by Prospective International Students to US Higher Education Institutions. See: http://www.nafsa.org/content/PublicPolicy/FortheMedia/appssurveyresults.pdf; (b) Survey for Foreign Students and Scholar Enrollment, Applications, and VISA Trends for Fall 2003. See: http://www.nafsa.org/content/PublicPolicy/FortheMedia/nafsasurveyhighlights1103.pdf.
[2] Boyer Commission on Educating Undergraduates in the Research University. See http://notes.cc.sunysb.edu/Pres/boyer.nsf.
[3] J. Tsapogas, The Role of Community Colleges in the Education of Recent Science and Engineering Graduates, InfoBrief, National Science Foundation, NSF 04-315. See http://www.nsf.gov/sbe/srs/.
[4] J. W. Moore, The Boyer report, *J. Chem. Educ.*, 1998, **75**, 935.
[5] R. A. McCray, R. L. DeHaan and J. A. Schuck, *Improving Undergraduate Instruction in Science, Technology, Engineering, and Mathematics: Report of a Workshop*, National Research Council, National Academy Press, Washington, DC, 2003.
[6] A. Jamieson, A. Curry and G. Martinez, School Enrollment in the United States – Social and Economic Characteristics of Students. See: http://www.census.gov/prod/2001pubs/p20-533.pdf.
[7] Science and Technology, Section 20, US Census Bureau, pp. 601–616. See: http://www.census.gov/prod/2001pubs/statab/sec20.pdf.
[8] L. Hudson, Demographic and attainment trends in postsecondary education. In *The Knowledge, Economy and Postsecondary Education* (eds A. P. Graham and N. Stacey), Committee on the Impact of the Changing Economy on the Education System, National Research Council, National Academy Press, Washington, DC, 2002, pp. 13–57.
[9] T. D. Snyder and C. M. Hoffman, *Digest of Education Statistics (NCES 2000-031)*, US Department of Education, National Center for Educational Statistics, Washington, DC, 2000, pp. 295–296.
[10] E. Seymour and N. M. Hewitt, *Talking about Leaving: Why Undergraduates Leave the Sciences*, Westview Press, Boulder, CO, 1997.
[11] C. M. Henry, Undergraduates in the research lab, *Chem. Eng. News*, 2004, **May 3**, 33.
[12] K. Bartlett, Towards a true community of scholars: undergraduate research in the modern university, *J. Mol. Struct. (THEOCHEM)*, 2003, **666/667**, 707–711.
[13] W. B. Wood and J. M. Gentile, Teaching in a research context, *Science*, 2003, **302**, 1510.
[14] M. P. Doyle, Academic excellence – the role of research, *J. Chem. Educ.*, 2002, **79**, 1038–1044.
[15] R. W. Murray, Undergraduate research secrets rediscovered, *J. Anal Chem.*, 2001, 237A.
[16] A. R. Hutchison and D. A. Atwood, Research with first- and second-year undergraduates: a new model for undergraduate inquiry at research universities, *J. Chem. Educ.*, 2002, **79**, 125–126.
[17] T. J. Wenzel, Undergraduate research: a capstone learning experience, *J. Anal. Chem.*, 2000, 547A–549A.

[18] N. C. Craig, The joys and trials of doing research with undergraduate, *J. Chem. Educ.*, 1999, **76**, 595–597.
[19] L. L. Kirk and L. F. Hanne, An alternate approach to teaching undergraduate research, *J. Chem. Educ.*, 1991, **68**, 839–841.
[20] M. D. Schuh and K. K. Karukstis, Getting started in research with undergraduates, *J. Chem. Educ.*, 2004, **81**, 322.
[21] K. K. Karukstis and T. J. Wenzel, Enhancing research in the chemical sciences at predominately undergraduate institutions, *J. Chem. Educ.*, 2004, **81**, 468–469.
[22] P. Dea, CUR 2000: the many facets of undergraduate research, *J. Chem. Educ.*, 2000, **77**, 432.
[23] T. E. Elgren, Undergraduate research fellowships undergo change, *J. Chem. Educ.*, 2000, **77**, 1113.
[24] K. K. Karukstit, A report on CUR: the many facets of undergraduate research, *J. Chem. Educ.*, 2000, **77**, 1388–1389.
[25] J. Slezak, Student inspired undergraduate research, *J. Chem. Educ.*, 1999, **76**, 1054–1055.
[26] J. A. Halstead, Creating undergraduate research opportunities in changing communities: CUR seventh national conference, *J. Chem. Educ.*, 1998, **75**, 407–408.
[27] G. Whipple-VanPatter, Research not foreign to two-year colleges, *J. Chem. Educ.*, 1998, **75**, 1210.
[28] J. A. Halstead, What is undergraduate research, *J. Chem. Educ.*, 1997, **74**, 1390–1391.
[29] J. G. Dunn and D. N. Phillips, Introducing second-year chemistry students to research work through mini-projects, *J. Chem. Educ.*, 1998, **75**, 866–869.
[30] J. F. Belliveau and G. P. O'Leary, Jr., Establishing an undergraduate research program, *J. Chem. Educ.*, 1983, **60**, 670–671.
[31] K. C. Cannon and G. R. Krow, Synthesis of complex natural products as a vehicle for student-centered, problem-based learning, *J. Chem. Educ.*, 1998, **75**, 1259–1260.
[32] V. Demczylo, J. Martinez, A. Rivero, E. Scoseria and J. L. Serra, Research projects for undergraduate students, *J. Chem. Educ.*, 1990, **67**, 948–950.
[33] T. R. Ruttledge, Organic chemistry lab as a research experience, *J. Chem. Educ.*, 1998, **75**, 1575–1577.
[34] G. B. Kharas, A new investigative sophomore organic laboratory involving individual research projects, *J. Chem. Educ.*, 1997, **74**, 829–831.
[35] R. E. Adelberger, Guest comment: undergraduate research – making physics interesting to all students, *Am. J. Phys.*, 1994, **62**, 872–873.
[36] D. A. Sabatini, *J. Prof. Issues Eng. Educ. Practice*, 1997, **123**, 98–102.
[37] J. C. Weisshaar, Frontier chemistry for freshmen, *J. Chem. Educ.*, 1994, **71**, 225–226.
[38] J. G. Dunn, R. I. Kagi and D. N. Phillips, Developing professional skills in a third-year course offered in western Australia, *J. Chem. Educ.*, 1998, **75**, 1313–1316.
[39] CUR website: http://www.cur.org/.
[40] NSF REU website: http://www.nsf.gov/home/crssprgm/reu/start.htm.
[41] NSF URC website: http://www.nsf.gov/pubs/2003/nsf03595/nsf03595.pdf.
[42] REU list website: http://www.nsf.gov/home/crssprgm/reu/reu_search.cfm.
[43] D. A. Waldow, C. B. Fryhle and J. C. Bock, *J. Chem. Educ.*, 1997, **74**, 441–442.
[44] URP website: http://www.science.duq.edu/opportunities/urp.html.
[45] S. A. O'Reilly, A. M. Wilson and B. Howes, Utilization of SciFinder scholar at an undergraduate institution, *J. Chem. Educ.*, 2002, **79**, 524–526.
[46] A. R. Bressette and G. W. Breton, Using writing to enhance the undergraduate research experience, *J. Chem. Educ.*, 2001, **78**, 1626–1627.
[47] A. M. Shachter, Integrating ethics in science into a summer undergraduate research program, *J. Chem. Educ.*, 2003, **80**, 507–512.
[48] L. M. Sweeting, Ethics in science for undergraduate students, *J. Chem. Educ.*, 1999, **76**, 369–372.

CHAPTER 15

Crossing the Line: Stochastic Models in the Chemistry Classroom

Michelle M. Francl

Department of Chemistry, Bryn Mawr College, 101 N. Merion Avenue, Bryn Mawr, PA 19010, USA

Contents

1. Introduction 215
2. Molecular dynamics 217
3. Stochastic methods 217
4. Conclusions 219
References 219

1. INTRODUCTION

"Physical Chemistry: The process by which all natural phenomena are reduced to $y = mx + b$" [1].

Browse through early mathematical reference texts for chemists, such as the classic 'Mathematical Preparation for Physical Chemistry' by Daniels [2], and it is clear that historically there is more than a little truth in this humorous definition. Several chapters are typically devoted in these tomes to graphical analyses of data with advice ranging from the selection of appropriate paper and scales, to how to most reliably extract slopes and intercepts. Early in the last century, computational chemists were not quantum mechanics or molecular dynamicists, but masters of curve fitting and error analysis [3].

Physical chemistry as a field has grown and diversified since these classic texts were written, and the physical chemistry curriculum has been challenged to grow and stretch along with it. While mathematical and computational techniques remain a mainstay of a student's experience in physical chemistry, both the necessary concepts and available tools have changed tremendously. Not only has the advent of inexpensive, widely available computing driven these changes, but the penetration of physical chemistry into other disciplines, ranging from medicine to environmental science, is a significant impetus for change as well.

In 2000, the Mathematical Association of America (MAA) convened a working group of chemists to consider the connections between the undergraduate mathematics and chemistry curricula [4]. The group noted the significant impact of

ANNUAL REPORTS IN COMPUTATIONAL CHEMISTRY, VOLUME 1
ISSN: 1574-1400 DOI 10.1016/S1574-1400(05)01015-7

technology on both curricula, commenting that it allowed chemists to take on questions of increased complexity. They identified nine uses of technology critical to chemists in general, of which I will highlight the first two as of particular interest to physical chemists: multivariate modeling and visualization, and iterative solutions. Six areas of mathematical expertise necessary to chemistry were acknowledged, four of which involve significant use of computers: multivariable relationships, numerical methods, visualization and data analysis. The chemists in the working group expressed concern about the impact of technology on the teaching of the base skills and concepts in these areas, as well as acknowledging that symbolic algebra programs are in common use in upper division courses such as physical chemistry.

More recently, Zielinski and Schwenz have considered the state of the physical chemistry curriculum in the new century [5]. They provide a comprehensive picture of recent pedagogical developments relevant to the teaching of physical chemistry. The choice of what to include in a first course in physical chemistry has grown more difficult as the range of the material has grown, and Zielinski and Schwenz advocate tailoring the courses to the interests and needs of students. They stress the need for physical chemistry students to have access to a robust computational resource.

One hopes that it is the rare student who now escapes physical chemistry without once using a computer to solve a problem: either a numerical problem or for a quantum chemical calculation. Symbolic programming environments such as Mathematica [6], Maple [7], Matlab [8] and Mathcad [9] are widely accessible to students and instructors, and more and more material is being developed for use in teaching physical chemistry [10–12]. The advent of GUI interfaces for quantum chemistry software has brought the use of programs such as AMPAC [13], GAMESS [14], Gaussian [15] and Spartan [16] into not only the physical chemistry classroom, but into general chemistry and organic chemistry as well. Current textbooks clearly incorporate both types of computational approaches into their problems, if not always into the body of the text, and ancillary materials using symbolic algebra environments are widely available [11].

It remains that the emphasis in most texts is on deterministic approaches, rather than stochastic models or dynamical simulations. A simple stochastic experiment undertaken in my office in which I randomly selected one recent physical chemistry text, opened to a chapter at random and determined the percentage of problems on a page that required students to fit data to a line, reveals that even current students might not disagree with the tongue-in-cheek definition of physical chemistry as a field given above [17]. Current practitioners certainly would take issue with this circumscribed view of computations in physical chemistry, as both molecular dynamics and stochastic methods such as Monte Carlo simulations are standard parts of the computational chemist's repertoire. The MAA working group on mathematics and chemistry [4] supported the notion that students must develop facility with common computational techniques, including both stochastic and deterministic approaches. Zielinski and Schwenz [5] advocate for introduction of stochastic methods into the curriculum for students on an engineering track. I would extend that emphasis and suggest that students destined for graduate work should also be exposed to at least the conceptual framework of stochastic algorithms and simple implementations of both Monte Carlo techniques and molecular dynamics. The accessibility of resources such as Mathematica and Mathcad makes it possible for students to 'get inside' these methods in ways not possible even a decade ago. Though many texts might not reflect this part of the field, if one looks beyond the texts and into the literature, one finds that stochastic methods and molecular dynamics are indeed finding their way into the physical chemistry curriculum.

2. MOLECULAR DYNAMICS

Lamberti *et al.* have developed a superb introduction to molecular dynamics for students (and for instructors who might not be intimately familiar with the workings of dynamics simulations) [18]. As well as providing a clear, easily readable and solid conceptual framework for these methods, the authors emphasize throughout the practical aspects of implementing a dynamics algorithm. Both simple Euler integration and Verlet's more sophisticated algorithm are laid out. Sample programs in FORTRAN and C, along with some suggested exercises for students, are available through JCEOnline [19]; the URL provided by the authors in the chapter is no longer accessible. The code provided could easily be ported to any of the symbolic algebra environments, and a laboratory exercise devoted to exploring these methodologies.

Cropper's *Mathematica Computer Programs for Physical Chemistry* [20] includes a treatment of molecular dynamics and includes CD programs for students to use and modify. The treatment here is also accessible to undergraduate students.

3. STOCHASTIC METHODS

Cropper's monograph also devotes space to stochastic methods, including covering Monte Carlo methods and techniques based on random walk algorithms. Applications to stochastic kinetics and random coil polymers are described. Again, the CD provides a starting point for working with the algorithms, rather than just reading about how they might work.

Bluestone describes a simple problem which can be solved using Monte Carlo methods [21]. Thermodynamic properties of a simple system with evenly spaced energy levels are computed using a Monte Carlo algorithm. The system is conceptually accessible to undergraduate physical chemistry students and applying Monte Carlo techniques emphasizes the underlying statistical nature of the observed thermodynamics.

Woller, while a teaching assistant at University of Nebraska, developed materials for an undergraduate physical chemistry lab introducing students to Monte Carlo methods and their use in molecular simulations [22]. Written in 1996, the page remains the second most common entry point into the University of Nebraska, Lincoln chemistry department web page (the most common entry is the departmental home page). The software which accompanies the lab is not available on the site, but the introduction is so easily readable by a physical chemistry student that I include it here. I have used it in class with an exercise to determine π to quickly convince students of the power of these methods.

More recently, Mira *et al.* provide a useful pedagogical framework comparing deterministic techniques for modeling chemical kinetic equations to stochastic approaches [23]. They point out that physical chemistry courses typically limit their discussion of chemical kinetics to deterministic methods, even though chemical reactions are inherently stochastic events. The limitations of deterministic methods are well known [24], particularly when describing processes such as nucleation or explosions and some biological reactions. Deterministic approaches can be algorithmically simpler; the simplest examples can be worked with pencil and paper, though even deterministic models can easily grow sufficiently complex to require extensive computational investment (e.g., solutions of coupled rate equations for atmospheric models). Two examples are used to highlight the differences in the approaches. Matlab is used to implement the algorithms, and the code is available at JCEOnline [19].

A more sophisticated piece of code, which uses stochastic approaches to model chemical kinetics, is available from IBM [25]. The code, designed to let experimentalists explore systems rapidly and develop mechanisms based on experimental data, can also be used in the classroom. It is available for several platforms, including Macintosh and Windows, and there is good documentation at the website as well. Instructors should expect to invest a bit of time in learning input methodologies, but the ability to see what happens when you alter a mechanism is worth the investment.

Genetic algorithms are another class of stochastic methods used frequently by computational chemists, as well as by chemical engineers and process chemists. Rowe and Colbourn cover the basics of neural networks, including fuzzy logic and genetic algorithms, highlighting the applications to industrial processes [26]. This would be a good starting point for a class about these often less familiar techniques. (Though many students will have heard about 'fuzzy logic' in the news, few are acquainted with the details.)

4. CONCLUSIONS

Physical chemistry is a dynamic and growing field; it is therefore a challenge to keep the undergraduate physical chemistry curriculum in concert with developments in the discipline, as well as to design courses that provide students with the necessary conceptual and practical background necessary to go on. Instructors should consider ways to expose students to both the practical aspects of molecular dynamics as well as to stochastic approaches for solving problems of interest to chemists. Materials accessible to undergraduate physical chemistry students are available which do not require substantial investment of faculty time to implement in the classroom. I look forward to the development of more materials to help instructors bridge the inevitable gap between physical chemistry as it is practiced and as it is taught.

REFERENCES

[1] The author of this bit of wisdom and wit appears to be lost in the mists of history, and perhaps wisely so.
[2] F. Daniels, *Mathematical Preparation for Physical Chemistry*, McGraw-Hill, New York, 1928.
[3] T. B. Crumpler and J. H. Yoe, *Chemical Computations and Errors*, Wiley, New York, 1940.
[4] N. C. Craig, Chemistry Report: MAA-CUPM Curriculum Foundations Workshop in Biology and Chemistry, *J. Chem. Educ.*, 2001, **78**, 582–586.
[5] T. J. Zielinski and R. W. Schwenz, Physical chemistry: a curriculum for 2004 and beyond, *Chem. Educator*, 2004, **9**, 108–121.
[6] *Mathematica*, Wolfram, Inc., Champaign, IL.
[7] *Maple*, Maplesoft, Waterloo, Ont., Canada.
[8] *Matlab*, The Mathworks, Inc., Natick, MA.
[9] *Mathcad*, Mathsoft, Cambridge, MA.
[10] See SymMath, a peer reviewed collection of symbolic algebra documents (in both Mathematica and Mathcad) curated by Zielinski at JCEOnline (http://jchemed.chem.wisc.edu/JCEDLib/SymMath/index.html). An earlier collection of Mathcad documents also assembled by Zielinski resides at http://bluehawk.monmouth.edu/~tzielins/mathcad/
[11] See for example: J. H. Noggle, *Physical Chemistry Using Mathcad*, Pike Creek Publishing, Newark, DE, 1997, or M. P. Cady and C. A. A. Trapp, *Mathcad Primer for Physical Chemistry*, W. H. Freeman and Company, New York, 2000.
[12] See for example: S. H. Young, J. D. Madura and A. Wierzbicki, Integration of numerical methods into the undergraduate physical chemistry curriculum using MATHCAD, *J. Chem. Educ.*, 1995, **72**, 606, or M. M. Francl, *Survival Guide for Physical Chemistry*, Physics Curriculum and Instruction, Inc., Lakeville, MN, 2001.
[13] *AMPAC*, Semichem, Inc., Kansas City, MO.
[14] GAMESS, http://www.msg.ameslab.gov/GAMESS/GAMESS.html. M. W. Schmidt, K. K. Baldridge, J. A. Boatz, S. T. Elbert, M. S. Gordon, J. H. Jensen, S. Koseki, N. Matsunaga, K. A. Nguyen, S. J. Su, T. L. Windus, M. Dupuis and J. A. Montgomery, *J. Comput. Chem.*, 1993, **14**, 1347–1363, A WebWare GUI interface is available for PC-GAMESS to subscribers from JCEOnline.

[15] *Gaussian*, Gaussian, Inc., Pittsburgh, PA.
[16] *Spartan*, Wavefunction, Inc., Irvine, CA.
[17] One in five problems on the page in question required students to do a line fit. The author acknowledges that the small sampling rate might have biased the results.
[18] V. E. Lamberti, L. D. Fosdick, E. R. Jessup and C. J. C. Schauble, A hands-on introduction to molecular dynamics, *J. Chem. Educ.*, 2002, **79**, 601–606.
[19] *JCEOnline: Journal of Chemical Education Online*. Available at http://jchemed.chem.wisc.edu/
[20] W. H. Cropper, *Mathematica Computer Programs for Physical Chemistry*, Springer, New York, 1998.
[21] S. Bluestone, A Monte Carlo simulation for a uniform ladder of energy levels: statistical thermodynamic properties, *J. Chem. Educ.*, 1995, **72**, 606.
[22] J. Woller, The Basics of Monte Carlo Simulations, www.wollernet.com/portfolio/writing/montecarlo/index.html
[23] J. Mira, C. M. Fernández and J. M. Urreaga, Two examples of deterministic versus stochastic modeling of chemical reactions, *J. Chem. Educ.*, 2003, **80**, 1488–1493.
[24] R. de Levie, Stochastics, the basis of chemical dynamics, *J. Chem. Educ.*, 2000, **77**, 771.
[25] See https://www.almaden.ibm.com/st/computational_science/ck/msim/index.shtml
[26] R. C. Rowe and E. A. Colbourn, Neural computing in product formation, *Chem. Educator*, 2003, **8**, 211–218.

CHAPTER 16

Simulation of Chemical Concepts, Systems and Processes Using Symbolic Computation Engines: From Computer-Assisted Problem-Solving Approach to Advanced Tools for Research

Jonathan Rittenhouse and Mihai Scarlete

Bishop's University, Lennoxville, Que., Canada J1M 1Z7a

Contents

1. Introduction 221
2. Creation of self-extracting databases 223
3. Storage capacity in symbolic/computational form 224
   3.1. Relationships between thermodynamic functions of state [11] 224
   3.2. Storage of the data in functional form 224
   3.3. Storage of quantified versions of chemical principles 225
4. Use of the graphing power of SCE for enhanced and accurate visualization of chemical concepts 225
   4.1. Visualization of wave functions, orbitals' phase and probability 225
5. Design of specialized procedures based on the rapidity and numerical computation power of the SCE 228
   5.1. Automatic procedure for the kinetic analysis of van't Hoff reactions 228
6. Emulation of professional software and advanced application-specific procedures 229
   6.1. Phase of molecular orbitals 229
   6.2. Modeling of molecules with high symmetry (maximum three parameters) 231
7. Development of tools for research activities requiring quantification of the description of the chemical system under scrutiny 231
   7.1. Process design using SCE 231
   7.2. Description of the oscillating systems involving transamination of poly(organo)silanes during polymer-source chemical vapor deposition (PS-CVD) 232
8. Conclusions 234
References 234

1. INTRODUCTION

On the basis of a continuous monitoring of the developments in symbolic computation engines (SCE) and their increasing use in pedagogical

ANNUAL REPORTS IN COMPUTATIONAL CHEMISTRY, VOLUME 1
ISSN: 1574-1400 DOI 10.1016/S1574-1400(05)01016-9

circumstances, and on several years of SCE-assisted classroom experience in the teaching of physical chemistry, we have identified four specific areas in which SCE have impacted the physical chemistry curriculum. We thus propose in this chapter a classification structured on the degree of integration of the SCE functions in the solution process. As we view it, of all the chemical branches physical chemistry is best positioned for the development of students' ability to convert qualitative concepts into quantitative descriptions of chemical systems. In this respect, the pedagogical use of SCE can be maximally beneficial to students' knowledge acquisition. We argue, then, that apart from the conceptual limits inherently associated with a logical description of any natural object, the accuracy of the description is ultimately linked to its appropriate numerical definition. In the past, in order to keep a proposed model unaltered by its mathematical description, the pace of integration of the quantitative treatment of chemical systems into the academic curriculum was more dependent on the required mathematical expertise of the student than on the conceptual difficulty itself. For example, coverage of molecular modeling, or of the activity of the normal modes of molecular vibration, was part of the graduate education because of the pedagogical response to the prohibitively laborious calculations required, rather than to the intellectual difficulty related to the clear understanding of the symmetry principles upon which both those fields are based.

Recently, as a response to the increased burden of mathematical algorithms necessary to solve chemical problems, SCE (or 'mathematical slaves') have imposed themselves as the pedagogical option currently favored. Thus, almost all current major textbooks in physical chemistry now include nonaltered numerical examples of the chemical concepts covered, based on the use of SCE. In addition, an increasing number of the practice problems proposed are designed directly for the use of the calculation/graphing power of SCE [1,2].

All major SCE manufacturers – Matematica™, MathCad™, Maple™, Matlab™ – designed pedagogically useful and practical tools to address the shortcomings of the traditional 'blackboard and chalk' approach: eliminating for example, the difficulty, experienced by a 'classic' teacher in trying to exemplify the normal modes of vibration of methane. However, even more crucial to the dramatically increasing use of SCE in teaching environments was when these SCEs embedded HTML-exporting procedures thus unleashing the pedagogic potential of the World Wide Web. Indeed, the most recent and even more exciting development in this area is the possibility offered by some SCEs to export to HTML not only text or figures/animations, but also interactive Java-based applets allowing 3D-manipulations of the SCE-objects.

Close observation of recent developments in SCE use demonstrates that it is now a common practice that the treatment of chemical problems is proposed in the applications package ('procedures') of the SCEs where dedicated servers offer to SCE users practically unlimited web space to propose new solutions to enhance

SCE-exposure benefits, leading, in turn, to an increased rate of incorporation of new subjects in the core package of procedures offered by the SCE libraries. The result of this sustained innovation in the last 2 years has led to an impressive collection of application-specific procedures, produced by large number of contributors covering many chemical aspects. However, till this time, only a limited number of attempts are known to us that have been designed to provide a consistent, complete SCE-assisted approach to cover all classical physical chemistry textbook chapters for the classroom [3–6]. We are aware, however, that such attempts are just the tip of the iceberg for a clear new trend in the teaching of science, and the teaching of physical chemistry in particular. The preferred format in these initiatives appears today to be a CD-ROM accompanied by an explanatory book, rather than a book accompanied by a CD-ROM, as was the case 5 years ago – a preference that we hold can be directly linked to the natural environment of the SCE, and to the inherent limitations of a hard copy to accommodate manipulations of 3D objects and animations.

As we stated above we have observed, following the increased level of SCE integration in the teaching of physical chemistry, four 'pilot areas' of new developments. They are:

- Creation of self-extracting, self-expanding databases (quantum chemistry [7–9], thermodynamics, spectroscopy).
- Use of the graphing power of SCE for enhanced and accurate visualization of chemical concepts.
- Design of specialized procedure based on the fast and powerful numerical computation engine of the SCE ('templates').
- Emulation of professional software and advanced dedicated applications (molecular modeling procedures [10]).
- Development of advanced modeling tools for research activities involving quantification of the description of the chemical systems under scrutiny.

In what follows we elaborate on our perception of the specificity of the SCE approach at each of the above-mentioned levels, each one illustrated with a selected example to serve for an application-gain analysis.

2. CREATION OF SELF-EXTRACTING DATABASES

The computational power of SCE coupled to storage facilities associated with a general computer-assisted environment can be used to create and expose the principles at the creation of physical chemistry databases. We have selected three independent examples, in the areas of thermodynamics, spectroscopy, and quantum chemistry, illustrating increasing degrees of the functionalities of SCE.

3. STORAGE CAPACITY IN SYMBOLIC/COMPUTATIONAL FORM

3.1. Relationships between thermodynamic functions of state [11]

The database is structured on the Bridgman table principle of the partial derivatives of state functions. The laborious, purely mathematical algorithm of the independent variable change of the functions of state is simply avoided. The capacity storage of the computer is the main function of the SCE used, together with the symbolic definition of the mathematical formulae. The use of the database created, simply requires a clear understanding of the definition of the functions of state, and thus allows focus on the physical interpretation of the derived relationship. For example, Table 1 presents three applications. Rows 1.1 and 1.2 show the commands and resulting output for two limiting cases, the straightforward definition of a basic quantity – C_p, and the output for a upper level relationship – respectively. Row 1.3 is an illustration of a practical application: the formula of the Joule–Thomson coefficient, $(\partial T/\partial P)_H$, commonly used as an example for the procedure of change of variables for thermodynamic state-functions, is extracted from a Bridgman library, and expressed as a function of state parameters.

3.2. Storage of the data in functional form

The example selected for this category is based on the consistency of the theoretical treatment of a problem with the experimental data, when the generalizing principle is not targeted. Due to the specificity of the large database required for the solution of various problems, thermodynamics is a fertile area for SCE-embedded applications designed for the storage of data in symbolic/numerical form adapted for SCE processing. Pedagogically oriented templates with limited model-complexity have been developed by educators, such as general templates for acid/base neutralization problems [12]. At the same time, using the

Table 1. Extraction of relationships between thermodynamic state functions using a Maple-assisted library based on Bridgman Tables

SCE input	SCE output
1.1. > Diff(H, T)[P] = dH_p/dT_p	$\left(\frac{\partial}{\partial T}H\right)_P = C_p$
1.2. > Diff(F, G)[H] = dF_h/dG_h	$\left(\frac{\partial}{\partial G}F\right)_H = \frac{dFH}{-V(C_p+S)+TS\left(\frac{\partial}{\partial T}V(T)\right)}$
1.3. > Diff(T, P)[H] = dT_h/dP_h	$\left(\frac{\partial}{\partial P}T\right)_H = -\frac{V-T\left(\frac{\partial}{\partial T}V(T)\right)}{C_p}$

same format, more elaborate templates have addressed the chemical equilibrium: determination of activity coefficients from various models for Gibbs free enthalpy [13], liquid/vapor phase equilibrium [14], and determination of chemical equilibrium in homogeneous chemical systems [15]. Since for all these calculations, a core entity of thermodynamic calculations is the C_p-function of temperature, an SCE-adapted database is under current elaboration. Such a database can be obtained through the collection and accuracy-sensitive weighted empirical interpolation of accessible data found in existent databases, such as International Critical Tables, Handbook of Physics and Chemistry, updated continuously with new experimental data. These values can be extracted and used for routine thermodynamic calculations, and the database, then, implemented in procedures used to calculate the chemical equilibrium in various systems, or to study a particular chemical reaction at various temperatures.

3.3. Storage of quantified versions of chemical principles

Quantification of second-order spin-coupling effects and magnetization proximity transfer in NMR has been proposed for COSY experiments, for AX, A_2MX J-resolved, and A_2MX 1D systems [16]. A somewhat different approach has been used in the creation of self-expanding database for use in spectroscopy. Here, we present a simple and efficient use of SCE in the creation of a library with direct application in the interpretation of mass spectra. The object of the project was to provide students with a library to calculate the intensities of the isotopic peaks in mass spectra. The net result has enhanced the consistency of the calculations and accuracy, reducing the very voluminous MS tables, necessary for the interpretation of a mass spectrum, to just a few command lines at the same time. This library has an output giving the exact position of the MIP, and the relative MIP, $M + 1$ and $M + 2$ isotopic peak intensities, as presented in Table 2. Apart from the creation and efficient handling of the library, the students are given the opportunity for appending various modules of less general applications. As a pedagogical issue, the comparison with the existent MS tables found in any mass spectrometry textbook allows students to better understand the significance and reliability of the isotopic peak counts in a mass spectrum.

4. USE OF THE GRAPHING POWER OF SCE FOR ENHANCED AND ACCURATE VISUALIZATION OF CHEMICAL CONCEPTS

4.1. Visualization of wave functions, orbitals' phase and probability

(N.B.: in the general case, the three-variable wavefunctions obtained require a visualization procedure impossible to use in textbook format, e.g., 3D animation.)

Table 2. SCE-assisted mass spectrum based on a self-expanding library predicting relative isotopic peaks intensities from the molecular formula

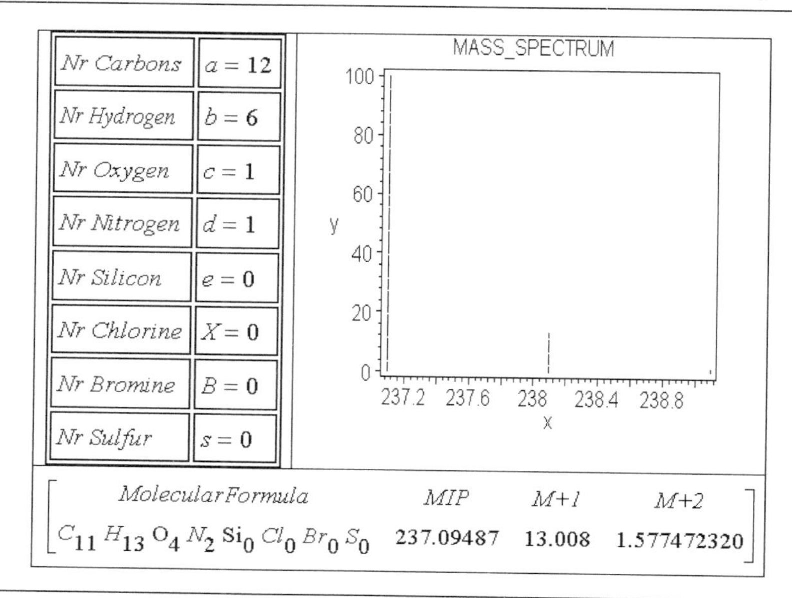

Introduction of molecular formula under tabular form, graphic representation of the predicted mass spectrum, tabular form for the molecular ion peak (MIP), and the isotopic peaks M + 1, M + 2.

Solving the Schrödinger equation for a specific quantum system, under various boundary conditions and/or symmetry constraints, is a fundamental component in any quantum chemistry course. Once the solutions are derived, the appropriate and efficient use of these solutions becomes the main focus for the student. The storage of the spherical harmonic solutions to the Schrödinger equation is a simple and useful tool for the teaching of quantum chemistry, very much like the use of crafted models used to represent the geometrical properties of molecules. The next step, consisting of the calculation of electronic probabilities, phase domains, or overlap integrals to cite only a few examples, is the focus of the chemist. The storage capacity of the SCE can be used to create the database of the spherical harmonic functions. These functions are then called as independent objects from the 'library', each being defined via the characteristic set of quantum numbers. Used already in the early stages of SCE integration, such applications rely on the accurate representation of a scientific model. The strength of the approach resides in the perfect correlation between the scientific model under scrutiny and its graphic representation. The results can then be used to correct current misrepresentations derived from the inaccuracy of common sense-derived 'drawing' of concepts, and to elaborate on the sources of inaccuracy. One example is the treatment of the geometrical representations of atomic orbitals for freshmen students, as presented in Table 3. The first row presents the command lines and graphical representations in spherical coordinates of the p_z wavefunctions (3.1.1), and of its square ('p_z-orbital', 3.1.2, respectively). Explanation of the general

Table 3. Accurate representations of the angular wavefunctions and atomic orbitals *via* automatic extraction from a self-expanding library

3.1.1 > plot3d(pz,_wavefunction, phi = 0...2 * Pi, theta = 0...Pi, coords = spherical, axes = box, scaling = constrained)

3.1.2. > plot3d(pz_wavefunction^2, phi = 0...2 * Pi, theta = 0...Pi, coords = spherical, axes = box, scaling = constrained)

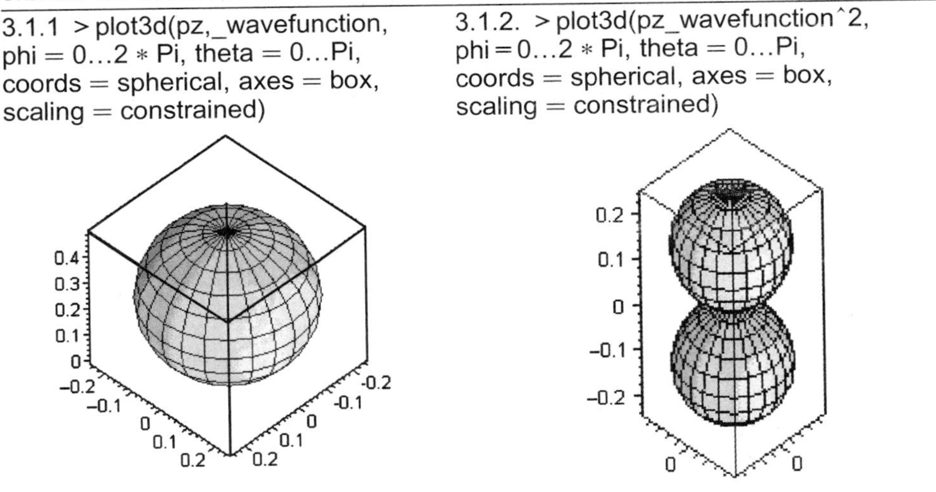

3.2.1. > plot3d(px_wavefunction^2 + py_wavefunction^2, phi = 0...2 * Pi, theta = 0...Pi, coords = spherical, axes = box, scaling = constrained)

3.2.2. > plot3d(px_wavefunction^2 + py_wavefunction^2 + pz_wavefunction^2, phi = 0...2 * Pi, theta = 0...Pi, coords = spherical, axes = box, scaling = constrained)

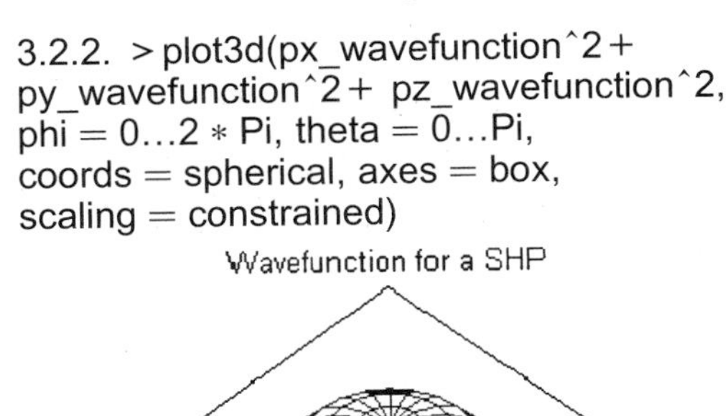

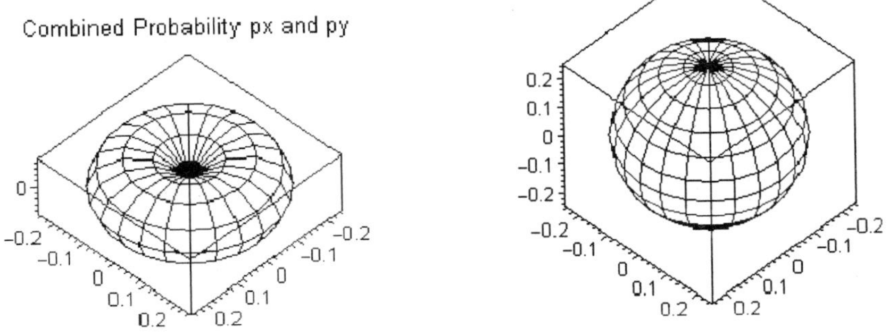

The generation of solutions to the Schrödinger equation applied to hydrogen-like systems is based on the appropriate selection of quantum number sets.

concept of 'spherical harmonics' becomes self-explanatory, as presented by the stepwise reconstruction presented in row 3.2 (2D mixing 3.2.1, and 3D spherical harmonics reconstruction – 3.2.2.)

5. DESIGN OF SPECIALIZED PROCEDURES BASED ON THE RAPIDITY AND NUMERICAL COMPUTATION POWER OF THE SCE

Obviously, the core-usage of SCE is targeting its main functionality, its computation power. The examples below belong to formal kinetics and quantum chemistry.

5.1. Automatic procedure for the kinetic analysis of van't Hoff reactions

We have chosen to illustrate this second level of integration of SCE, where the storing capacity is complemented by dedicated procedures to identify the solution of a problem in formal chemical kinetics. Similar attempts integrating Microsoft Excel Solver [17–21], or other computational programs, [22–24] accompanied the simultaneous development of similar Maple-assisted procedures. A more accurate solution, including statistical weights of the experimental points has been proposed by Zielinski and Allendoerfer [25]. The approach under our scrutiny is commonly called a 'template approach' because the procedure can be generalized to obtain solutions for a defined class of problems, rather than offering a solution to a singular problem. The design of such a template is based on a modular approach to the problem resolution: data collection, selection of the kinetic model from a finite series of possible models, sequential correlation of the experimental data with all kinetic models considered, followed by the evaluation of the degree of correlation between the experimental data and the models.

The selection of the model with the highest correlation coefficient represents the solution to the problem. In this section we present such a template (Table 4), designed to find automatically the kinetic parameters of a van't Hoff reaction – the experimental rate constant and the order of reaction (row 4.5) based on the maximization of the correlation coefficient (column 3 in table element 4.4.2), with respect to the experimental data introduced in vector-form (row 4.1, 4.2, and 4.3). The computational process can be visualized inside the SCE file in real time, *via* a graphical interface (not shown here, since the visualization is based on a 2D animation procedure). Numerical intermediate results before selection of the maximum correlation coefficient can be optionally displayed, showing the correspondent fit to a proposed reaction order (column 2 in 4.4.2). Column I in 4.4.2 is just an internal counter of the procedure, indicating the loop sequence.

Table 4. Automatic calculation of the parameters of a van't Hoff reaction

4.1. Product concentration := [0.260, 0.680, 1.020, 1.470, 1.690]

4.2. Time := [3600, 10,800, 18,000, 32,400, 43,200]

4.3. Initial reactant concentration := 2.162

4.4.1. Output of the procedure for maximization of the correlation coefficient function of the reaction order *R*;
Column 1 = nr. crt.;
Column 2 = order;
Column 3 = *R*-value

4.4.2.

$$\begin{bmatrix} 1 & -2.00 & 0.88161837 \\ 2 & -1.75 & 0.89570465 \\ . & . & . \\ 12 & 0.75 & 0.99886447 \\ 13 & 1.00 & 0.99998805 \\ 14 & 1.25 & 0.99890481 \\ . & . & . \\ 28 & 4.75 & 0.86881441 \\ 29 & 5.00 & 0.85950195 \end{bmatrix}$$

4.5.

$$\begin{bmatrix} \text{Correlation coefficient} & \text{Experimental order rxn} & \text{Rate constant} \\ 0.99998805 & 1.00 & \left[0.000035196716\frac{1}{s}\right] \end{bmatrix}$$

6. EMULATION OF PROFESSIONAL SOFTWARE AND ADVANCED APPLICATION-SPECIFIC PROCEDURES

The emulation of molecular modeling procedures has been particularly targeted in SCE-assisted physical chemistry. These procedures are all almost easily corroborated with appropriate graphic user interfaces due to the integrated graphing capabilities with the SCE. Various applications include calculations of the properties of molecular orbitals (potential energy surfaces, iso-surfaces, property maps), structural optimization (equilibrium geometry, conformation analysis, geometry of transition states), and vibrational modeling. The following two examples have been selected to exemplify this section.

6.1. Phase of molecular orbitals

Although the effects of the orbital overlap can be understood by direct visualization of the phase of the orbitals, the phase distribution within hybrids, or extended molecular orbitals, is not straightforward. In this case, a fast and affordable mixing procedure (LCAO, or even more evolved evaluation methods) is required for the visualization of the orbital-set basis. The example provided in Table 5 is an example of fast evaluation of the phase distribution in the sp^3d^2 set of hybrids.

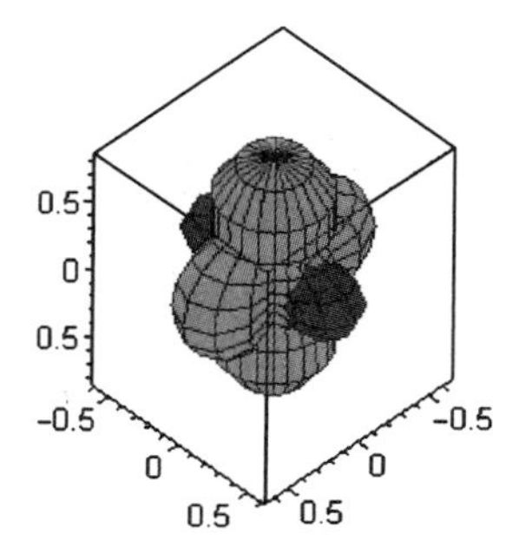

The figure immediately above presents the results of the Maple input (example based on a self-extracting specific library of hydrogen-like wavefunctions): >hyb1: = 1/sqrt(6) * WFs + 1/sqrt(2) * WFpz + 1/sqrt(3) * WFdz;: p1: = plot3d (hyb6, phi = 0...2 * Pi, theta = 0...2 * Pi, coords = spherical, axes = frame, scaling = CONSTRAINED, axes = box, color = [signum(hyb1),0,1]): etc...display({p1,p2...p6}).

Table 5. Representation of phase distribution in sp^3d^2 hybrid-set

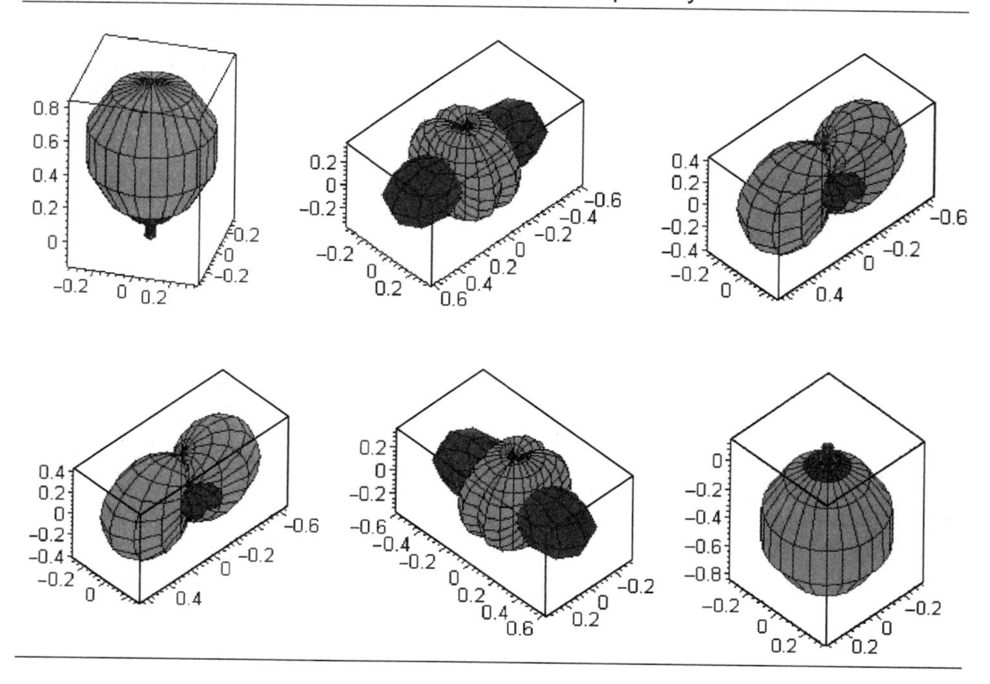

Table 6. Selected Newton–Raphson output for geometry optimization of D_{3d}-C_2H_6 *vs.* C–C–H bending angle (column 3), and six equal C–H bond lengths (column 2)

Bond length	Bond length	Bending angle	Nr count
1.126475049	1.126475049	109.3382661	1
. . .	. . .	. . .	. . .
1.240993905	1.240993905	109.3382661	4
1.241000000	1.241000000	109.3382661	5

6.2. Modeling of molecules with high symmetry (maximum three parameters)

Table 6 shows the Maple output of a procedure designed to optimize the geometry of D_{3d}-ethane by minimizing the first-order perturbation of Gibbs free enthalpy, using the Newton–Raphson convergence. Minimization of the number of parameters is based on the constraint of an exact D_{3d} symmetry of the optimized configuration of the molecule.

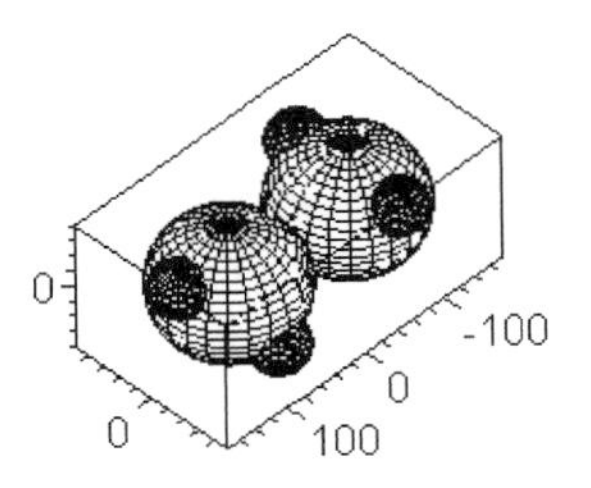

Observe that, if the two parameters chosen are C–C–H bending angle and C–H length, convergence to the 10th digit is obtained after only five iterations, while the bending angle is resolved within the first loop, as shown in Table 6.

7. DEVELOPMENT OF TOOLS FOR RESEARCH ACTIVITIES REQUIRING QUANTIFICATION OF THE DESCRIPTION OF THE CHEMICAL SYSTEM UNDER SCRUTINY

7.1. Process design using SCE

The borders between teaching, research, and even industrial practice, may become diffuse in cases when real data are used in a highly accurate model made possible by the expanded limits of the numeric analysis. We have selected for this presentation one example featuring an integrated technological/commercial analysis of an in-house synthesis of acetic anhydride *via* acetone cracking [26]. Although designed for educational purposes, the model presented below has been adopted by industry. The analysis, based on Douglas' methodology [27,28], involves the modeling of a process integrating the kinetic behavior of six simultaneous reactions with ketene intermediate, a thermodynamic analysis, and specific financial aspects of the process. The model made extensive use of SCE-embedded packages, such as fitting experimental data, numerical solutions of non linear systems of algebraic equations, and unconstrained linear optimization. The model has a modular structure: input of kinetic and thermodynamic chemical data, batch/continuous process design, product recovery by liquid separation from the anhydride column, sensitivity analysis of project economics. The result of the model is a graph of the 'overall economic potential' *vs.* 'acetone conversion'.

7.2. Description of the oscillating systems involving transamination of poly(organo)silanes during polymer-source chemical vapor deposition (PS-CVD)

PS-CVD [29] is intended to introduce a combinatorial approach for the experimental design of ceramic thin films for semiconductor and optoelectronic devices *via* synthesis of semiconductor thin layers with oscillating properties. The method was developed in our group for the synthesis of thin layers on large surfaces, where chemical vapor deposition is not applicable, such as large, irregular or dielectric substrates. The method uses the oscillating chemical system generated in the gas phases *via* controlled fragmentation of polysilanes exposed to specific thermal and/or chemical conditions. One such complex set of chemical reactions is presented below, involving reactions with ammonia. The system produces oscillations of the gaseous silazane intermediates, appearing as a result of an autocatalytic transamination step in the fragmentation of polysilanes. An oscillating regime of the targeted intermediate species (polycarbosilazane) is induced by the chemical system. Four chemical reactions are selected as the analysis core of the system. This time-based oscillating regime is transformed into spatial oscillations *via* local perturbation of the reaction medium. The analysis of this perturbation was possible by the design of a SCE-based procedure, involving three correlated modules. The first involved the analysis of the oscillating chemical system (a series of four reversible reactions), parameterized with respect to the relative concentrations of the species in the system and a set of eight rate constants; the graphical output of the solution of the system of eight differential equations associated with the four reversible reactions is presented in Row 7.1. The behavior of the system within a perturbed gaseous phase contained in a 3D space involved modeling of the flow of the intermediate and was obtained *via* an in-house Maple-assisted visualization of a CFX-imported solution for general fluid

Table 7. Simulation of the PS-CVD oscillatory system

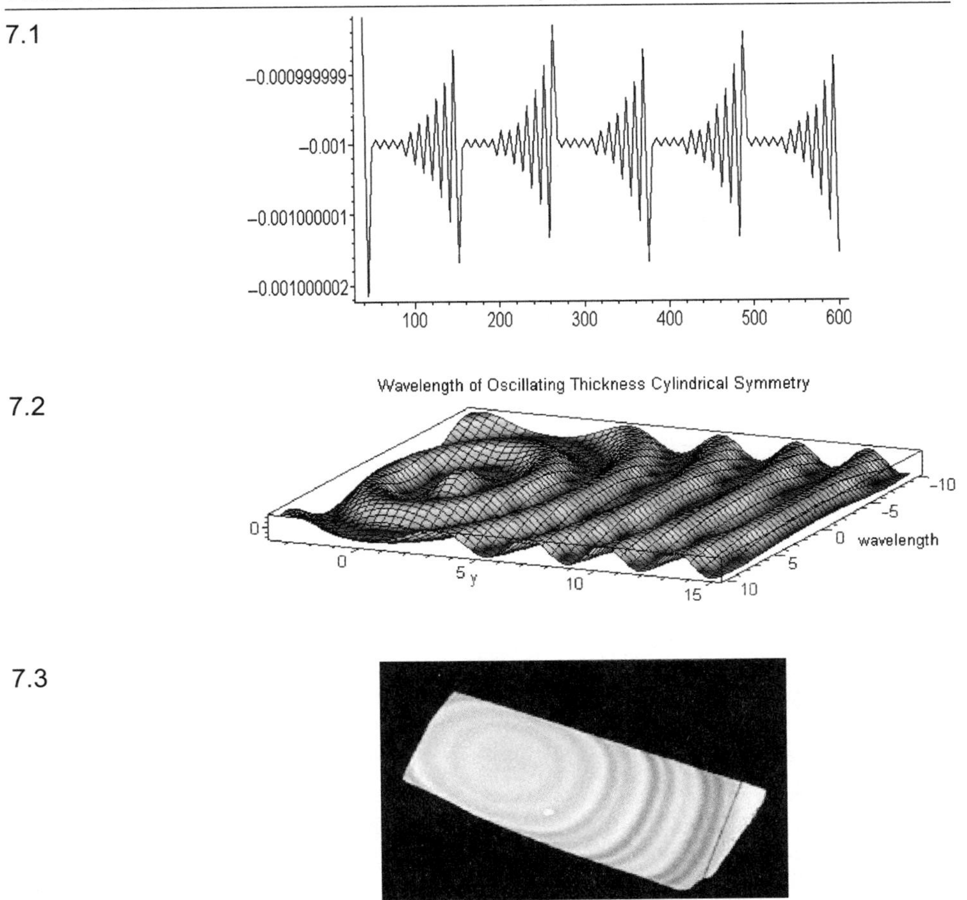

dynamics based on mixed mass/heat transfer in a Navier–Stokes environment (Table 7).

Two other SCE-based procedures have been designed: a second one for the modeling of the local perturbation, parameterized with respect to intensity of the perturbation and localization in the system, and, finally, a third one, involving the introduction of border conditions defining the cylindrical symmetry of the actual experimental setup and prediction of the interference pattern based on the oscillation of the production of the intermediate. At the end of the analysis, a selection process based on a trial-and-error procedure was used to determine the most probable set of parameters required to maximize the correlation of the general solution to the oscillating system – Row 7.2 – with the experimental result presented in Row 7.3.

8. CONCLUSIONS

At the end of the 18th century, the scientist A. Quetelet made the then radical observation that “we may judge the degree of perfection to which a science has arrived by the facility with which it may be submitted to calculation”. As time has shown, such an observation has held true when applied, for example, to the early 20th century physics of microparticles that led to the full theoretical development of quantum mechanics. Nor does one need Quetelet’s courage today to observe that, within all chemical branches – organic, inorganic, general, or biochemistry – physical chemistry is best positioned to develop a student’s ability to convert qualitative concepts into quantitative descriptions of chemical systems. As we hope this chapter has demonstrated, SCE-assisted pedagogy can make the understanding and study of physical chemistry approach ever closer the state of ‘perfection’ that Quetelet posited over 300 years ago and open doors to new possibilities that neither he nor we can fully predict.

REFERENCES

[1] P. Atkins, *Physical Chemistry*, W.H. Freeman, New York, 1997.
[2] R. A. Alberty and R. J. Silbey, *Physical Chemistry*, Wiley, New York, 1996.
[3] J. H. Noggle, *Physical Chemistry Using Matematica*, CRC Press, Boca Raton, FL, 1996.
[4] G. Woodbury, *Physical Chemistry*, Brookes/Cole, Florence, KY, 1997.
[5] M. Scarlete, *WWW Living-Book of Physical Chemistry*, BU Press, Lennoxville, Canada, 1999.
[6] M. M. Francl, *Survival Guide for Physical Chemistry*, E. Maley, Lakeville, MN, 2001.
[7] M. Horbatsch, *Quantum Mechanics Using Maple*, Springer, Berlin, 1995.
[8] D. Harrington, A Maple package for computing and graphing hydrogen and Slater-type orbitals, Maplet Maple Applications Center, 2002, **7**, 13. URL: http://www.maplesoft.com
[9] M. Komma, Hydrogen orbitals, Maplet Maple Applications Center, 2002, **12**, 26. URL: http://www.maplesoft.com
[10] F. M. Fernandez, *Introduction to Perturbation Theory in Quantum Mechanics*, CRC Press, Boca Raton, FL, 2001.
[11] M. Scarlete, *Maple-Assisted Phenomenological Thermodynamics*, Bishop’s University, e-Press, Lennoxville, Canada, 2001.
[12] W. A. Hoffmann, Acid–base equilibria, Maplet Maple Applications Center, 2001, **6**, 31. URL: http://www.maplesoft.com
[13] R. Taylor, Activity coefficients in binary systems, Maplet Maple Applications Center, 2001. URL: http://www.maplesoft.com
[14] J. Kitchin, Calculation of the bubble point pressure for a mixture using the Peng Robinson equation of state and simple mixing rules, Maplet Maple Applications Center, 2000, **15**, 4. URL: http://www.maplesoft.com
[15] J. Grotendorst and J. Dornseiffer, Calculation of chemical equilibrium compositions, Maplet Research Center Juelich, 2002.
[16] R. P. F. Kanters, COSY spectrum of a Homonuclear AX Spinsystem, 2001; Homonuclear A2MX J-resolved spectrum with hetero X, Maplet Maple Applications

Center, 2001; Homonuclear A2MX one-dimensional spectrum, Maplet Maple Applications Center, 2001, **12**, 11. URL: http://www.maplesoft.com
[17] P. Denton, Analysis of first order kinetics using Microsoft Excel Solver, *J. Chem. Educ.*, 2000, **77** (11), 1525–1526.
[18] D. C. Harris, Nonlinear least squares curve fitting with Microsoft Excel Solver, *J. Chem. Educ.*, 1998, **75**, 119–122.
[19] E. Vitz, Conceptualizing kinetics with curve fitting, *J. Chem. Educ.*, 1998, **75** (11), 1661–1664.
[20] M. Iannone, Using Excel Solver: an addendum to the HCl infrared spectrum, *J. Chem. Educ.*, 1998, **75**, 1188–1189.
[21] J. Machuca-Herrera, Nonlinear curve fitting using spreadsheets, *J. Chem. Educ.*, 1997, **74**, 1525–1526.
[22] R. D. Poshusta, *Comput. Phys.*, 1991, **5** (2), 248–252.
[23] T. G. Copeland, Kinetic analysis using computers, *J. Chem. Educ.*, 1984, **61**, 778–780.
[24] P. J. Moore, Analysis of kinetic data for a first-order reaction with unknown initial and final readings by the method of non-linear least squares, *Chem. Soc. Faraday Trans. J.*, 1972, **68**, 1890–1893.
[25] T. J. Zielinski and R. D. Allendoerfer, Least square fitting of non-linear data in the undergraduate laboratory, *J. Chem. Educ.*, 1997, **74**, 1001–1008.
[26] C. Richardson, S. Sopher, R. Petak, A. Linninger, Conceptual process design and numerical computing techniques, Birkhäuser, Dordrecht, The Netherlands, 1995, Vol. 2 & 3, pp. 3–14.
[27] J. M. Douglas, *Conceptual Design of Chemical Processes*, McGraw-Hill, New York, 1988.
[28] J. M. Douglas, A hierarchical decision procedure for process synthesis, *AIChE J.*, 1985, 353–360.
[29] S. Martic, J. Campbell, M. Bercu, C. Aktik and M. Scarlete, Spectroscopic analysis and semiconductor properties of amorphous thin films containing silicon–carbon–nitrogen deposited via a polymeric route, *Can. J. Anal. Sci. Spectrosc.*, 2003, **48** (1), 77–84.

Section 6
Emerging Science

Section Editor: Ralph Wheeler
Department of Chemistry & Biochemistry
University of Oklahoma
620 Parrington Oval
Room 208
Norman, OK 73019
USA

CHAPTER 17

The Challenges in Developing Molecular Simulations of Fluid Properties for Industrial Applications

Raymond D. Mountain and Anne M. Chaka

Computational Chemistry Group, Physical and Chemical Properties Division, Chemical Science and Technology Laboratory, National Institute of Standards and Technology, Gaithersburg, MD 20899-8380, USA

Contents

1. Introduction 239
2. The intermolecular potential function barrier 241
3. The sampling barrier 242
4. Challenges/opportunities 243
References 243

1. INTRODUCTION

Molecular simulation methods, both molecular dynamics and Monte Carlo, and computer speeds have developed to the point where it is possible to envision these methods as being able to provide reliable estimates of thermal properties (equation of state, vapor pressure, viscosity to name a few) of industrially interesting fluids. These methods are considered to be one of the 'enabling' technologies of computational chemistry that are expected to facilitate the application of chemical science knowledge of condensed phase properties in the chemical industry for conditions where experimental data are sparse [1]. To date, these methods have been used to implement the connections statistical mechanics provides between the thermal properties and the underlying intermolecular interactions of the molecules of the fluid [2]. This is accomplished by sampling the phase space of the system and using the coordinates of the configurations so generated to calculate various fluid properties. There are several books available that describe simulation methods and the reader is referred to them for details on the technique [3–5].

There are two barriers to realizing the vision of robust fluid thermophysical property predictions from molecular simulations. The first barrier is the need for 'accurate' interaction potentials (or force fields) between the molecules that can be evaluated 'cheaply' since millions of energy and force determinations are

ANNUAL REPORTS IN COMPUTATIONAL CHEMISTRY, VOLUME 1
ISSN: 1574-1400 DOI 10.1016/S1574-1400(05)01017-0

needed for even a small system simulation. The second barrier is the need for algorithms that can efficiently generate statistically significant estimates of fluid properties. These barriers and various ways to try to overcome are examined in the following sections. The terms 'accurate' and 'cheaply' will depend on the specific problems of industrial interest. Data and property information are most likely to be available on commodity materials, but industrial competition requires fast and flexible means to obtain data on novel materials, mixtures, and formulations under a wide range of conditions. For the vast majority of applications, particularly those involving mixtures and complex systems (such as drug–protein interactions or polymer nanocomposites), evaluated property data simply do not exist and are difficult, time-consuming, or expensive to obtain. For example, commercial laboratory rates for measuring vapor–liquid equilibria for two state points of a binary mixture are of the order of $30–40k. Hence, industry is looking for a way to supply massive amounts of data with reliable uncertainty limits *on demand*. Predictive modeling and simulation have the potential to help meet this demand.

The Council for Chemical Research's Vision 2020 [6] states that the desired target characteristics for a virtual measurement system for chemical and physical properties are as follows: problem setup requires less than 2 h, completion time is less than 2 days, cost including labor is less than $1000 per simulation, and that it is usable by a non-specialist, i.e., someone who cannot make a full-time career out of molecular simulation. Unfortunately, we are a long way from meeting this goal, particularly in the area of molecular simulations.

All fluid properties are ultimately determined by the electronic structure of the chemical species present, but effectively linking this quantum world to the macroscopic properties of dense fluids, such as density, solubility, viscosity, boiling point, has never been done. Forging this link in a rigorous manner requires exploitation of the quantum wavefunction to obtain an accurate description of the intermolecular forces that determine condensed phase properties, integrated with a proper treatment of statistical mechanics in a molecular simulation framework to capture the dynamic nature of a fluid. Statistical mechanical methods using molecular simulations are reliable and successful in determining properties of simple classical model systems. Descriptions of intermolecular interactions responsible for condensed phase properties are currently limited to fits of experimental data. The reliability of the resulting potential models is unknown and they cannot be systematically transferred, extended, or improved. This problem was clearly demonstrated recently by the results of the 'First Industrial Fluid Properties Simulation Challenge' [7]. This challenge [8] was organized by several US chemical companies and coordinated by NIST as a first step to show quantitatively that "...a lack of validation of different methods and of reliable comparison studies was a major limitation to industrial application of atomistic scale (fluid properties) simulation."

The 'Challenge' results underscore the notion that modeling *inter*molecular forces remains an art form rather than a robust, scientific methodology. In contrast, *intra*molecular interactions (bonding, structure) can be readily described by quantum mechanics for small, isolated gas-phase molecules. The total energy of the weaker *inter*molecular interactions can be accurately calculated for pairs of molecules, but this requires the highest level, most computationally intensive theoretical methods. Hence, it is not possible to simply increase the number of molecules in the system to extend quantum mechanics to the condensed phase for fluids. A doubling of computer power every 18 months is not sufficient to overcome the formal scaling of n^7 (where n = number of electrons) necessary for accurate calculation of intermolecular forces [9]. At a fundamental level, it is the polarizability and electrostatic interactions resulting from the electronic structure of a molecule that determine how it interacts with its environment and the resultant properties of the fluid. These interactions are two of the most difficult aspects of the bulk phase to treat accurately, because they arise out of both short- and long-range interactions, respond dynamically to a changing environment, and are a true multi-body effect. The force fields in common use for simulations that have been developed from static binary interactions have limited fundamental validity.

2. THE INTERMOLECULAR POTENTIAL FUNCTION BARRIER

In molecular simulations, the potential functions describing the intermolecular interactions determine how the phase space of the system is sampled as well as being used to evaluate the physical properties of the system. This means that the extent to which the computed properties agree with experimental values for those quantities depends critically on how well the potential functions represent the true intermolecular interactions. The early simulation research used model potentials, such as the Lennard-Jones 12-6 potential, with only passing intent that the results might correspond to a physical fluid [10].

Although such simple model potentials are not capable of representing real physical fluids, these models were valuable for developing physical understanding of the way intermolecular interactions influence various properties. For simulations to be useful for industrial purposes, it is necessary that the intermolecular potentials be able to generate properties that are approximately congruent with the fluids of interest under the appropriate state conditions of temperature, pressure, and composition. The challenge is one of learning how to do this so that the potentials are predictive rather than just descriptive. Most of the model potentials currently in use are obtained by a combination of quantum chemistry calculations plus an empirical component where selected properties are used to adjust the potentials. Since this is more an art than science, the predictions from the model potentials must be compared with measured physical properties in order to validate the model.

The situation is worse for mixtures. Frequently, empirical, or semi-empirical combining rules are used to generate potential parameters for the unlike molecule interactions from the potentials for the pure components. This introduces yet another level of uncertainty. Because most industrial applications involve mixtures, the real challenge is not just the generation of models for like molecules, but also calls for the development of methods to produce good interactions between unlike molecules. The combining rules are reasonable when the two species are chemically similar, but the results contain large, unknown uncertainties.

3. THE SAMPLING BARRIER

In order for a molecular simulation to have value in most industrial settings, it must be possible to obtain answers 'promptly'. The saying that problems should be solved 'over coffee', 'over lunch', or 'overnight' sets a typical time scale for providing answers. There are (at least) two ways to speed up simulations so that such methods could become able to fit into an industrially acceptable performance level.

The first is to make use of faster computers and simulation algorithms than can make efficient use of parallel computing environments. Molecular dynamics algorithms have been developed that can scale nearly linearly with the number of processors [11–13]. In principle, this is also the case for Monte Carlo algorithms. Molecular dynamics and Monte Carlo simulations are used to obtain thermodynamic properties and can do this well once the potentials are specified. Also, there is a simple method that can be used to monitor a simulation and indicate when a statistically adequate sample has been generated. It makes use of the ergodic requirement that time averages of properties associated with individual molecules converge to the per molecule average of the entire sample [14]. It applies equally well to Monte Carlo simulations when time average is replaced by cumulative average obtained during the random walk in phase space that occurs during the simulation [15].

Transport properties, such as the thermal conductivity and the viscosity, are more difficult to obtain from a molecular dynamics simulation because the usual linear response approach, also known as the Green-Kubo formulation of transport coefficients, requires a large number of statistically independent time origins in order to obtain reliable estimates. In practice, the time required to obtain a viscosity coefficient is at least an order of magnitude longer than is needed to obtain thermodynamic properties. This has led to the development of non-equilibrium methods where the system is 'sheared' to mimic more closely the experimental approach to measure the viscosity [16]. Even so, this method requires a number of simulations at different shear rates so that an extrapolation to low shear rates can be made with some confidence. The result is that the time needed to obtain a viscosity coefficient remains long. While much less effort has been devoted to determining the thermal conductivity, the difficulties are the same as for viscosity.

Both the linear response approach and the non-equilibrium approach use the same basic assumption. A gradient is applied and the resulting current is calculated with the transport coefficient being the ratio of the two quantities. Recently, a suggestion was made to reverse this by imposing a current and calculating the resulting gradient [17]. This basic idea has been used to determine the thermal conductivity of a fluid [18] and also the thermal diffusion in a mixture [19]. This approach is not yet mature, but may make it possible to obtain transport coefficients with a reasonable time cost.

Another alternative approach to estimating the viscosity is to simulate a well-characterized, transient flow and to infer the viscosity from the decay of the momentum current [20]. The initial results obtained are encouraging. The equivalent scheme for estimating the thermal conductivity has not been reported. The idea of inferring the viscosity from the decay of a current is not new, but this method appears to be much more efficient than the ones based on the decay of the transverse momentum current correlation function [21,22].

4. CHALLENGES/OPPORTUNITIES

For the barrier of obtaining accurate intermolecular potentials, the challenge is to develop quantum mechanical methods that accurately calculate intermolecular forces in the condensed phase. Secondly, one needs to describe those forces in a readily evaluated computational method with sufficient accuracy to capture the underlying physics.

For the sampling barrier, the short-term task is to critically evaluate the novel non-equilibrium methods mentioned previously [17–20]. For the longer term, the challenge is to develop user-friendly simulation packages so that the learning curve cost of using simulations is reduced to a level that encourages industrial use.

REFERENCES

[1] T. Thompson (ed.), *Chemical Industry of the Future, Technology Roadmap for Computational Chemistry*, 1999, http://www.ccrhq.org/vision/roadmaps/complete.html

[2] D. A. McQuarrie, *Statistical Mechanics*, Harper and Row, New York, 1976.

[3] M. P. Allen and D. J. Tildesley, *Computer Simulation of Liquids*, Clarendon Press, Oxford, 1987.
[4] D. C. Rapaport, *The Art of Molecular Dynamics Simulation*, Cambridge University Press, New York, 1985.
[5] D. Frenkel and B. Smit, *Understanding Molecular Simulation: From Algorithms to Applications*, 2nd edition, Academic Press, New York, 2002.
[6] http://www.chemicalvision2020.org/techvision.html
[7] Results were reported at the 2002 Fall Meeting of the AIChE and soon will be published in *Fluid Phase Equilibria*. http://www.cstl.nist.gov/FluidSimulationChallenge/
[8] *Thermophysical Properties of Industrial Fluids by Molecular Simulations* (June 18–19, 2001), Gaithersburg, MD. See Ref. [7] for a link to the web site.
[9] M. Head-Gordon, Quantum chemistry and molecular processes, *J. Phys. Chem.*, 1996, **100**, 13213–13215.
[10] A. Rahman, Correlations in the motion of atoms in liquid argon, *Phys. Rev.*, 1964, **138**, A405–A411.
[11] R. K. Kalia, S. de Leeuw, A. Nakano and P. Vashishta, Molecular-dynamics simulations of Coulombic systems on distributed-memory MIMD machines, *Comput. Phys. Commun.*, 1993, **74**, 316–326.
[12] A. Nakano, P. Vashishta and R. K. Kalia, Parallel multiple time-step molecular dynamics with three-body interaction, *Comput. Phys. Commun.*, 1993, **77**, 303–312.
[13] S. Plimption, Fast parallel algorithms for short-range molecular dynamics, *J. Comput. Phys.*, 1995, **117**, 1–19.
[14] D. Thirumalai and R. D. Mountain, Activated dynamics, loss of ergodicity, and transport in supercooled liquids, *Phys. Rev. E*, 1993, **47**, 479–489.
[15] R. D. Mountain and D. Thirumalai, Quantative measure of efficiency of Monte Carlo simulations, *Physica A*, 1994, **210**, 453–460.
[16] P. T. Cummings and D. J. Evans, Nonequilibrium molecular dynamics approaches to transport properties and non-Newtonian fluid rheology, *Ind. Eng. Chem. Res.*, 1992, **31**, 1237–1252.
[17] F. Muller-Plathe, Reversing the perturbation in nonequilibrium molecular dynamics: an easy way to calculate the shear viscosity, *Phys. Rev. E*, 1999, **59**, 4894–4898.
[18] F. Muller-Plathe, A simple nonequilibrium molecular dynamics method for calculating the thermal conductivity, *J. Chem. Phys.*, 1997, **106**, 6082–6085.
[19] D. Reith and F. Muller-Plathe, On the nature of thermal diffusion in binary Lennard-Jones liquids, *J. Chem. Phys.*, 2000, **112**, 2436–2443.
[20] G. Arya, E. J. Maginn and H.-C. Chang, Efficient viscosity estimation from molecular dynamics simulation via momentum impulse relaxation, *J. Chem. Phys.*, 2000, **113**, 2079–2087.
[21] R. D. Mountain, Molecular dynamics study of liquid rubidium, *Phys. Rev. A*, 1982, **26**, 2859–2868.
[22] B. J. Palmer, Transverse-current autocorrelation-function calculations of the shear viscosity for molecular liquids, *Phys. Rev. E*, 1994, **49**, 359–366.

CHAPTER 18

Computationally Assisted Protein Design

Sheldon Park and Jeffery G. Saven

Department of Chemistry, University of Pennsylvania, 231 S. 34th Street, Philadelphia, PA 19104, USA

Contents

1. Introduction 245
2. Computational protein design 246
   2.1. Target structure 246
   2.2. Degrees of freedom 246
   2.3. Energy function 246
   2.4. Solvation and patterning 247
   2.5. Search methods 247
3. Computationally designed proteins 248
4. Conclusion 250
Acknowledgement 251
References 251

1. INTRODUCTION

Protein design refers to the identification of amino acid sequences that fold to predetermined structures. In some cases, structure itself may also be an element of design. *De novo* designed proteins have the potential to serve as novel therapeutics, catalysts, biomaterials, and molecular scaffolds. In addition, protein design tests our understanding of folding and structure–function relationships, since the biological functions of proteins are usually contingent on their forming unique, well-defined three-dimensional structures.

Two different experimental approaches have advanced *de novo* protein design. Building upon trends observed in the known structures of proteins, simple structural motifs common to proteins may be assembled to form whole proteins or protein complexes. This hierarchical protein design [2] has been successful in designing proteins with repeating building blocks such as helix bundles and coiled coils [3,4]. On the other hand, partially random protein libraries with diversities $>10^5$ may be generated, from which variants with desired characteristics may be selected using a high throughput screen [5]. Catalytic antibodies and phage display demonstrate the power and versatility of combinatorial approaches to protein engineering [6,7], which are appropriate for cases where we have incomplete knowledge about the determinants of structure and/or function.

ANNUAL REPORTS IN COMPUTATIONAL CHEMISTRY, VOLUME 1
ISSN: 1574-1400 DOI 10.1016/S1574-1400(05)01018-2

Despite their notable successes, both hierarchical and combinatorial approaches to protein design are troubled by a number of issues, including the complexity of protein structures, the accuracy of the energy functions used for sequence–structure compatibility, and the exponentially large number of possible sequences. This review summarizes the recent accomplishments in computational protein design (CPD), focusing on efforts where large numbers of simultaneous mutations are identified. CPD allows sequences to be rapidly screened or characterized *in silico*, with far-reaching implications on how proteins may be discovered and crafted in the future.

2. COMPUTATIONAL PROTEIN DESIGN

CPD involves several fundamental elements, summarized in this section [8].

2.1. Target structure

The choice of the target structure is a key decision made during CPD. Most redesign efforts start with a high-resolution structure obtained through X-ray crystallography or solution NMR, sometimes with modifications to reflect specific design goals [1]. The reuse of an existing structure need not limit functional diversity that may be achieved, since nature often produces proteins with vastly different functionalities using the same protein fold [9]. Novel tertiary structures may also be modeled to obtain new structures and topologies [1,10].

2.2. Degrees of freedom

Two types of degrees of freedom that must be simultaneously optimized during CPD are the amino acid identities and their side chain conformations. Not all amino acids are required to create functional proteins in the laboratory [11], and the use of prepatterning or a reduced alphabet can vastly simplify design. From statistical analyses of high-resolution structures, combinations of low-energy side chain torsion angles are preferentially populated, yielding discrete sets of rotamers. Rotamer approximations reduce the degrees of freedom available to each amino acid and greatly accelerate the computation [12].

2.3. Energy function

Protein sequence–structure compatibility is evaluated using physicochemical potentials. The physical potentials are usually optimized independently of the protein design process itself, and several well-characterized atomistic potentials are available (e.g., Amber [13], CHARMM [14], Gromos [15]). Most potentials have terms involving bond lengths, bond angles, and dihedral angles, as well as terms accounting for the van der Waals, electrostatic and H-bonding interactions. Often only the two-body, noncovalent terms are explicitly evaluated during CPD, since bond lengths, bond angles, and dihedral angles are determined by the rotamers. The energy is compared among candidate sequences. Reduced, database-derived potentials that address structural propensities and do not include atomistic detail may also be incorporated in sequence design [16].

2.4. Solvation and patterning

Since hydrophobic effects provide a main driving force behind protein folding [17], it is important to include terms that describe the solvation preferences of the amino acids. The use of explicit solvent during design is computationally prohibitive, and solvation effects are usually modeled with a phenomenological free energy term that quantifies the hydrophobicity or solvent exposure of the side chain [18]. Another way of accounting for solvation is through hydrophobic patterning [19] to ensure that nonpolar residues are preferentially buried in the interior while polar residues are exposed to the solvent. The patterning of the sequence may also be influenced by the secondary structure propensities among amino acids, i.e., preferences for α-helix [20–22] and β-sheet [23–25].

2.5. Search methods

Sequence search methods based on Monte Carlo (MC) algorithms are intuitive and straightforward to implement [26–28]. A random search through the sequence space may be computationally inefficient. Several MC-based algorithms have been developed to bias the trial moves and increase the acceptance probability, including MC with quenching, biased MC [29], and mean field biased MC [30]. Simulated annealing (SA) allows a rapid convergence to low energy states [31], though these need not be global optima [32].

Genetic algorithms (GAs) are inherently parallel algorithms that use a population of candidate solutions to arrive at optimal solutions through mutations, crossovers and natural selection [33]. Being a stochastic algorithm, GAs are able to optimize a solution on a rugged fitness landscape such as the free energy landscapes encountered during protein design.

Pruning and elimination methods such as dead-end elimination (DEE) identify the global minimum energy sequence for pairwise potentials [34]. The algorithm systematically compares rotamer pairs at each residue position and discards one when its lowest energy state is higher than the highest energy state attainable with the other rotamer. This leads to a narrowing of the search space during the computation. While significant challenges remain due to the exponentially growing computation time with protein size and diversity, the method has been instrumental in many CPD projects.

Rather than identify particular sequences directly, statistical methods estimate the site-specific amino acid probabilities for sequences folding to a target structure [35,36]. Modeled on the concept of entropy maximization in statistical mechanics, the algorithm defines an effective entropy as a function of the individual amino acid probabilities. Maximization subject to desired energetic and functional constraints yields the site-specific probabilities of the amino acids. Since the entire sequence space can be characterized using this method, the probabilistic approach to protein design can easily address large systems that are too large for direct sequence search. Furthermore, the method is versatile enough to identify optimal and suboptimal sequences, and provides information that may be readily used to guide the construction of combinatorial experiments [35,36].

3. COMPUTATIONALLY DESIGNED PROTEINS

An increasing number of computationally designed proteins have been reported in the literature, and here we mention a few successes. Broadly classified into two types, design efforts have emphasized (a) achieving well-structured folded states using large-scale sequence variability or (b) engineering novel functional properties, in some cases achieving both.

The early efforts in CPD mainly focused on hydrophobic core packing and developing algorithms to allow multiple mutations to be simultaneously introduced within the protein interior [37–40]. The quality of the designed core was then tested using biochemical, structural, and functional assays. Later algorithms extended the scope of design to include nonglobular proteins. Proteins with repeating α-helical secondary structural elements were among the favorite design targets, and several authors reported the design of helix bundles including two or more helices [41,42]. Backbone degrees of freedom were also introduced in some cases to relax the structure as different side chains are tested [43,44]. Energy landscape methods have also been applied to the design of a 47-residue three-helix bundle [16].

Computational design helped construct model β proteins containing two or three strand antiparallel β-sheets. For example, one computational design, Betanova, features a three-stranded β-sheet with four residues per strand and exhibits cooperative folding and unfolding [45]. The protein was designed in an iterative process from which a sequence compatible with the target backbone structure was selected based on β-hairpin stability, amino acid β-sheet propensities, statistical preferences for interstrand residue pairs, and side chain rotamer conformations. More recently, the WW-domain, a 35-residue β protein, has been computationally redesigned [44].

Mayo and coworkers have used a design scheme that applies DEE to find globally optimal sequences for a given structure. A designed ββα-motif protein, resembling the tertiary structure of a zinc finger DNA binding module, folds stably without the requisite Zn^{2+} metal ion [46]. The same algorithm was also used to redesign rubredoxin so that the molecule can stably fold without iron [47]. Recognizing the importance of binary patterning, the Mayo group automated the patterning of hydrophobicity in a structure, and used the method to design a thermophilic variant of the engrailed homeodomain [48].

Recently, a 97-residue α/β protein Top7 with a novel fold was successfully designed and its structure determined [10]. Thus, a globular protein fold not found in nature is physically possible, and the design extends the realization of nonnatural protein structures beyond a previously designed right-handed helical coiled coil [49]. The iterative cycle of sequence design and backbone optimization using MC minimization algorithms was critical to the success. The crystal structure of Top7 at 2.5 Å resolution has the designed topology with 1.17 Å RMSD over all backbone atoms. The similarity between the designed and predicted structure validates the energy function used, which had been partially parameterized using known protein structures and sequences.

CPD has been used to engineer new functionalities into proteins, e.g., metal or ligand affinity. The protein design algorithm DEZYMER was used to introduce targeted mutations in thioredoxin, thereby evolving a tetrahedral tetrathiolate iron center [50]. The designed protein, which forms a 1:1 monomeric complex with Fe(III), supports multiple cycles of oxidation and reduction. Working with five periplasmic binding proteins from *E. coli*, a series of new proteins with designed binding sites were engineered to recognize ligands with a wide range of chemical properties in terms of molecular shape (polar, aliphatic, and aromatic), chirality, functional groups (nitro, hydroxyl, and carboxylate), internal flexibility, charge, and water solubility [51]. When some of the receptors were tested for ligand specificity, the designed receptors exhibited chiral stereospecificity as well as sensitivity to the presence of various functional groups on the ligand.

Probabilistic protein design has been used to arrive at the sequence of a 114-residue four helix bundle with a diiron center (DFsc), where the structure, sequence, and function were designed *de novo*. The protein possesses a well-ordered interior as evident in the 1D and 2D NMR spectra (Fig. 1) [1]. Modeled after a previously designed antiparallel dimer, the target template was generated by redesigning the loops to connect the helices. During the sequence calculation, the amino acid identities of a subset of 26 residues were constrained, where these

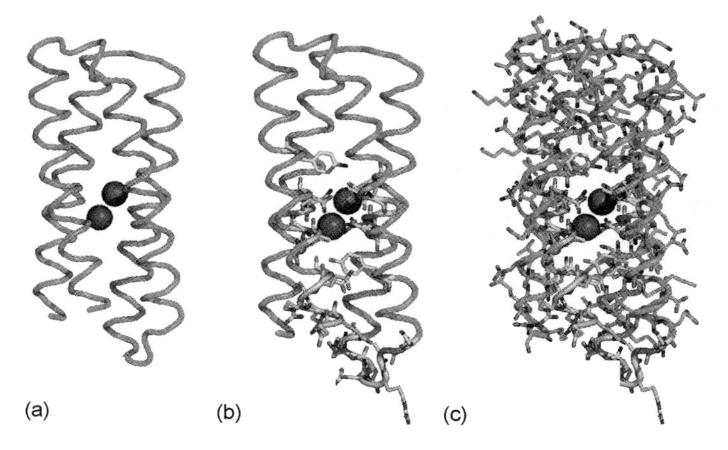

Fig. 1. Design of 114-residue metalloprotein. (a) Target four-helix backbone with dinuclear metal center. (b) Structure with fixed residues that provide metal binding, helix initiation, active site accessibility, and an interhelical turn. (c) Computationally determine the identities of the remaining 88 amino acids [1].

residues participate in metal binding, facilitate access to the active site, initiate a helix, or form a turn sequence. The remaining 88 residues were identified using a statistical computational assisted design strategy (*scads*). The parallel calculation of the site-specific amino acid probabilities at all positions allowed a large sequence space to be efficiently searched. The designed protein exhibits catalytic activity against known peroxidase substrates.

Scads has also been used to redesign an integral membrane protein to make it water soluble. To that end, 28% of the residues from the highly hydrophobic transmembrane region of KcsA were simultaneously mutated to hydrophilic residues [52]. The water-soluble variant was expressed in high yield and possessed the expected helical content, tetrameric oligomerization state, and toxin binding properties of the wild-type membrane soluble form. Such methods open a new route to better understanding membrane proteins, which are notoriously difficult to express and structurally characterize.

4. CONCLUSION

The many degrees of freedom involving both sequence and local (side chain) structure lead to the combinatorial complexity of protein sequence design. Through quantitative analysis of the sequence–structure relationship, CPD promises to play a critical role in furthering our understanding of proteins and in discovering *de novo* proteins with novel properties.

ACKNOWLEDGEMENT

J.G.S. is a Cottrell Scholar of Research Corporation, and the authors acknowledge support from the NIH (GM 61267) and NSF (DMR 00-79909).

REFERENCES

[1] J. R. Calhoun, H. Kono, S. Lahr, W. Wang, W. F. DeGrado and J. G. Saven, Computational design and characterization of a monomeric helical dinuclear metalloprotein, *J. Mol. Biol.*, 2003, **334**, 1101–1115.
[2] J. W. Bryson, S. F. Betz, H. S. Lu, D. J. Suich, H. X. Zhou, K. T. O'Neil and W. F. DeGrado, Protein design: a hierarchic approach, *Science*, 1995, **270**, 935–941.
[3] L. Regan and W. F. DeGrado, Characterization of a helical protein designed from first principles, *Science*, 1988, **241**, 976–978.
[4] P. B. Harbury, T. Zhang, P. S. Kim and T. Alber, A switch between two-, three-, and four-stranded coiled coils in GCN4 leucine zipper mutants, *Science*, 1993, **262**, 1401–1407.
[5] G. MacBeath, P. Kast and D. Hilvert, Redesigning enzyme topology by directed evolution, *Science*, 1998, **279**, 1958–1961.
[6] S. J. Pollack, J. W. Jacobs and P. G. Schultz, Selective chemical catalysis by an antibody, *Science*, 1986, **234**, 1570–1573.
[7] R. H. Hoess, Protein design and phage display, *Chem. Rev.*, 2001, **101**, 3205–3218.
[8] J. G. Saven, Designing protein energy landscapes, *Chem. Rev.*, 2001, **101**, 3113–3130.
[9] N. Nagano, C. A. Orengo and J. M. Thornton, One fold with many functions: the evolutionary relationships between TIM barrel families based on their sequences, structures and functions, *J. Mol. Biol.*, 2002, **321**, 741–765.
[10] B. Kuhlman, G. Dantas, G. C. Ireton, G. Varani, B. L. Stoddard and D. Baker, Design of a novel globular protein fold with atomic-level accuracy, *Science*, 2003, **302**, 1364–1368.
[11] S. Akanuma, T. Kigawa and S. Yokoyama, Combinatorial mutagenesis to restrict amino acid usage in an enzyme to a reduced set, *Proc. Natl Acad. Sci. USA*, 2002, **99**, 13549–13553.
[12] R. L. Dunbrack, Jr., Rotamer libraries in the 21st century, *Curr. Opin. Struct. Biol.*, 2002, **12**, 431–440.
[13] S. J. Weiner, P. A. Kollman, D. A. Case, U. C. Singh, C. Ghio, G. Alagona, S. Profeta and P. Weiner, A new force-field for molecular mechanical simulation of nucleic-acids and proteins, *J. Am. Chem. Soc.*, 1984, **106**, 765–784.
[14] B. R. Brooks, R. E. Bruccoleri, B. D. Olafson, D. J. States, S. Swaminathan and M. Karplus, Charmm – a program for macromolecular energy minimization, and dynamics calculations, *J. Comput. Chem.*, 1983, **4**, 187–217.
[15] J. Hermans, H. J. C. Berendsen, W. F. Vangunsteren and J. P. M. Postma, A consistent empirical potential for water–protein interactions, *Biopolymers*, 1984, **23**, 1513–1518.
[16] W. Jin, O. Kambara, H. Sasakawa, A. Tamura and S. Takada, De novo design of foldable proteins with smooth folding funnel: automated negative design and experimental verification, *Struct. Fold. Des.*, 2003, **11**, 581–590.
[17] A. R. Fersht, *Structure and Mechanism in Protein Science*, Freeman, New York, 1999.
[18] D. Eisenberg and A. McLachlan, Solvation energy in protein folding and binding, *Nature*, 1986, **319**, 199–203.

[19] S. Kamtekar, J. M. Schiffer, H. Y. Xiong, J. M. Babik and M. H. Hecht, Protein design by binary patterning of polar and nonpolar amino-acids, *Science*, 1993, **262**, 1680–1685.
[20] K. T. O'Neil and W. F. Degrado, A thermodynamic scale for the helix-forming tendencies of the commonly occurring amino-acids, *Science*, 1990, **250**, 646–651.
[21] A. Chakrabartty, T. Kortemme and R. L. Baldwin, Helix propensities of the amino acids measured in alanine-based peptides without helix-stabilizing side-chain interactions, *Protein Sci.*, 1994, **3**, 843–852.
[22] P. C. Lyu, M. I. Liff, L. A. Marky and N. R. Kallenbach, Side chain contributions to the stability of alpha-helical structure in peptides, *Science*, 1990, **250**, 669–673.
[23] C. K. Smith, J. M. Withka and L. Regan, A thermodynamic scale for the beta-sheet forming tendencies of the amino acids, *Biochemistry*, 1994, **33**, 5510–5517.
[24] C. A. Kim and J. M. Berg, Thermodynamic beta-sheet propensities measured using a zinc-finger host peptide, *Nature*, 1993, **362**, 267–270.
[25] D. L. Minor, Jr. and P. S. Kim, Measurement of the beta-sheet-forming propensities of amino acids, *Nature*, 1994, **367**, 660–663.
[26] E. I. Shakhnovich and A. M. Gutin, A new approach to the design of stable proteins, *Protein Eng.*, 1993, **6**, 793–800.
[27] H. W. Hellinga, The construction of metal centers in proteins by rational design, *Fold. Des.*, 1998, **3**, R1–R8.
[28] B. Kuhlman and D. Baker, Native protein sequences are close to optimal for their structures, *Proc. Natl Acad. Sci. USA*, 2000, **97**, 10383–10388.
[29] A. P. Cootes, P. M. G. Curmi and A. E. Torda, Biased Monte Carlo optimization of protein sequences, *J. Chem. Phys.*, 2000, **113**, 2489–2496.
[30] J. Zou and J. G. Saven, Using self-consistent fields to bias Monte Carlo methods with applications to designing and sampling protein sequences, *J. Chem. Phys.*, 2003, **118**, 3843–3854.
[31] S. Kirkpatrick, C. D. Gelatt and M. P. Vecchi, Optimization by simulated annealing, *Science*, 1983, **220**, 671–680.
[32] C. A. Voigt, D. B. Gordon and S. L. Mayo, Trading accuracy for speed: a quantitative comparison of search algorithms in protein sequence design, *J. Mol. Biol.*, 2000, **299**, 789–803.
[33] J. H. Holland, *Adaptation in Natural and Artificial Systems: An Introductory Analysis with Applications to Biology, Control, and Artificial Intelligence*, University of Michigan Press, Ann Arbor, MI, 1975.
[34] J. Desmet, M. de Maeyer, B. Hazes and I. Lasters, The dead-end elimination theorem and its use in protein side-chain positioning, *Nature*, 1992, **356**, 539–542.
[35] J. M. Zou and J. G. Saven, Statistical theory of combinatorial libraries of folding proteins: energetic discrimination of a target structure, *J. Mol. Biol.*, 2000, **296**, 281–294.
[36] H. Kono and J. G. Saven, Statistical theory for protein combinatorial libraries. Packing interactions, backbone flexibility, and the sequence variability of a main-chain structure, *J. Mol. Biol.*, 2001, **306**, 607–628.
[37] J. W. Ponder and F. M. Richards, Tertiary templates for proteins. Use of packing criteria in the enumeration of allowed sequences for different structural classes, *J. Mol. Biol.*, 1987, **193**, 775–791.
[38] P. E. Correa, The building of protein structures from alpha-carbon coordinates, *Proteins*, 1990, **7**, 366–377.
[39] C. Lee and S. Subbiah, Prediction of protein side-chain conformation by packing optimization, *J. Mol. Biol.*, 1991, **217**, 373–388.
[40] L. Holm and C. Sander, Database algorithm for generating protein backbone and side-chain co-ordinates from a C alpha trace application to model building and detection of co-ordinate errors, *J. Mol. Biol.*, 1991, **218**, 183–194.

[41] J. W. Bryson, J. R. Desjarlais, T. M. Handel and W. F. DeGrado, From coiled coils to small globular proteins: design of a native-like three-helix bundle, *Protein Sci.*, 1998, **7**, 1404–1414.
[42] A. E. Keating, V. N. Malashkevich, B. Tidor and P. S. Kim, Side-chain repacking calculations for predicting structures and stabilities of heterodimeric coiled coils, *Proc. Natl Acad. Sci. USA*, 2001, **98**, 14825–14830.
[43] P. B. Harbury, B. Tidor and P. S. Kim, Repacking protein cores with backbone freedom: structure prediction for coiled coils, *Proc. Natl Acad. Sci. USA*, 1995, **92**, 8408–8412.
[44] C. M. Kraemer-Pecore, J. T. Lecomte and J. R. Desjarlais, A de novo redesign of the WW domain, *Protein Sci.*, 2003, **12**, 2194–2205.
[45] T. Kortemme, M. Ramirez-Alvarado and L. Serrano, Design of a 20-amino acid, three-stranded beta-sheet protein, *Science*, 1998, **281**, 253–256.
[46] B. I. Dahiyat and S. L. Mayo, De novo protein design: fully automated sequence selection, *Science*, 1997, **278**, 82–87.
[47] P. Strop and S. L. Mayo, Rubredoxin variant folds without iron, *J. Am. Chem. Soc.*, 1999, **121**, 2341–2345.
[48] M. L. Connolly, Solvent-accessible surfaces of proteins and nucleic acids, *Science*, 1983, **221**, 709–713.
[49] P. B. Harbury, J. J. Plecs, B. Tidor, T. Alber and P. S. Kim, High-resolution protein design with backbone freedom, *Science*, 1998, **282**, 1462–1467.
[50] C. D. Coldren, H. W. Hellinga and J. P. Caradonna, The rational design and construction of a cuboidal iron–sulfur protein, *Proc. Natl Acad. Sci. USA*, 1997, **94**, 6635–6640.
[51] L. L. Looger, M. A. Dwyer, J. J. Smith and H. W. Hellinga, Computational design of receptor and sensor proteins with novel functions, *Nature*, 2003, **423**, 185–190.
[52] A. M. Slovic, C. M. Summa, J. D. Lear and W. F. DeGrado, Computational design of a water-soluble analog of phospholamban, *Protein Sci.*, 2003, **12**, 337–348.

SUBJECT INDEX

A

ab initio modeling 187–8
ab initio thermochemical methods 33, 37, 45
absorption
  intestinal 137–8
  see also ADMET properties
active transport 139–40
acyl carrier protein synthase (AcpS) 179
adenosine triphosphate (ATP) site recognition 187–8
adiabatic approximations 20, 25, 27
ADMET properties
  active transport 139–40
  aqueous solubility 135–7, 162
  blood–brain barrier permeation 140–2
  computational prediction 133–51
  cytochrome P450 interactions 143–4
  drug discovery 159–62
  efflux by P-glycoprotein 140, 160–1
  intestinal absorption 137–8
  intestinal permeability 134–5, 161
  metabolic stability 142–3, 162
  oral bioavailability 134, 138–9, 159–60
  plasma protein binding 142
  toxicity 144
AGC group of kinases 196
agrochemicals 163
AMBER force fields 92, 94–7, 99, 119–21
angular wavefunctions 225–8
aqueous solubility 135–7, 162
atomic orbital representations 225–8
atomistic simulation
  boundary conditions 80
  experimental agreement 77–8
  force fields 77, 79, 80–2
  methodological advances 79
  nucleic acids 75–89
  predictive insights 78–9
  sampling limitations 80–2
ATP *see* adenosine triphosphate
AUTODOCK 122–3

B

B3LYP functional 32, 48–50
back-propagation neural networks (BPNN) 136–7
base pair opening 77
basis sets 13–15, 32–3
Betanova 248–9
Bethe–Salpeter equation 27
binding affinities 78
binding free energy
  calculating 114–19
  protein–ligand interactions 113–30
  scoring functions 119–26
bio-molecular simulation
  atomistic simulation 75–82
  nonequilibrium approaches 108
  protein force fields 91–102
  protein–ligand interactions 113–30
  water models 59–74
bioavailability 134, 138–9, 159–60
blood–brain barrier permeation 140–2, 160–1
bond breaking
  configuration interaction 51
  coupled cluster methods 52–3
  generalized valence bond method 47–8
  Hartree–Fock theory 46, 48–51
  multireference methods 51–3
  perturbation theory 51–2
  potential energy surface 54
  quantum mechanics 45–56

self-consistent field methods 46–7, 53
spin-flip methods 53
Born–Oppenheimer approximation 3, 54
boundary conditions 80
Boyer Commission 206–7
BPNN *see* back-propagation neural networks
Bridgman tables 224

C

CAMK group of kinases 186, 196
Carnegie Foundation 206–7
casein kinase 2 (CK2) 197
Casida's equations 21–2, 25
CASSCF *see* complete-active-space self-consistent field
CBS-*n* methods 36–7
CC *see* coupled cluster
CD *see* circular dichroism
CDKs *see* cyclin-dependent kinases
central nervous system (CNS) drugs 160–1
charge transfer (CT) 26
CHARMM force fields 77, 79, 92–5, 97–9, 119–20
chemical vapor deposition (CVD) 232–3
CI *see* configurational interaction
circular dichroism (CD) spectra 22–4
cluster-based computing 113
CMAP *see* correction maps
CMGC group of kinases 186, 192–4
CNS *see* central nervous system
complete-active-space self-consistent field (CASSCF) method 47, 53
compound equity 171
computational protein design (CPD) 245–53
degrees of freedom 246
energy function 246–7
examples 248–50
search methods 247–8
solvation and patterning 247
target structures 246
computational thermochemistry
ab initio methods 33, 37, 45
CBS-*n* methods 36–7
density functional theory 32–3
empirical corrections 34–6
explicitly correlated methods 39
G1, G2, G3 theory 34–6
hybrid extrapolation/correction 36–7
isodesmic/isogyric reactions 34
nonempirical extrapolation 37–9
quantum mechanics 31–43
semi-empirical methods 31–2
Weizmann-*n* theory 37–9
configurational interaction (CI) 9–10, 48, 51
conformational flexibility 173
consensus approaches 145
correction maps (CMAP) 95, 96, 98
correlation methods 8–11
Council for Chemical Research 240
Council on Undergraduate Research (CUR) 206–7, 208
coupled cluster (CC) methods 10–11, 37–40, 48–50, 52–3
CPD *see* computational protein design
CT *see* charge transfer
CUR *see* Council on Undergraduate Research
current density 27
CVD *see* chemical vapor deposition
cyclin-dependent kinases (CDKs) 186, 192–4
cytochrome P450 interactions 143–4

D

D&C *see* divide and conquer
DA *see* discriminant analysis
databases
drug-likeness 155–6
ligand-based screening 172–5

self-extracting 223, 225
symbolic computation engines 223, 224–5
de novo protein design 245
dead-end elimination (DEE) 247–8, 249
degrees of freedom 246
density functional theory (DFT)
bond breaking 48–9
computational thermochemistry 32–3
protein–ligand interactions 116
state of the art 4, 11–12, 13–15
time-dependent 20–30
DEZYMER algorithm 249
DFT *see* density functional theory
discriminant analysis (DA) 138
distributed computing 113
distribution *see* ADMET properties
divide and conquer (D&C) algorithm 116–17
DOCK program 173–4, 177, 178, 189
docking 79, 114, 119, 121, 155, 169, 172–4, 178, 189–96
drug discovery 155–68
agrochemicals 163
aqueous solubility 162
chemistry quality 157
CNS drugs 160–1
databases 155–6
drug-likeness 155–7
intestinal permeability 161
lead-likeness 159
metabolic stability 162
oral drug activity 159–60
positive desirable chemistry filters 158–9
promiscuous compounds 162–3
drug-likeness 155–7

E

education
research-based experiences 205–14
stochastic models 215–20
symbolic computation engines 221–35
efflux by P-glycoprotein 140, 160–1
electron correlation methods 8–11
electronic Schrödinger equation 3–15
empirical force fields 91–102
empirical scoring functions 122–3
energy function 246–7
Ewald summation techniques 59, 62, 75
exact exchange 26–7
excited state structure/dynamics 24
excretion *see* ADMET properties
extended systems 26

F

FEP *see* free energy perturbation
FlexX 173, 178, 189
FLO99 178
Florida Memorial College 212
fluctuation theorem 109
fluid properties 239–44
focal-point approach 39
force fields
molecular simulations 239–40
nucleic acids 77, 79, 80–2
protein–ligand interactions 116, 119–21
proteins 91–102
structure-based lead optimization 177
fragment positioning 175–7
free energy 96, 103–11, 113–30
free energy perturbation (FEP) 104, 106
fuzzy logic 218

G

G-Score 123
G1, G2, G3 theory 34–6
generalized gradient approximation (GGA) 12

generalized valence bond (GVB) method 47–8
global matrices 116–17
graphical representations 225–8, 232–3
GROMOS force fields 97
GVB *see* generalized valence bond

H

Hartree–Fock (HF) method 4, 5–11, 13–15, 20–1, 46, 48–51
Hellmann–Feynman theorem 21
hierarchical protein design 245
high-throughput screening (HTS) 171–2
Hohenberg–Kohn (HK) theorem 11, 20
homodesmotic reactions 34
homology models 170, 188–9
HTS *see* high-throughput screening
hybridization, structure-based 191–2
hydration free energies 103

I

intermolecular potential functions 241–2
intestinal absorption 137–8
intestinal permeability 134–5, 161
isodesmic/isogyric reactions 34

J

Jarzynski relationship 103, 104–10

K

kinome targeting 185–202
  applications 192–7
  ATP site recognition 187–8
  homology models 188–9
  kinase family 186–7
  methodology 188–92
  selectivity 190–1
  structure-based hybridization 191–2
  virtual screening 189–90
knowledge-based scoring functions 123–5
Kohn–Sham (KS) equations 11, 20–2, 25

L

lead optimization *see* structure-based lead optimization
lead-likeness 159
Lennard–Jones (LJ) potential 93–4, 116, 121
LES *see* locally enhanced sampling
library enumeration 178
ligand binding 103
ligand-based screening 172–5, 178–9
linear interaction energy 117
LJ *see* Lennard–Jones
local spin density approximation 11–12
locally enhanced sampling (LES) 79
LUDI scoring function 123, 173

M

many-body perturbation theory 10
Maple 228, 230–2
master equations 115–16, 119–20
Mathematical Association of America 215–16
MCSCF *see* multi-configurational self-consistent field
MCSS program 173–4, 177
MD *see* molecular dynamics
metabolic stability 142–3, 162
  see also ADMET properties
MLR *see* multiple linear regression
MM *see* molecular mechanics
molecular dynamics (MD) simulation 75–8, 217, 239, 242
molecular mechanics (MM) 119, 120–2
molecular modeling 59–130
  atomistic simulation of nucleic acids 75–89
  free energy 103–11, 113–30
  nonequilibrium approaches 103–11

protein force fields 91–102
protein–ligand interactions 113–30
water models 59–74
TIP4P 62–4, 69–72
TIP4P-Ew 64–5, 69–72
TIP5P 65–7, 69–72
TIP5P-E 67–72
molecular orbital representation 229–31
molecular simulations 177–8, 239–44
Møller–Plesset form 10, 48–50
Monte Carlo methods 216–18, 239, 242, 247–8
multi-configurational self-consistent field (MCSCF) method 9–10, 46–7
multiple excitations 25
multiple linear regression (MLR) 136
multireference methods 51–3

N

National Science Foundation (NSF) 206–7, 209
nonequilibrium approaches
computational uses 109
experimental applications 108
free energy calculations 103–11
Jarzynski relationship 103, 104–10
theoretical developments 108–9
NSF *see* National Science Foundation
nucleic acids 75–89
nuisance compounds 162–3, 190

O

OPLS/AA force fields 92–4, 97
oral bioavailability 134, 138–9, 159–60
oral drug activity 159–60
orbital representations 225–8, 229–31
oscillating systems 232–3

P

P-glycoprotein 140, 160–1
parallel computing 242
PARAM force fields 97
partial least squares (PLS) analysis 134–5, 138
patterning 247
PB *see* Poisson–Boltzmann
PDB *see* Protein Data Bank
permeability, intestinal 134–5, 161
perturbation theory 10, 51–2
PES *see* potential energy surface
pharmaceutical chemicals
ADMET properties 133–51
drug discovery 155–68
structure-based lead optimization 169–83
virtual screening protocols 114, 120, 125
pharmacophore models 172–4
PhDOCK 173–4, 177
physical chemistry 215–17
plasma protein binding (PPB) 142
PLS *see* partial least squares
PMFScore 124–5
Poisson–Boltzmann (PB) equation 117–19, 120–2
polymer-source chemical vapor deposition (PS-CVD) 232–3
poly(organo)silanes 232–3
positive desirable chemistry filters 158–9
potential energy surface (PES) 3–4, 54
potential functions 241–2
PPB *see* plasma protein binding
predictive modeling 133–51, 240
privileged structures 158
probabilistic protein design 249–50
problem-solving templates 228
process design 231–2
promiscuous compounds 162–3, 190
Protein Data Bank (PDB) 113, 117, 123–4
protein design 245–53
degrees of freedom 246
energy function 246–7
examples 248–50
search methods 247–8

solvation and patterning 247
target structures 246
protein force fields 91–102
condensed-phase 94–6
free energies of aqueous solvation 96
gas-phase 94–6
optimization 96–9
united-atom 97
protein kinases *see* kinome targeting
protein–ligand interactions 113–30
PS-CVD *see* polymer-source chemical vapor deposition

Q

QSAR/QSPR models 133–51
quantum–classical enzymatic calculations 103
quantum mechanics 3–56
basis sets 13–15, 32–3
bond breaking 45–56
computational thermochemistry 31–43
configurational interaction 9–10, 48, 51
coupled cluster methods 10–11, 37–40, 48–50, 52–3
density functional theory 4, 11–12, 13–15, 32–3, 48–9
electron correlation methods 8–11
generalized valence bond method 47–8
Hartree–Fock method 4, 5–11, 13–15, 20–1, 46, 48–51
perturbation theory 10, 51–2
potential energy surface 3–4, 54
self-consistent field methods 6–8, 9–10, 37, 46–7, 53
semi-empirical methods 12–13, 15
symbolic computation engines 225–8
time-dependent density functional theory 20–30
quasi-static (QS) transformations 105, 133–51

R

RASSCF *see* restricted-active-space self-consistent field
re-parameterizations 59, 60–1, 67, 72
Research Experiences for Undergraduates (REU) 209
research institutions 205–14
restrained electrostatic potential 92–3
restricted Hartree–Fock (RHF) 46, 48–50
restricted-active-space self-consistent field (RASSCF) method 47
REU *see* Research Experiences for Undergraduates
RHF *see* restricted Hartree–Fock
Roothaan–Hall equations 6–8
Runge–Gross theorem 27

S

sampling barriers 242–3
SAR *see* structure–activity relationships
scads 250
scaling methods 6–8
Schrödinger equation 3–15
scoring functions 119–26
self-consistent field (SCF) methods 6–10, 37, 46–7, 53
self-consistent reaction field (SCRF) 118, 121
self-extracting databases 223, 225
semi-empirical methods 12–13, 15, 31–2
signal trafficking *see* kinome targeting
solubility 135–7
solvation 117–19, 247
spin-flip methods 53
statistical computational assisted design strategy (scads) 250
stochastic models 215–20
storage capacity 224–5
structure–activity relationships (SAR) 91, 133–51
structure-based drug design 114, 120, 125

structure-based hybridization 191–2
structure-based lead optimization 169–83
  application to specific targets 179
  compound equity 171
  discovery 171–5
  fragment positioning 175–7
  high-throughput screening 171–2
  library enumeration 178
  ligand–target complex evaluation 178–9
  modification 175–9
  molecular simulation 177–8
  structure visualization 175
  virtual screening 169, 172–5
support vector machines 137, 145
symbolic computation engines (SCE) 221–35
  advanced application-specific procedures 229–31
  computation power 228–9
  emulation of professional software 229–31
  graphical representations 225–8, 232–3
  process design 231–2
  quantification 225, 231–3
  self-extracting databases 223
  specialized procedures 228–9
  storage capacity 224–5

T

target structures 246
TDDFT *see* time-dependent density functional theory
template approach 228–9
thermal conductivity 242–3
thermochemistry, computational 31–43
thermodynamics
  integration method 104
  nonequilibrium approaches 103–11
  protein–ligand interactions 113–30
  symbolic computation engines 224–5
  water models 59–72
time-dependent density functional theory (TDDFT) 20–30
  computational aspects 21–2
  developments 26–8
  electronic excitations 20–1
  exact exchange 26–7
  performance 22–4
  qualitative limitations 25–6
time-dependent Hamiltonian operators 104
TIP4P 62–4, 69–72
TIP4P-Ew 64–5, 69–72
TIP5P 65–7, 69–72
TIP5P-E 67–72
TKL *see* tyrosine kinase-like
TKs *see* tyrosine kinases
Top7 249
toxicity 144, 190
  see also ADMET properties
transamination 232–3
transferable intermolecular potential (TIP) water molecules 59–74
two-electron integrals 6–7, 12–13
tyrosine kinase-like (TKL) group of kinases 186, 196–7
tyrosine kinases (TKs) 186, 194–5

U

UHF *see* unrestricted Hartree–Fock
undergraduate research 205–14
Undergraduate Research Programs (URPs) 208, 209–12
united-atom protein force fields 97
university research 205–14
unrestricted Hartree–Fock (UHF) 46, 50–1
URPs *see* Undergraduate Research Programs

V

van't Hoff reactions 228–9

vertical excitation 22–4
virtual screening 169, 172–5, 189–90
  high throughput 120
  protocols 114, 120, 125
viscosity 242–3
visualization 175, 225–8, 232–3

W
water models 59–74
  bio-molecular simulation 59–61
  five-site 65–72
  four-site 62–5, 69–72
  methods 61–2
  TIP4P 62–4, 69–72
  TIP4P-Ew 64–5, 69–72
  TIP5P 65–7, 69–72
  TIP5P-E 67–72
wavefunctions 225–8
Weizmann-*n* theory 37–9

X
XScore 123

Z
Z-factor equation 22